National Conference on Recent Advances in Fuel Cells and Solar Energy 2019

IOP Conference Series: Earth and Environmental Science
Volume 312

Karaikal, India
11 - 12 May 2019

Editors:

Sendhil Kumar Natarajan
P. Karthikeyan

ISBN: 978-1-5108-9651-2
ISSN: 1755-1307

Printed from e-media with permission by:

Curran Associates, Inc.
57 Morehouse Lane
Red Hook, NY 12571

Some format issues inherent in the e-media version may also appear in this print version.

Institute of Physics
Dirac House, Temple Back
Bristol BS1 6BE UK

Phone: 44 1 17 929 7481
Fax: 44 1 17 920 0979

techtracking@iop.org

Additional copies of this publication are available from:

Curran Associates, Inc.
57 Morehouse Lane
Red Hook, NY 12571 USA
Phone: 845-758-0400
Fax: 845-758-2633
Email: curran@proceedings.com
Web: www.proceedings.com

TABLE OF CONTENTS

DESIGN AND DEVELOPMENT OF DUAL AXIS SUN TRACKING SYSTEM FOR FLOATING PV PLANT 1
Sendhil Kumar Natarajan, Aman Kumar, Raqeeb Mohamed, R Rathna, Swaraj Mondal, Subbarama Kousik Suraparaju

NUSSELT NUMBER AND FRICTION FACTOR BEHAVIOUR OF CIRCULAR WAVY TRANSVERSE RIBS ARTIFICIALLY ROUGHENED SOLAR AIR HEATER 10
Sheshang S Chandel, Ajay Verma, Maheandera Prabu Paulraj, Santosh Kumar Sahu

EXPERIMENTAL ANALYSIS OF THERMOELECTRIC DEVICE TO PRODUCE COOLING EFFECT USING SOLAR ENERGY AND PHASE CHANGING MATERIAL 21
P. Lakshmi Prasanna, Dr. K Dharma Reddy

OPTIMISATION OF PARAMETERS IN THERMAL ENERGY STORAGE SYSTEM BY ENHANCING HEAT TRANSFER IN PHASE CHANGE MATERIAL 28
V. Sravani, Dr. K. Dharma Reddy

PRODUCTION OF METHANOL FROM BIOGAS USING METHANOTROPHS FOR THE APPLICATION IN A MICROBIAL FUEL CELL 34
Alok Kumar, Anand Ramanathan

DEVELOPMENT OF COMPUTER AIDED MODELLING AND OPTIMIZATION OF MICROWAVE PYROLYSIS OF BIOMASS BY USING ASPEN PLUS 41
M Dineshkumar, Banagiri Shrikar, Anand Ramanathan

PERFORMANCE ANALYSIS OF SOLAR PHOTOVOLTAIC PANEL AT KARAIKAL WEATHER CONDITIONS 48
Sendhil Kumar Natarajan, Subbarama Kousik Suraparaju, E Elavarasan, K Arjun Singh

PERFORMANCE EVALUATION OF COMPOUND PARABOLIC CONCENTRATOR WITH EVACUATED TUBE HEAT PIPE 55
P. Vijayakumar, G. Kumaresan, S. Anish kumar, U. Mohamed Faizal, G.R. Vijay Chandran, K.S. Vishnu Adharsh

DEVELOPMENT OF SOLAR OVEN EMPLOYED WITH PARABOLIC CONCENTRATOR 70
P. Vijayakumar, G. Kumaresan, S. Anish kumar, S. Sudhagar, G.R. Vijay Chandran, K.S. Vishnu Adharsh

PERFORMANCE ANALYSIS OF SOLAR ASSISTED VAPOUR JET REFRIGERATION SYSTEM WITH REGENERATOR (CRMC METHOD) 79
Prakash P Sathiya, A Kalaisselvane

EFFECT OF RF/DC MAGNETRON SPUTTERING PROCESS PARAMETERS FOR IMPROVING EFFICIENCY OF SOLAR PV CELL 92
Natarajan Purushothaman, R Elansezhian, A V Raviprakash

A SUCCINCT REVIEW ON FUEL CELLS 100
Subbarama Kousik Suraparaju, Sendhil Kumar Natarajan, P Karthikeyan

A SHORT REVIEW ON RECENT TRENDS AND APPLICATIONS OF THERMOELECTRIC GENERATORS 108
Subbarama Kousik Suraparaju, Gudapati Kartheek, Gummalla Venkata Sunil Reddy, Sendhil Kumar Natarajan

OPTIMAL ENERGY SCHEDULING OF RENEWABLE ENERGY SOURCES IN SMART GRID USING CUCKOO OPTIMIZATION ALGORITHM WITH ENHANCED LOCAL SEARCH 116
S Antonyraj, G Giftson Samuel

ASPEN HYSYS SIMULATION OF BIOMASS PYROLYSIS FOR THE PRODUCTION OF METHANOL 123
M Arul, M Dinesh Kumar, Anand Ramanathan

LIFE CYCLE ASSESSMENT ON BIOFUEL PRODUCTION FROM BIOMASS GASIFICATION AND SYNGAS FERMENTATION 128
R. Muthudineshkumar, R. Anand

ENERGY AND LIFE CYCLE ASSESSMENT OF SOLAR ASSISTED MICROWAVE PYROLYSIS OF WASTE BIOMASS 135
A. Santhoshkumar, Anand Ramanathan

PRODUCTION OF BIODIESEL FROM AEGLE MARMELOS CORREA SEED OIL FOR FUEL CELL APPLICATION 146
Vinoth Thangarasu, Anand Ramanathan

THERMAL PERFORMANCE OF A SALT GRADIENT NON-CONVECTIVE SOLAR POND IN SUBTROPICAL REGION CLIMATIC CONDITIONS 152
Sendhil Kumar Natarajan, Susant Kumar Sahu, Arjun Singh K

NUMERICAL STUDY OF NATURAL CONVECTION IN FLAT RECEIVER WITH AND WITHOUT SECONDARY REFLECTOR FOR SOLAR PARABOLIC DISH SYSTEM 163
Sendhil kumar Natarajan, B Raviteja, D Sri Harshavardhan, G Hari Prasath, Subbarama Kousik Suraparaju

EXPERIMENTAL INVESTIGATION OF DRYING POTATO FOR KARAIKAL CLIMATIC CONDITION 173
Sendhil Kumar Natarajan, E Elavarasan

PRODUCTION OF BIO-OIL FROM SOAPNUT SEED BY MICROWAVE PYROLYSIS 180
A. Mathiarasu, M. Pugazhvadivu

PERFORMANCE ASSESSMENT OF A SUSTAINABLE SOLAR AIR HEATER (SAH) BY USING DIFFERENT PROFILES OF ABSORBER PLATES 188
Singuru Rajesh, Vandrasi Vinay Kumar, Jana Sai Ram

DESIGN AND FABRICATE A SOLAR TORREFACTION UNIT TO IMPROVE THE SUGARCANE BAGASSE QUALITY FOR PYROLYSIS REACTION 199
R Nagarajan, S Murugavelh

THREE DIMENSIONAL NUMERICAL VALIDATION AND INVESTIGATION ON AIR COOLING SYSTEM OF LI-ION BATTERY USED IN HYBRID ELECTRIC VEHICLES 206
Bhargav Ganesh Chekuri, Priyanka Satya Subhasree Vaskuri, S Shiva Kumar Chary, Raj R Karuppa Thundil

DESIGN AND ANALYSIS OF POWER CONVERTORS FOR ELECTRIC VEHICLE APPLICATIONS 213
Safare Sudeep Babu, S Shiva Kumar Chary, Syed Muhammad Abdullah, D Elangovan, Raj R Thundil Karuppa

NUMERICAL VALIDATION AND STUDY OF PARTICULATE FLOW IN CYCLONE SEPARATOR USING COMMERCIAL COMPUTATIONAL FLUID DYNAMICS CODE 225
Mayuresh Mutha, Vineet Katkamwar, Bageerathan Thirunavukkarasu, Raj R Thundil Karuppa, P Sivamurugan

PSO BASED DISPATCH AND CONTROL OF HYBRID POWER SYSTEM WITH PV-DIESEL GENERATOR- A PROTOTYPE 237
K Bharathi Kannan, E Elavarasan, Aavala Yashoda, Sovana Kumari, C H Ram JethMalani

HELIOTROPISM OF SOLAR PV USING THE 4-BAR MECHANISM ANALYSIS 247
Anuj Bimal Desai, Sendhil Kumar Natarajan

TECHNO-ECONOMIC ANALYSIS OF SOLAR TRIGENERATION SYSTEM 252
B Anand, S Murugavelh

BIO OIL PRODUCTION FROM AGRO WASTE RESIDUE: THERMOCHEMICAL CONVERSION TO IMPROVE OIL QUALITY 257
K Midhun Prasad, K Murugavelh

A NUMERICAL STUDY OF FLUID FLOW AND PRESSURE DROP ANALYSIS OF CATALYTIC CONVERTER USING COMMERCIAL CFD CODE 265
Aniket Patil, Mayuresh Takale, Bageerathan Thirunavukkarasu, Raj R Thundil Karuppa

A REVIEW ON COMPUTATIONAL FLUID DYNAMICS ANALYSIS ON GREENHOUSE DRYER 275
Sendhil Kumar Natarajan, E Elavarasan

Author Index

IOP Conference Series
Earth and Environmental Science

National Conference on Recent Advances in Fuel Cells and Solar Energy-2019

N-CRAFS 2019

11th and 12th May, 2019

Organized By

DEPARTMENT OF MECHANICAL ENGINEERING,

National Institute of Technology Puducherry, Karaikal.

Sponsors

VCollab

RENEWABLE ENERGY AGENCY OF PONDICHERRY

ENERGY IS LIFE BEE CONSERVE IT

Greenserve Energy Managment Solutions

GREENPRO ENERGY TECHNOLOGIES

IOP Conf. Series: Earth and Environmental Science **312** (2019) 011001 doi:10.1088/1755-1315/312/1/011001

Editors:

1. **Dr. Sendhil Kumar Natarajan,**
   Assistant Professor and Head,
   Department of Mechanical Engineering,
   National Institute of Technology Puducherry,
   Thiruvettakudy, Karaikal – 609 609
   UT of Puducherry, India.
   Email: sendhil80@nitpy.ac.in, drsendhil1980iitmuk@gmail.com
   Phone: 04368 265230, Ext: 212
   Mob: +91 9597541522

2. **Dr. P. Karthikeyan**
   Professor,
   Dept. of Automobile Engineering,
   PSG College of Technology,
   Pelamedu, Coimbatore, Tamilnadu -641 004, INDIA.
   Email: apk@aut.psgtech.ac.in, apkarthipsg@gmail.com
   Mobile: +91-9443682803; +919488850019

IOP Conf. Series: Earth and Environmental Science **312** (2019) 011001 doi:10.1088/1755-1315/312/1/011001

National Conference on Recent Advances in Fuel Cells and Solar Energy (N-CRAFS '19)
11th-12th May 2019
Department of Mechanical Engineering, NIT Puducherry, Karaikal

MESSAGE FROM THE DIRECTOR

It is an immense pleasure to know that the Department of Mechanical Engineering, National Institute of Technology Puducherry, Karaikal is organizing the National conference on "Recent Advances in Fuel cell and Solar Energy (N-CRAFS - 2019) during 11th - 12th May, 2019. Being a centrally funded institute and an institute of national importance, NIT Py has been actively involved in teaching, research, consultancy and many funding projects. The institute many programs for the benefit of academics and industries.

In this regard, I congratulate the HOD, faculty and staff members as well as students of the Mechanical Engineering Department. This event would definitely benefit the participants coming from various institutes and organizations.

I wish for the grand success of the conference.

Prof. K. Sanakaranarayanasamy

Director, NIT Puducherry.

IOP Conf. Series: Earth and Environmental Science **312** (2019) 011001 doi:10.1088/1755-1315/312/1/011001

National Conference on Recent Advances in Fuel Cells and Solar Energy (N-CRAFS '19)
11th-12th May 2019
Department of Mechanical Engineering, NIT Puducherry, Karaikal

MESSAGE FROM THE CHAIRMAN

It is our pleasure to organize the National conference on "Recent Advances in Fuel cell and Solar Energy (N-CRAFS - 2019) during 11th - 12th May, 2019. We are extremely happy in receiving good response for this program. I feel it as a good opportunity of everyone to share their ideas in this common platform. I congratulate all our organizing and hospitality committee involved in taking the responsibilities. My sincere thanks to all the participants who have contributed a lot in success of this event. I hope all the academicians; research scholars and academicians will be greatly benefited from this conference.

Chairman

N-CRAFS – 2019 and

HoD - Mechanical Engineering, NIT Puducherry.

IOP Conf. Series: Earth and Environmental Science **312** (2019) 011001 doi:10.1088/1755-1315/312/1/011001

National Conference on Recent Advances in Fuel Cells and Solar Energy (N-CRAFS '19)
11th-12th May 2019
Department of Mechanical Engineering, NIT Puducherry, Karaikal

About Institute

National Institute of Technology, Puducherry (NITPY) situated in 258 acres near village Poovam in Thiruvettakudy, Karaikal, is one of the thirty-one National Institutes of Technology. NIT Puducherry was sanctioned by the Government of India in 2009, as part of the Eleventh Five-Year Plan (2007–2012).

Notwithstanding the tender age, NITPY is committed to produce effective and responsible technocrats who have the ability to serve the nation on its journey to growth and prosperity. In this process, students have been admitted from the academic year 2010-11 onwards through JEE Mains with 50% of seats reserved for the candidates having residence/native at Puducherry, Karaikal, Mahe, Yanam and the remaining 50% seats are filled on all India basis. The institute provides keen guidance in terms of experienced and expert faculties as well as facilities to sculpt the bright minds into professionally balanced individuals, capable enough to take up challenges of nature, of society and of the marketplace. Presently, 38 faculty members along with 36 non-teaching staff are employed for an overall strength of 586 students.

About Department

Department of Mechanical Engineering was established in July 2014 with an undergraduate (B.Tech) Program. At present, the department also offers Ph. D programme. Dedicated faculty members with expertise in diverse domains of mechanical engineering such as Computer Aided Design, Ergonomics at Work Place, Industrial Safety, Power Plant Structural Analysis, Natural Fibre Composites, Energy Engineering, Renewable Energy, Internal Combustion Engines, Thermal Engineering, Fluid Mechanics, which contributes to the department knowledge hub. Likewise, the non-teaching staff are sound in technical skills and help to strengthen the student's theoretical understanding through lab practices.

Department puts a great effort to help women gain a sense that they belong to the Mechanical Engineering discipline. It also wants to see more women opting to pursue Mechanical Engineering. It facilitates active industrial projects for students such as IGCAR Kalppakam, Pondicherry Power Corporation Limited Karaikal, etc. With this, the students are made aware of current technologies, innovations and the necessity of industrial collaborations as well as addressing the real-world problems. Our department also encourages our students to form a technical/arts/innovation club, develop entrepreneur skills, participate in sports meet, engage in cultural activities and other co-curricular activities to improve their professional skills and groom them as a well rounded off personalities to provide the required leadership skills in fulfilling Make In India manthra to compete with the fastest growing country with available resources. Our department emphasizing on active interdisciplinary research, initiate collaboration with foreign universities and drawing external sources of funding from various funding agencies such as DST-SERB, DST-Inspire Faculty Fellowship, DST-UKIERI with Loughbourgh University, London,UK.

IOP Conf. Series: Earth and Environmental Science **312** (2019) 011001 doi:10.1088/1755-1315/312/1/011001

National Conference on Recent Advances in Fuel Cells and Solar Energy (N-CRAFS '19)
11th-12th May 2019
Department of Mechanical Engineering, NIT Puducherry, Karaikal

DEPARTMENT PROFILE

Faculties in the Department

Prof. K. Sankaranarayanasamy
Professor and Director
Areas of Interests:
Ergonomics at Work place, Industrial Safety, Laser Material Processing, Power plant Structure analysis, Natural Fiber Composites

Dr. Natarajan Sendhil Kumar
Asst. Professor & Head
Areas of Interests:
Solar Thermal, Concentrating Photovoltaic, Integration of High Temperature Solar Thermal and CPV, Heat Transfer and Fluid Flow, Jet Ejectors, Waste Frying Oils, Fuel Cells.

Dr. N. M. Sivaram
Asst. Professor
Areas of Interests:
Production Engineering, Industrial Safety Engineering

Dr. Johnney Mertens
Asst. Professor
Areas of Interests:
Polymer Composites, Material Characterization, Gear

Dr. B M V A Raju
Asst. Professor
Areas of Interests:
Design for Manufacturing & Assembly (DFMA), Environmental Conscious Manufacturing, Assembly Automation, Rapid Prototyping

DST - INSPIRE FACULTY

Dr. Vadivukkarasan M

Areas of Interests:
Liquid Spray Atomization, Hydrodynamic Instability, Thin Liquid films and Interfacial Phenomenon

National Conference on Recent Advances in Fuel Cells and Solar Energy (N-CRAFS '19)
11th-12th May 2019
Department of Mechanical Engineering, NIT Puducherry, Karaikal

Keynote Lecture – 1

Dr. Rui Chen

Professor of Low Carbon Engineering,

Loughborough University, United Kingdom.

BIO SKETCH

Professor Rui Chen, graduated in 1985 from Tianjin University, obtained the PhD at Loughborough in 1997, started academic career in 1999 as a lecturer in Thermofluids and obtained the full professor and chair in Low Carbon Power Engineering in 2009. He was the deputy head of the Department of Aeronautical and Automotive Engineering for research during 2010-2014 and the Head of Thermofluids and Dynamics Research Theme at Loughborough University in 2015.

He has over 30 years' experience of academic research and industrial development and was elected the fellow of Institute of Mechanical Engineers (IMechE) in 2007. He is a member of science board of UK Hydrogen and Fuel Cells Hub and Energy Storage Hub. His research covers both modelling and experimental aspects of energy technologies in hydrogen and fuel cell technologies, avionic thermal management, and combustion kinetics.

He has published over 160 academic research papers and has been awarded the Professional Engineering Publishing Award by the Institute of Mechanical Engineers (IMechE) for the best paper in 2006 and the Third-Place winner of the 2013 Dr Bernard S. Baker Award for Fuel Cell Research. He has led numerous projects, secured over £6M research fund as the principle investigator and further £37M as a co-investigator, in collaborations with many leading industries include Ford, Mahle Powertrains, AVL, JLR, Intelligent Energy, Lotus engineering, Romax Technologies, BAE Systems, Airbus and Boeing etc.

IOP Conf. Series: Earth and Environmental Science **312** (2019) 011001 doi:10.1088/1755-1315/312/1/011001

National Conference on Recent Advances in Fuel Cells and Solar Energy (N-CRAFS '19)
11th-12th May 2019
Department of Mechanical Engineering, NIT Puducherry, Karaikal

Keynote Lecture – 2

Dr. P. Karthikeyan
Professor, Department of Automobile Engineering, PSG Tech., Coimbatore, Tamilnadu.

BIO SKETCH

Dr. P. Karthikeyan, Professor in the Dept. of Automobile Engineering at PSG College of Technology, Coimbatore, INDIA has significantly contributed to fuel cell research in India. He received his PhD Degree at Indian Institute of Technology Madras (IIT-M), Chennai on 2008. After successful completion of his PhD Degree at Indian Institute of Technology Madras (IIT-M), he was received BOYSCAST fellowship (2009-10) from DST, Government of India for conducting Advanced Research Training in the area of "Fuel Cells - Water Management in PEM Fuel Cells" in USA. In part of his research work on Post - Doctoral Fellowship, he also visited Fuel Cell Dynamics and Diagnostics Laboratory, Pennsylvania State University, Pennsylvania, USA and Electro Chemical Engineering and Membrane Science Lab at School of Engineering, Vanderbilt University, Nashville, Tennessee, USA during 2009-10. He had also visited High Temperature Super and Semi-Conductor Laboratory, University of Houston, Houston, Texas & College of Engineering, University of Berkeley, San Francisco, California and Stanford University, San Francisco, USA. Recently, he visited CINVESTAV-IPN, Saltillo, Mexico on 2013 & 2015 under Indo-Mexican Bilateral Project. He received various National and International Research Projects sponsored by various government funded agencies such as DST, DST-UKIERI, AICTE-RPS, DST-CONACYT and DST-EPSRC summing up to the tune of 3.5 crores and submitted the proposal in the range of 13.5 crores; also Head for Research activity in Fuel cell Technology @ PSG CT, Coimbatore. He had authored more than 44 research papers in reputed national and international journal, 19 peer reviewed international conferences, and filed 2 technical patents (Invention Disclosure No.: D2010-65). Also, set up fuel cell laboratory in University of Toledo, Toledo, OHIO, USA & PSGCT, Coimbatore, India. He had research coordination with various universities such as the Centre for Research and Advanced Studies (CINVESTAV-IPN), Unidad Saltillo-Mexico, University of Toledo, OHIO, USA, Loughborough University, Loughborough, Leicestershire UK and Curtin University, Australia. His research area includes, thermal and water management issues on scaling and stacking up of fuel cell systems, studies on PEMFCs with porous inserts on conventional flow field, MEAs with platinum alloy catalysts and non-noble metal catalysts using Nano-materials, optimization of flow channel design and various operating parameters for performance enhancement of PEMFCs, development of indigenous fuel cell stacks for automobile and stationary applications, solving various issues for system development and system integration, design and development of concentrating photovoltaic (CPV) for solar power generation, and design and development of solar energy aided electrolyzer for hydrogen production.

IOP Conf. Series: Earth and Environmental Science **312** (2019) 011001 doi:10.1088/1755-1315/312/1/011001

National Conference on Recent Advances in Fuel Cells and Solar Energy (N-CRAFS '19)
11th-12th May 2019
Department of Mechanical Engineering, NIT Puducherry, Karaikal

Keynote Lecture – 3

Dr. Ravi Kumar K

Assistant Professor, Centre for Energy Studies,

Indian Institute of Technology Delhi, New Delhi, India.

BIO SKETCH

Dr. Ravi Kumar received his Ph.D. from Indian Institute of Technology (IIT) Madras in 2012. At IIT Madras, he worked on investigation of energy efficient receiver for line-focus concentrating solar power (CSP) systems such as parabolic trough collector and linear Fresnel reflector for process heating and power generation. He also worked on solar field design, feasibility study of large scale CSP plants and multi-stage solar desalination system and so on.
After his PhD, he worked as a postdoctoral fellow at ABB Corporate Research Center, Bangalore for a year. Then, he joined as an Associate Scientist in the same organization on July 2013 and promoted as Scientist in 2014. His research work at ABB Corporate Research Center includes solar aided thermal power generation, solar heat for process heating applications and multi-physics modelling of secondary steel making (continuous casting) process & measurement devices. Since November 2016, he has been part of Centre for Energy Studies, Indian Institute of Technology Delhi as an Assistant Professor.

Research Interests:

Dr. Ravi Kumar focused on design and development of efficient and cost effective solar thermal energy systems for various applications like process heating, power generation, desalination and so on. He also focused on development of thermal energy storage technologies for medium and high temperature applications. His research area includes:

- Solar thermal power generation
- Solar process heat applications
- Efficient receivers for CSP systems
- Energy storage
- Hydrogen storage and
- Forecasting of solar radiation.

IOP Conf. Series: Earth and Environmental Science **312** (2019) 011001 doi:10.1088/1755-1315/312/1/011001

National Conference on Recent Advances in Fuel Cells and Solar Energy (N-CRAFS '19)
11th-12th May 2019
Department of Mechanical Engineering, NIT Puducherry, Karaikal

Keynote Lecture- 4

Dr. Sendhil Kumar Natarajan

Assistant Professor & Head, Department of Mechanical Engineering

National Institute of Technology Puducherry, Karaikal.

BIOSKETCH

Dr. Sendhil Kumar Natarajan, is working as an Assistant Professor & Head in the Department of Mechanical Engineering, NIT Puducherry. He did B.Tech in Mechanical Engineering and M.Tech., in Energy Technology in Pondicherry Engineering College in the year of 2001 and 2003. Later, he joined as a Research scholar, at IIT Madras, under Half Time Teaching and Research Assistantship (HTRA Scheme – Full time), MHRD scholarship and obtained his Ph.D. in area of solar energy in the year of 2008. After his Ph.D., he joined as Senior R&D Engineer, at CSEM-UAE, Ras Al Khaimah, UAE. Later, he joined as a Post-Doctoral Research Fellowship at Heriot-Watt University, Edinburgh, Scotland, United Kingdom under European Union Project. He got young faculty award and outstanding researcher award in the field of Solar Thermal. He published almost 16 International Journals papers in SCI and 30 National and International conferences in the area of energy. Two of his research publications, cited as hottest article in the science direct. Apart from this, three of his International journals about receivers got more than 100 citations. Total citations of 767. He has two funded projects. One is from DST-SERB as a PI for Concentrating Solar Systems, and other one is from DST-UKIERI as a Co-PI for fuel cell research. His research area is Solar Thermal system and CSP system, CPV, Integration of high temperature solar thermal systems, Heat transfer and fluid flow and Fuel cells.

IOP Conf. Series: Earth and Environmental Science **312** (2019) 011001 doi:10.1088/1755-1315/312/1/011001

National Conference on Recent Advances in Fuel Cells and Solar Energy (N-CRAFS '19)
11th-12th May 2019
Department of Mechanical Engineering, NIT Puducherry, Karaikal

SESSION CHAIR

Dr. R. THUNDIL KARUPPA RAJ

Dr. R Thundil Karuppa Raj has completed his doctoral degree from I C Engines laboratory of Mechanical Department of IIT Madras under the guidance of Prof. V.Ganesan during June 2007. He has his master level degree graduation from NIT Trichy in the area of Design and Production of Thermal Power Equipments in First Class with Distinction. He is basically a mechanical engineering graduate from Bharathiar University during May 1998. He also has completed his Master of Business Administration from Madurai Kamaraj University with first class.

He has nearly 21 years of working experience which includes 13 and half years in teaching, 4 years in research and 3 and half year in industry. He worked as Technical Leader in the CFD division of Mechanical Engineering Services of Robert Bosch Engineering and Business Solutions Limited, Coimbatore. Right now, he is working as Professor and Head of the department of Automotive Engineering associated with the School of Mechanical Engineering, VIT, Vellore. He has also served as the Programme Chair for B.Tech (Mechanical with Specialization in Energy Engineering) programme and presented the programme for ABET, USA and NAAC accreditation during his tenure.

To his research interest, he has published research articles in high impact factor Thomson Reuters journals like Fuel, Energy, Applied Thermal Engineering, International Journal of Heat and Mass Transfer, Energy Conversion and Management, ASME Journal of Thermal Sciences and Engineering Applications, International Journal of Hydrogen Energy, Renewable Energy etc. He has nearly 90 publications which include more than 50 Scopus indexed international journals, 30 international conferences and 10 national conferences. He has produced five Ph.D. students in the area of Internal Combustion Engines using both experimental and numerical techniques and currently guiding 6 more Ph.D. students. He is serving as reviewer for many international peer reviewed scopus indexed journals. He is editorial member in the journals of Applied Sciences, Trends in Applied Sciences Research, Singapore Journal of Scientific Research. He is also a fellow membership of International Science Congress association, Society of Automotive Engineers, Association of Energy Engineers. He has produced youtube lectures for the subject Engineering Thermodynamics. His area of interest is thermodynamics, thermal and combustion engineering using commercial CFD tools.

IOP Conf. Series: Earth and Environmental Science **312** (2019) 011002 doi:10.1088/1755-1315/312/1/011002

National Conference on Recent Advances in Fuel Cells and Solar Energy (N-CRAFS '19)
11th-12th May 2019
Department of Mechanical Engineering, NIT Puducherry, Karaikal

Sponsors

STATE DESIGNATED AGENCY
RENEWABLE ENERGY AGENCY, PUDUCHERRY (REAP) (U.T. of Puducherry)

To promote the concept of Renewable Energy Sources and to implement the schemes on Renewable Energy in the Union Territory of Puducherry, Renewable Energy Agency, Puducherry was established also it is Society Registration Act in December 2005. Renewable Energy Agency, Puducherry is the State Nodal Agency of Ministry of New and Renewable Energy, New Delhi and State Designated Agency of Bureau of Energy Efficiency, Ministry of Power, Government of India, New Delhi. Besides it is the implementing agency for various Renewable Energy and Energy Conservation projects in the Union Territory of Puducherry.

VCollab

Visual Collaboration Technologies Inc, VCollab, USA, (www.vcollab.com), founded in year 2000, head quartered in Troy, Michigan. The Company with a vision to provide cutting edge Visual Collaboration Solutions for the Manufacturing Industry. VCollab also has offices in Dallas, Texas, and Bangalore, India with worldwide distribution network.

VCollab has rich & diverse experiences in the CAD/CAM/CAE disciplines and expertise in 3D visualization, Virtual Reality & Collaboration Technologies. VCollab much early identified the main bottleneck in CAD/CAM/CAE, PLM, SLM & VR effectiveness is cross-discipline Visualization & Collaboration and need for a visual collaboration platform for the product development using Engineering Simulation data. Also, the number of simulation solvers and associated formats, integrated HPC infrastructures, large size of the simulation results files are making it very difficult to process, share and communicate the large amounts of product performance data across global centers.

VCollab provides Solutions in CAD/CAM/CAE Visualization, Presentation, Documentation & Collaboration through the VCollab suite of Products which includes VCollab Pro, VMoveCAD, VMove CAE, VCollab Presenter etc. The Products helps to maximize the utilization of the CAD & CAE investments and streamline the 3D product performance data communications across the global centers, suppliers and cross functional teams of the product development companies.

The VCollab solution today is known for its unique capabilities to improve the productivity of simulation teams and also to improve the collaboration between simulation experts and other engineers involved in the product development process, making access to product performance data more available whenever needed.

Major global leading automotive OEMs and Aero OEMs are the early adopters of VCollab technology & to name a few the clientele includes General Motors, Ford, Toyota, Suzuki, Chrysler, Boeing, Airbus, Honeywell, Pratt & Whitney, Thales, Ashok Leyland, DRDO Defense Research labs (ADA/ADE/NAL/RCI/RDE/ASL etc., Scientific labs, US, Canadian, Japanese, Indian Academic Institutions including IIT M, PSG, MSR & more.

Greenserve Energy
Managment Solutions

GREENPRO
ENERGY TECHNOLOGIES

IOP Conf. Series: Earth and Environmental Science **312** (2019) 011003 doi:10.1088/1755-1315/312/1/011003

National Conference on Recent Advances in Fuel Cells and Solar Energy (N-CRAFS '19)
11th-12th May 2019
Department of Mechanical Engineering, NIT Puducherry, Karaikal

List of Committees

Chief Patron

Dr.K.Sankaranarayanasamy

Director

National Institute of Technology Puducherry, Karaikal

Patron

Dr.G.Seetharaman

Registrar

National Institute of Technology Puducherry, Karaikal

Chairman

Dr.N.Sendhil Kumar

Assistant Professor and Head

Department of Mechanical Engineering,

National Institute of Technology Puducherry, Karaikal

Organizing Committee

1. Dr.N.M.Sivaram, Assistant Professor, NIT Puducherry
2. Dr.A.Johnney Mertens, Assistant Professor, NIT Puducherry
3. Dr.M.V.A.Raju Bahubalendruni, Assistant Professor, NIT Puducherry
4. Dr.M.Vadivukkarasan, DST INSPIRE Faculty, NIT Puducherry
5. Dr.Narendran Rajagopalan, Assistant Professor, NIT Puducherry
6. Dr.S.Madappa, Assistant Professor, NIT Puducherry
7. Dr.S.Babu, Students Activity and Sports Officer, NIT Puducherry

IOP Conf. Series: Earth and Environmental Science **312** (2019) 011003 doi:10.1088/1755-1315/312/1/011003

National Conference on Recent Advances in Fuel Cells and Solar Energy (N-CRAFS '19)
11th-12th May 2019
Department of Mechanical Engineering, NIT Puducherry, Karaikal

National Advisory Committee

1. Prof. G.N. Tiwari, Rtd. Professor, IIT Delhi, New Delhii
2. Prof. K.S. Reddy, IIT Madras, Chennai
3. Prof. S. Iniyan, Anna University, CEG campus, Chennai.
4. Prof. S. Murugan, NIT-Rourkela.
5. Dr.Raj Ganesh S.Pala, IIT Kanpur
6. Dr.A.S.K.Sinha, IIT(BHU), Varanasi
7. Dr.Om Prakash Singh, , IIT(BHU), Varanasi
8. Dr.Anil Kumar Emadabathuni, IIT Tirupati, Tirupati
9. Dr.K.Ravi Kumar, IIT Delhi, New Delhi
10. Dr.Suresh M V J J, GE Power, Bangalore
11. Dr.Devanuri Jaya Krishna, NIT Warangal
12. Dr.Tangellapalli Srinivas, NIT Jalandhar
13. Dr.P.Karthikeyan, PSG Tech., Coimbatore
14. Dr.D.Ragupathy, NIT Puducherry
15. Dr.R.Anand, NIT Trichy
16. Dr.Narendra Laxman Gajbhiye, NIT Bhopal
17. Dr.Arumuga Perumal D, NIT Surathkal
18. Dr.Veershetty Gumtapure, NIT Surathkal
19. Dr.Nabin Sarmah, University of Tezpur, Assam
20. Dr.A.Sreekumar, CGET, Pondicherry University
21. Dr. Ranjan Das, IIT Ropar.

IOP Conf. Series: Earth and Environmental Science **312** (2019) 011004 doi:10.1088/1755-1315/312/1/011004

National Conference on Recent Advances in Fuel Cells and Solar Energy (N-CRAFS '19)
11th-12th May 2019
Department of Mechanical Engineering, NIT Puducherry, Karaikal

CONFERENCE PHOTOGRAPHS

IOP Conf. Series: Earth and Environmental Science **312** (2019) 011004 doi:10.1088/1755-1315/312/1/011004

National Conference on Recent Advances in Fuel Cells and Solar Energy (N-CRAFS '19)
11th-12th May 2019
Department of Mechanical Engineering, NIT Puducherry, Karaikal

IOP Conf. Series: Earth and Environmental Science **312** (2019) 011005 doi:10.1088/1755-1315/312/1/011005

Peer review statement

All papers published in this volume of *IOP Conference Series: Earth and Environmental Science* have been peer reviewed through processes administered by the proceedings Editors. Reviews were conducted by expert referees to the professional and scientific standards expected of a proceedings journal published by IOP Publishing.

Design and development of dual axis sun tracking system for floating PV plant

Sendhil Kumar Natarajan*, Aman Kumar, Raqeeb Mohamed, Rathna R, Swaraj Mondal and Subbarama Kousik Suraparaju

Department of Mechanical Engineering,
National Institute of Technology – Puducherry, Karaikal 609609
U.T of Puducherry, INDIA.

Email: sendhil80@nitpy.ac.in, drsendhil1980iitmuk@gmail.com

Abstract. Increase in need of non-renewable energy sources has directed to the installation of PV plants in many areas. But considering the availability of land and the charges, the PV plant can be installed on water bodies like dams, reservoirs, lake, waste water treatment plants, canals etc. Floating type PV panels has a number of advantages compared to the land-based PV plant; its efficiency can be increased by effective cooling and tracking mechanism. The installation in water bodies offers additional advantages to the aquatic environment like reducing the evaporation rate of the water body, reduces unwanted plant growth in water body; improving the water quality. In this paper, detailed description of floating and the rotating structures in the plant. For higher efficiency the solar panel should be tracking in two axes, that is, tracking of azimuthal and altitude axis. Hence, dual axis tracking system is adopted and the mechanism is explained. Torque calculations are made for selection of correct stepper motor and hybrid linear actuator. Different materials for the platform are compared. A prototype made of wood was design and developed in NIT Puducherry.

Keywords: Solar, Photovoltaics, Floating, Dual axis tracking.

1. Introduction

In the current scenario, the renewable energy sources are substituting the non-renewable energy sources and also fulfilling the energy demands all over the world. Among all the available renewable energy sources, solar energy is the most predominant and most promising energy alternative due to its ubiquity and also it is freely available all over the world. Solar thermal and Solar Photovoltaics (PV) are the applications of the solar energy where solar PV is the most common application of solar energy. Many developments have been evolved int eh realm of solar PV cells from early 1800s to present day first [1–7]. The establishment of solar PV panels has the prerequisite difficulty of more land which will always be an expensive. The immediate solution for this economic glitch is installing the solar PV on the water body which can reduce the acquirement of land with high price. Floating solar PV has its own advantages when compared to ground mounted solar PV or roof mounted solar PV panels like more efficiency because of cooling effect of water. Also, this floating solar PV enables the decrease in water evaporation from the water bodies and prevents the water pollution by restricting the formation of algae etc. The platforms used for floating solar PVs are completely recyclable and withstand to effect of ultraviolet rays, corrosion etc., by using high-density polyethylene. So, with these advantages floating solar PV systems became very logical substitute for existing solar PV systems by utilizing the waterbodies and also these systems are economically viable compared to other solar PV systems [8–12]. The floating solar PV systems can be more efficient if they are equipped with solar tracing system. This solar tracking system enables the floating PV to find the Sun's position and tilt according to it. These tracking and tilting can be done by various methods. Also, this tracking of different types like single axis and double axis tracking with open loop and closed loop. These systems are mainly done by MATLAB or Arduino programming to track the exact position of sun [13–22]. In the current paper, we designed a floating solar PV system with some selection criteria for materials and tracked the sun for double axis i.e, azimuthal and elevation tracking system with the Arduino programming.

2. Floating PV

The floating PV installed on as floating system on the water bodies is the current trending technology in solar PV system which reduces the economic difficulty of installing solar PV system on the expensive and valuable land. This floating system arrangement consists of floating and rotating platforms which solar PV panels are mounted and connected to tracking system through cables.

i. **Floating platform** – A floating platform should have adequate buoyancy enough to float itself as well as with a substantial load. It should be designed in such a way that it accommodates no. of solar cells connected in series and parallel combination and also along with tracking system.
ii. **Rotating platform** – It is basically placed above the floating platform; the PV panels are actually placed in this and sits on the latter with rollers. This has gear at its circumference and it is rotated by motor.
iii. **Mooring system** – A mooring system typically indicates to any perpetual structure to which a container may be protected. In this floating solar PV system, this mooring system safeguards the panels from turning or floating away by keeping in the same position.
iv. **Solar PV module** – The crystalline PV modules have proved to be efficient compared to the amorphous, so we are using crystalline type in our project. The commercially available panels have aluminum frames, which on years starts to corrode so alternatives such as polymers can be adopted.
v. **Cables and Connectors** – The generated electricity from the floating system is transferred to land through cables and connectors. In this case, we used floating cbles for transfer of generated electricity.

2.1. Design of floating platform.

The primary requirements of a floating platform are:

1. Efficient – the consumption of power by the motor is reduced by use of worm gear drive and due to high transmission ratio, providing high torque with the use of frictionless bearing the efficiency can be increased.
2. Economical – by adopting plastic gears the cost is reduced compared to the steel frames.
3. Low manufacturing Cost – simplicity of the design and the low cost for the manufacturing process.
4. Maintenance Free – plastic gears with the self-lubricating property and corrosion resistant makes it maintenance free
5. Less Installation Resources – modular structure and easy assembling of components make it portable to transfer and also requires less man power and assembling equipment.
6. Lesser Resistance towards wind force.

We have chosen hexagonal structure because, when arranging the hexagons together the empty spaces or gap is nil and compared to other shapes the material required in this case is less. This reduces the overall cost of the project. In the platform slots are made in each side for the wind flow throughout the wind can destroy the stability of the plant.

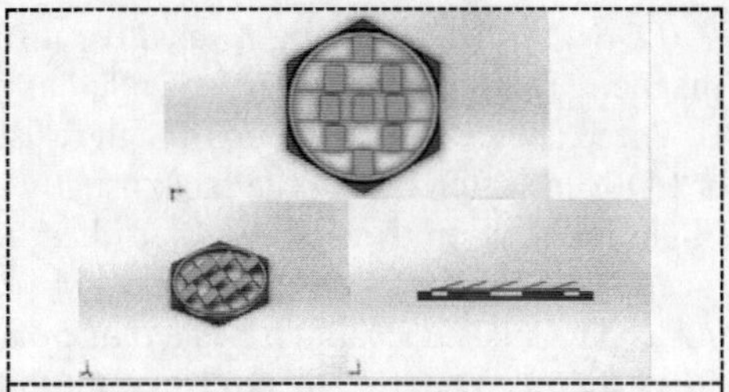

Figure1. Structure of the platform

2.2. Design of rotating platform.

The rotating platform is placed above the floating platform. It is joined with rollers in the bottom for the easy movement. It has spur gear tooth profile in its circumference which is in turn rotated by another spur gear with smaller radius which is connected in parallel with stepper motor using shafts. It has another structure fixed to it using a hybrid linear actuator and the panel are kept in this structure. The panels are kept in such a way that the shadow of one panel does not fall on the other.

2.3. *Arduino Program for Sun Tracking Mechanism.*

There are two types of tracking,

1. The Sun's position based on the collected data can be fed to the Arduino and the panel position is predetermined.
2. The other one is by using LDR (Light Dependent Resistor), the sun's position can be calculated instantly and this can be fed to Arduino.

Pin Diagram of Arduino: Arduino Uno is the representative illustration of Arduino board. It comprises of ATmega328- a 28 pin microcontrollers. This is made up of 14 digital input/output pins of which 6 can be used as Pulse Width Modulation (PWM) outputs, 6 analog inputs, a 16 MHz crystal oscillator, a Universal Serial Bus (USB) connection, a power jack, an In-Circuit Serial Programming (ICSP) header, and a reset button

Power Jack: It is powered either by pc through an USB or by an external source such as battery or adapter.

Digital Inputs: It comprises of 14 digital inputs/output (I/O) pins, in which each pin provide or take up current of 40mA. Some of them have special functions such as pins 0 and 1, which act as Rx and Tx respectively. For serial communication, pins 2 and 3-which are external interrupts and subsequently pins 3,5,6,9,11 provides PWM output and pin 13 where LED is connected.

Analog inputs: It has 6 analog input/output pins, in which each pin is able to provide a resolution of 10 bits.

ARef: It offers reference to the analog inputs

Reset: It is used to reset the microcontroller.

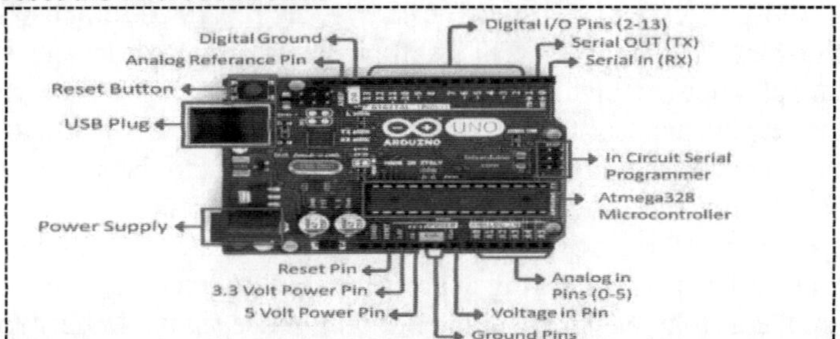

Figure2. Arduino Pin Diagram
(http://oyoana.com/wiki/index.php/Intro_to_the_arduinoooo)

2.3.1. Light Dependent Resistor (LDR): An LDR is a light sensitive instrument whose resistivity is a function of the incident electromagnetic radiation and it is also called as photo conductive cells, photo conductors, or simply photocells. These devices are made up of high resistive semiconductor materials. There are various indications for LDR, in which one of the most usually used symbols is shown in the figure below. The arrow shows light incident on the device. As said earlier, LDRs are light dependent devices whose resistance is declined when light falls on them and amplified when there is no light or in darkness. This resistance is called as dark resistance and it has a range of 1012 Ω. If a constant voltage is applied and magnitude of incident light is increased then the current starts increasing.

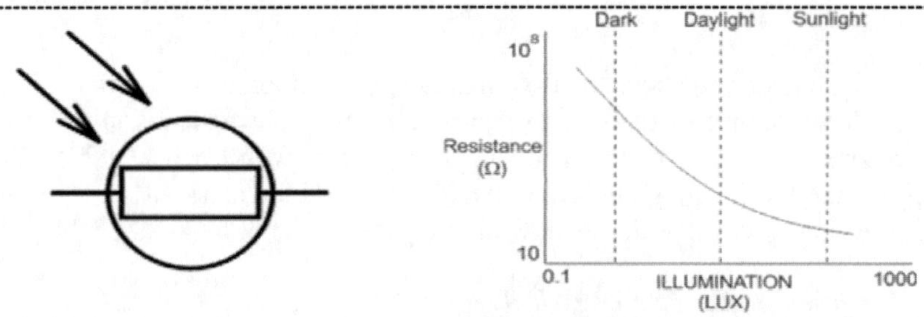

Figure3. LDR and graph shows resistance vs. illumination curve for a particular LDR.

(Source: https://www.electrical4u.com/light-dependent-resistor-ldr-working-principle-of-ldr)

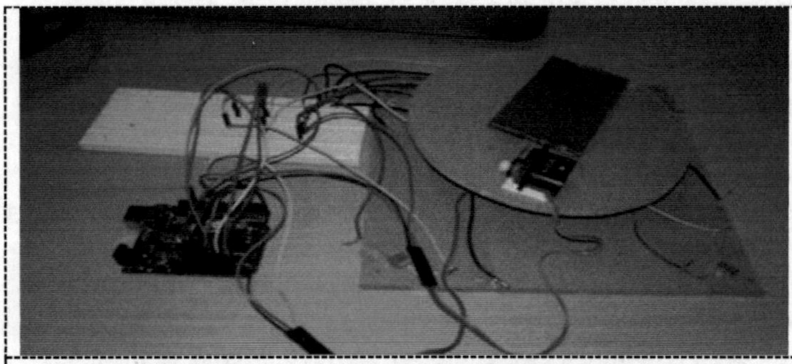

Figure4. Circuit connection for the program

2.4. Working of dual axis sun tracking mechanism

Initially the panel is fixed approximately at 11° (since the latitude of Karaikal is 10.9254° N). The sun's position is tracked using LDR sensors. Here we have employed dual axis tracking system, azimuthal and elevation tracking. 4 LDR sensors are used. The sensors are separated from each other, so that they are hidden from each other. They are placed in small piece of cardboard on four sides and separated using cardboard. As we all know as the light falls on LDR the resistance starts decreasing, the value of resistance from all the sensors are sent to the Arduino and the program is made in such a way that it compares the data and generates an output. The Arduino sends signal to the stepper motor. The stepper motor is connected with the driver gear. This driver gear drives the rotating platform which itself is a spur gear. As the stepper motor turns one step it rotates the driver gear to a particular angle and this in turn rotates the rotating platform. This tracking enables the plant to track sun every hour, that is, tracks the sun as it moves from east to west. The Arduino not only sends signal to the stepper motor but also to the hybrid linear actuator. The linear actuator connects the rotating platform and the base where the panels are kept. Hence according to the signal, the linear actuator lifts the base upwards. We all know that, the sun's declination angle varies from -23.5° to 23.5° no longer the sun path remains the same so, this elevation tracking can track the sun on daily basis.

IOP Conf. Series: Earth and Environmental Science **312** (2019) 012001 doi:10.1088/1755-1315/312/1/012001

2.5. Torque calculation

For azimuthal tracking of PV panel, we are adopting worm gears.
Radius of spur gear (R) = 0.3 m
Mass of the total rotating platform (m) = 30 kg
Torque required rotate the worm gear (T) = I × α
Moment of Inertia,
$I = \frac{1}{2} \times m \times R \times R$
$= \frac{1}{2} \times 30 \times 0.3 \times 0.3$
$= 1.35\ m^2$
To rotate 180° in 12 hours, ω = 0.25 rad/sec
$\alpha = \omega / t$
$= 0.25 / 1$
$= 0.25\ rad/sec^2$
Required torque,
$T = 1.35 \times 0.25$
$= 0.3375$ N-m
Selection of stepper motor based on 0.3375 N-m torque

2.6. Selection of stepper motor

Stepper motor selected is **OMHT17-275** (High Torque Stepper Motor)
FRAME SIZE: NEMA 17
STEP ANGLE: 1.8°

2.6.1. Features and Benefits:

- High Torque
- Cost effective
- Up to 0.44 N-m Holding Torque

Table 1. Standard Motor Specifications.

Step Angle (°)	1.8
Frame Size	NEMA 17
Body Length (m)	0.048
Current (A)	1.7
Holding Torque (N-m)	0.44
Resistance (Ω)	1.7
Voltage (V)	2.8
Rotor Inertia (g-cm^2)	82
Number of leads	8
Connection	Parallel
Weight (kg)	0.357
Branding	Power Step High Torque

Maximum power rating of the stepper motor = V × I

$= 2.8 \times 1.7$
$\mathbf{= 4.76W}$

2.6.2. Torque Calculation for Hybrid Linear Actuator.

For elevation axis tracking hybrid linear actuators connected to motor are,

IOP Conf. Series: Earth and Environmental Science **312** (2019) 012001 doi:10.1088/1755-1315/312/1/012001

$$F = \frac{Q(2\pi\mu r - P)}{(2\pi r + \mu P)}$$

Where,
μ = Frictional coefficient
r = Pitch circle radius of lead screw
Q = Perpendicular load acting
P = Pitch of lead screw
F = Perpendicular force acting
Expected weight of the rotating platform = 30 kg (including wind force)
Q = 15 kg = 147N
μ = 0.35
r = 2.97mm

$$F = \frac{147[(2 \times 3.14 \times 0.35 \times 0.00297) - 0.000305]}{[2 \times 3.14 \times 0.00297) + 0.35 \times 0.000305]}$$

F = 162.833N
Torque = 5.478 kg-cm

2.7. Selection of Material for The Platform
A comparison is made between the plastic materials and material is selected. The three materials opted for comparison were:
1. POLYETHER ETHER KETONE (PEEK)
2. NYLON 6/6 GF-30
3. HIGH DENSITY POLYETHYLENE (HDPE)

Table2. Comparison of Properties

PHYSICAL PROPERTIES			
Property	**PEEK**	**HDPE**	**NYLON 6,6**
Density (g/cc)	1.49	0.94-1.0	1.36
Moisture Absorption at Equilibrium (%)	0.2	0.01-0.03	1.6
MECHANICAL PROPERTIES			
Tensile Strength at Break (MPa)	156	113	140
Elongation at Break (%)	3.0	1.0-2.9	5.0
Flexural Strength (MPa)	250	215	315
Flexural Modulus (GPa)	10.0	11.7	10.8

2.8. Advantages of High-Density Polyethylene
Based on the above comparison high density polyethylene (HDPE) is selected for the below mentioned reason.
1. Super buoyancy (lighter than water) – The density is in the range of 930 to 970 kg/m³ which makes HDPE lighter than water.
2. Zero corrosion – Oxygen, salt and water have no effect on HDPE. It is not affected by the ravages of time. HDPE offers zero corrosion and zero maintenance.
3. Resistance to marine growth – no anti fouling needed.
4. High impact resistance – Its viscous and elastic characteristics prevent it from deformation. It is tough and flexible.
5. Black strength (UV resistant) – One can use black HDPE because of its resistance to Ultraviolet (UV) rays. Black HDPE is mixed with carbon, which avoids the material to turn out to be as brittle due to climatic conditions.
6. Production flexibility – Can be constructed by plate welding, giving great flexibility.

7. Easy of repair.
8. Low carbon footprint – The carbon footprints of HDPE making is 5 times lesser than Aluminium in terms of durability. HDPE is indestructible; it may stay up to 40 years.
9. Durability: 100% recyclable.

2.9. Selection of Reinforcement

Reinforcing the structure increases the physical properties. There is much reinforcement available in the market; some of them are discussed below.

i. **Carbon – fiber reinforcement**
   a) Very light results in light weight structures
   b) Impact strength
   c) Expensive, the price will be decreased when there is advancement in the production technology
   d) Thermal expansion is zero
   e) Temperature resistant
   f) Corrosion resistant
   g) Very stiff and brittle
   h) High strength to weight ratio

ii. **Natural – fiber reinforcement**
   a) Low density
   b) Less expensive and Biodegradable
   c) Tensile modulus and flexural modulus
   d) Renewable resources
   e) Moisture absorption is more
   f) Coupling between natural fiber and polymer is difficult

iii. **Glass – fiber reinforcement**
   a) High modulus of elasticity and Higher elongation at break
   b) It is cheap
   c) Good chemical resistant to acids and solvents
   d) Electrically insulated
   e) Relatively Higher temperature resistant
   f) Low moisture absorption
   g) Higher strength to weight ratio
   h) It's not stiff, but this can overcome by designing sandwich structure, light weight core can be sandwiched between two glass skins and then the end is stiff.

Based on the above comparison on properties and reliability glass - fiber reinforcement is used.

2.10. Design of Gears

Spur gears are selected and their design is as follows:
Pressure angle is chosen as 20° (mostly preferred for plastic gears)
Pitch Diameter of the gear, $D = 60$ cm
Taking the no. of teeth, $N = 120$
Diametral pitch $= N / D$
$= 120 / 60$
$= 2$ / cm
Addendum $= 1 / P$
$= 1 / 2 = 0.5$cm
Whole depth = addendum + dedendum
$h_t = 2.33 / P$
$= 2.33 / 2 = 1.165$ cm
Standard circular tooth thickness,
$C_t = \pi / 2P = 0.785$ cm

IOP Conf. Series: Earth and Environmental Science **312** (2019) 012001 doi:10.1088/1755-1315/312/1/012001

Root diameter, dr = D - 2 h_t
=60 – (2 * 1.165)
=57.67 cm
Module, m = D / N
=0.5 cm
Tangential Force = 2000× T/d
= 2000 × 0.3375 / 0.6
=1.125 kN = 11036 kgf
Calculation for pinion
Pitch diameter = 6 cm
Pressure angle = 20°
No. of teeth = 12

$$\text{Bending Stress } (\sigma_b) = \frac{F}{(m \times y \times b)} = \frac{1125}{(30 \times 0.7658 \times 3.5)} = 13.99 \text{ N/m}$$

By considering all the design calculations, using selected materials and components driven by Arduino program, the floating PV is developed as shown in figure,

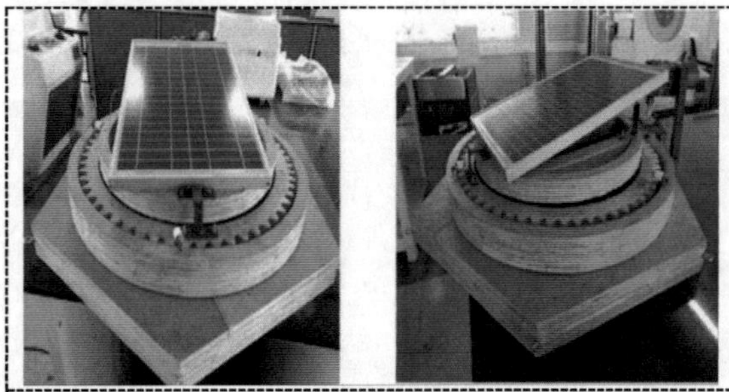

Figure5. Developed prototype in front and side views

Figure6. Panel in lifted position

3. Conclusions

1. From the several proposed design for floating platform, the hexagonal structure was found to be efficient, cost effective and easy installation.
2. Dual-axis sun tracking mechanism was explained, the azimuthal tracking was controlled by stepper motor and the elevation tracking was controlled by hybrid linear actuator.
3. The selection of the stepper motor and hybrid linear actuator was made by calculating the torque produced by the structure.
4. An Arduino program was developed to control Arduino which in turn controls the stepper motor and the hybrid linear actuator.
5. From Polyether ether ketone, nylon 6,6 and High-density polyethylene (HDPE) materials, HDPE was found to reliable with 10% of glass fiber reinforcement as it increases the strength of the structure.
6. The gear parameters of spur gear were calculated and the design was made according to the calculation.
7. Prototype with optimum tracking was fabricated.

4. Future Scope

1. The efficiency of the floating PV plant can be compared with the land-based PV plant experimentally for some duration and results and graphs can be made based on the observations made.

IOP Conf. Series: Earth and Environmental Science **312** (2019) 012001 doi:10.1088/1755-1315/312/1/012001

2. The original model with high density polyethylene can be made and tested.
3. The project can further be expanded to make it work in saltwater lagoons where waves would be present by using dampers or secondary reflectors.

References

[1] History of Solar Cells https://www.solarpowerauthority.com/a-history-of-solar-cells/

[2] Loferski J J 1993 The First Forty Years: *Prog. PHOTOVOLTAICS Res. Appl.* **1** 67–78

[3] Shirland F A 1966 THE HISTORY , DESIGN , FABRICATION AND PERFORMANCE OF CdS THIN FILM SOLAR CELLS *Adv. Energy Convers.* **6** 201–22

[4] Green M A 2005 Silicon Photovoltaic Modules : *Prog. PHOTOVOLTAICS Res. Appl.* **13** 447–55

[5] BERNÈDE J C 2008 ORGANIC PHOTOVOLTAIC CELLS : HISTORY , PRINCIPLE AND TECHNIQUES *J. Chil. Chem. Soc* **53** 1549–64

[6] Parida B, Iniyan S and Goic R 2011 A review of solar photovoltaic technologies *Renew. Sustain. Energy Rev.* **15** 1625–36

[7] Grätzel M 2003 Dye-sensitized solar cells *J. Photochem. Photobiol. C Photochem. Rev.* **4** 145–53

[8] Cazzaniga R, Cicu M, Rosa-clot M, Rosa-clot P, Tina G M and Ventura C 2017 Floating photovoltaic plants : Performance analysis and design solutions *Renew. Sustain. Energy Rev.* 1–12

[9] Lee Y, Joo H and Yoon S 2014 Design and installation of floating type photovoltaic energy generation system using FRP members *Sol. ENERGY* **108** 13–27

[10] Trapani K and Millar D L 2014 The thin film flexible floating PV (T3F-PV) array : The concept and development of the prototype *Renew. Energy* **71** 43–50

[11] Sahu A, Yadav N and Sudhakar K 2016 Floating photovoltaic power plant : A review *Renew. Sustain. Energy Rev.* **66** 815–24

[12] Ho C J, Chou W and Lai C 2015 Thermal and electrical performance of a water-surface floating PV integrated with a water-saturated MEPCM layer *Energy Convers. Manag.* **89** 862–72

[13] Ramya P and Ananth R 2016 The Implementation of Solar Tracker Using Arduino With Servomotor *Int. Res. J. Eng. Technol.* **3** 2395–56

[14] Kaur T, Mahajan S, Verma S, Priyanka and Gambhir J 2017 Arduino based low cost active dual axis solar tracker *1st IEEE Int. Conf. Power Electron. Intell. Control Energy Syst. ICPEICES 2016* 2–6

[15] Pradeep K P J, Reddy K S P, Mouli C C and Raju K N 2014 Development of Dual-Axis Solar Tracking using Arduino with Lab VIEW *Int. J. Eng. Res. Appl.* **17** 3–6

[16] Amaize P, Uzairue S and Sanni T F 2018 Arduino Based Solar Tracking System For Energy Improvement Of Pv Solar Panel . *Proc. Int. Conf. Ind. Eng. Oper. Manag. Washingt. DC, USA, Sept. 27-29, 2018* 2469–78

[17] Mishra J, Thakur R and Deep A 2017 Arduino based Dual Axis Smart Solar Tracker *Int. J. Adv. Eng. Manag. Sci.* **3** 532–5

[18] Nadkarni N S, Raju B, Fernandes M X and Kimberly P 2017 Solar Tracking System Using Arduino *Int. J. Sci. Res. Dev.* **5** 2321–0613

[19] Lu S, Dai R, Zhang G and Wang Q 2018 Investigation of Street Lamp With Automatic Solar Tracking System *J. Sol. Energy Eng.* **140** 061002

[20] Rambhowan Y and Oree V 2014 Improving the dual-axis solar tracking system efficiency via drive power consumption optimization *Appl. Sol. Energy* **50** 74–80

[21] Ferdaus R A, Mohammed M A, Rahman S, Salehin S and Mannan M A 2014 Energy Efficient Hybrid Dual Axis Solar Tracking System *J. Renew. Energy* **2014** 1–12

[22] Racharla S and Rajan K 2017 Solar tracking system–a review *Int. J. Sustain. Eng.* **10** 72–81

Nusselt number and friction factor behaviour of circular wavy transverse ribs artificially roughened solar air heater

Sheshang S Chandel[1], Ajay Verma[2], Maheandera Prabu Paulraj[3], Santosh Kumar Sahu[4]

[1,2]Dept of Mechanical Engineering, Shri Shankaracharya Group of Institutions, Bhilai- 490020, C.G, India.

[3,4]Fluid Mechanics and Machinery Laboratory, Discipline of Mechanical Engineering, Indian Institute of Technology Indore, Simrol- 453552, Indore, Madhya Pradesh, India.

E-mail: csheshang1611@gmail.com

Abstract. The present study deals about the fluid flow and heat transfer behaviours of circular wavy wall transverse rib artificially roughened SAH. The circular wavy ribs are used in the solar air (SAH) heater with different undulation (N) and constant amplitude (A) below absorbing plate at testing zone for the present numerical investigation. Heat transfer behaviour varies about number of undulation with the intention that the optimum undulation of circular wavy wall transverse rib is necessary for better performance of SAH. The FLUENT v 14.5 is used to solve the governing equations (mass, momentum and energy). The RNG (k- ε) model is utilized to capture the turbulent effects and also the SIMPLE algorithm is used to couple the pressure and velocity. The geometrical parameter, pitch (p/e) ranges from 7.14 to 14.28 and relative roughened height (e/D) ranges from 0.021 to 0.042 are taken. The Reynolds numbers (Re) is varied between 3800 and 18000 for the constant heat (q) flux of plate 1000 W/m2. The results, Nusselt number (Nu_r/Nu_s) enhancement, and friction (f_r/f_s) factor performance are presented.

Key words: solar air heater, circular wavy wall, CFD, friction factor, Nusselt number

1. Introduction

Solar air (SAH) heater is the low cost heat transfer mechanism which absorbs heat from sun rays and transfers as thermal energy. Solar air (SAH) heaters have a plate placed at upper surfaces of duct which takes temperature from aluminium plate and bottom surfaces are taken insulated to avoid the heat losses. The duct is enclosed with inlet and outlet. The air flows inside the duct takes the heat energy from the aluminium absorbing plate and transfer to a required place by using electric blower or fan. Now days, solar air (SAH) heater are mainly using for space heating application, process heating application, ventilation applications and many more. Various turbulence creators in the terms of different roughened surfaces, rib, fins and vortex rings are used for enhancing overall heat transfer reported by Yadav et al. [1]. Prasad and Saini [2] conducted the experiment by using circular wire which are placed close to the hot test shield and captured in to flow pattern near to the wall. And they found positive results in thermal performance with penalty in pressure drop. Aharwal et al. [3] introduce inclined ribs geometries and reports enhancement of Nusselt (Nu) number for angle of (α) attack effects. The fluent software help to predict the effect inside solar air heater (SAH) channel and

IOP Conf. Series: Earth and Environmental Science **312** (2019) 012002 doi:10.1088/1755-1315/312/1/012002

a range of two-dimensional artificially roughened geometries by the Chaube et al. [4] and A S Yadav [5]. The results shown on their studies are related to thermal performance in different shapes and parameter of roughness by using commercial computational (CFD) software FLUENT. S Kumar and R P Saini [6] used three-dimensional computational domain chosen for the study of fluid and heat characteristics of arc shape ribs. Hatami et a. [7] uses the cavity of circular wavy shape for natural convection and their results was positive for local Nusselts (Nu) number. In this present investigation, the wavy ribs roughened surface is taken for solar air drying applications. The wavy roughened surface is placed near the absorbing (aluminium) plate (upper wall of testing section). The forced turbulent effects are captured by CFD code (fluent version14.5).

2. Problem definition

The two-dimensional rectangular SAH duct is considered for study the variations in flow pattern and heat transfer of circular wavy ribs artificially roughened solar air heaters. In present investigation the wavy circular ribs are introduced as a roughened surface and placed on a bottom of the absorbing plate. The flow parameters are taken similar as A S Yadav and J L Bhagoria [8] for predicting the behaviours of working fluid flow. Entrance section and exit section are taken $5\sqrt{WH}$ and $2.5\sqrt{WH}$ recommended by ASHRE Standard 93-2003[9]. The parameters of duct for entrance section length (L_1) 225mm, tested section length (L_2) 115mm and length (L_3) 115mm for exit section are taken with 100mm width and 20mm height. Computational domain with circular wavy walls is show in figure 1. Aluminium absorbing plate of 0.5 mm thick is placed at top wall of testing zone with GI wavy circle ribs. The pitch ranges between two ribs are taken as constant at 10mm and the three values are considered for rib height 0.7mm, 1.0mm and 1.4mm for analysis. The circular wavy rib function is taken from M Hatami, D Song and D Jing [7] and the function is $r = r_{in} + Acos(N(\zeta))$. Here, r_{in} base radius of circle, r true inner radius of circle, A is amplitude taken as constant of 0.2mm and N is number of undulation which is in range from 4-12. The ζ taken as rotational angle. The fluid and material properties are taken at room temperature. List of properties of aluminium and air are mentioned in Table 1 which are using in present study.

3. CFD Simulation

Computational Fluid (CFD) Dynamic code fluent (version 14.5) is used to capture the flow behaviours of fluid and characters of heat transfer for 2D artificially roughened solar air collecter; by using finite volume (FVM) scheme, general Navior-Stokes equations (mass and momentum) and governing equation for energy are solved. Steady condition is considered for present investigation, assuming fully developed and turbulence effect. The properties of present material and fluid are taken as constant throughout the investigation. The no radiation effect and no slip at wall are assumed for analysis. The FLUENT v14.5 (CFD code) is used in present numerical investigation. The 2-D geometry is developed by using design modular of ANSYS; mesh generation is done by using ANSYS ICEM HEX tool which is present at ANSYS Workbench 14.5v and then this meshing model are used for further simulation process. The assemble desktop is used for simulating and the configuration, processor- Intel(R) Xeon(R) processor, Ram-32 GB, NVIDIA Quadro K620- with Professional 64-bit has been used for the present study.

3.1 Grid Generation

The different cell shapes with different element size are generated and tested to capture the turbulent effects on present study and are mentioned in figure 2. The triangle mesh, triangle and square combined mesh and square mesh are generated for proper capturing flow behaviour of fluid on computational domain.

Table 1 Thermo-physical properties used for computational analysis.

Properties	Working fluid (air)	Absorber plate (aluminum)
Density, ‘ρ’ (kg m^{-3})	1.225	2719
Specific heat, ‘Cp’ (J kg^{-1} K^{-1})	1006.43	871
Viscosity, ‘μ’ (N m^{-2})	1.7894e-05	-
Thermal conductivity, ‘k’ (W m^{-1} K^{-1})	0.0242	202.4

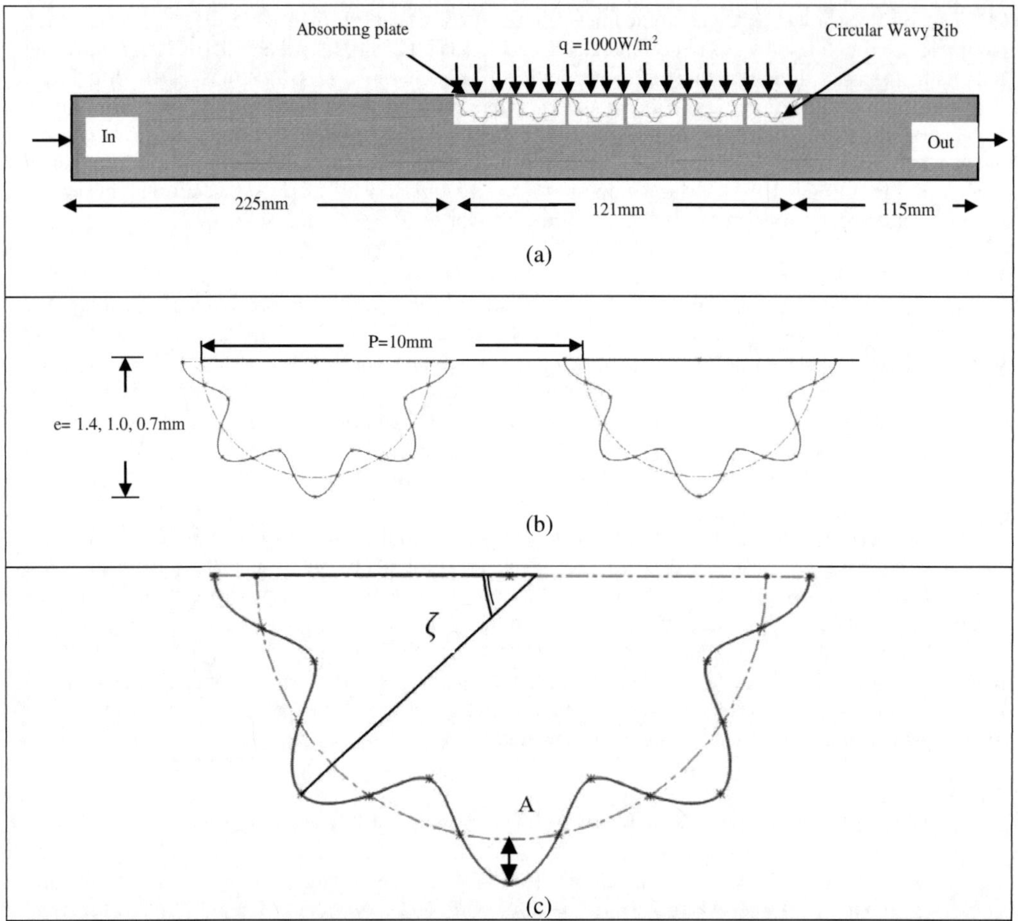

Figure 1. Artificially roughened SAH (circular wavy ribs),
(a) Schematic diagram of computational domain, (b) Circular wavy rib with pitch 10, (c) Circular wavy rib with amplitude (A=0.2mm) and angle of rotation (ζ=45°).

Table 2. Grid test for present analysis

Mesh Size (mm)	No. of elements	Nu	% error	f	% error
0.26	137435	102.86	_	0.012072	_
0.25	145190	104.64	1.70	0.012396	4.13
0.24	160534	106.37	1.62	0.012611	2.61
0.23	175256	108.07	1.57	0.012721	1.70
0.22	191542	108.85	0.75	0.012883	0.86
0.21	208676	110.88	1.08	0.012072	1.25

IOP Conf. Series: Earth and Environmental Science **312** (2019) 012002 doi:10.1088/1755-1315/312/1/012002

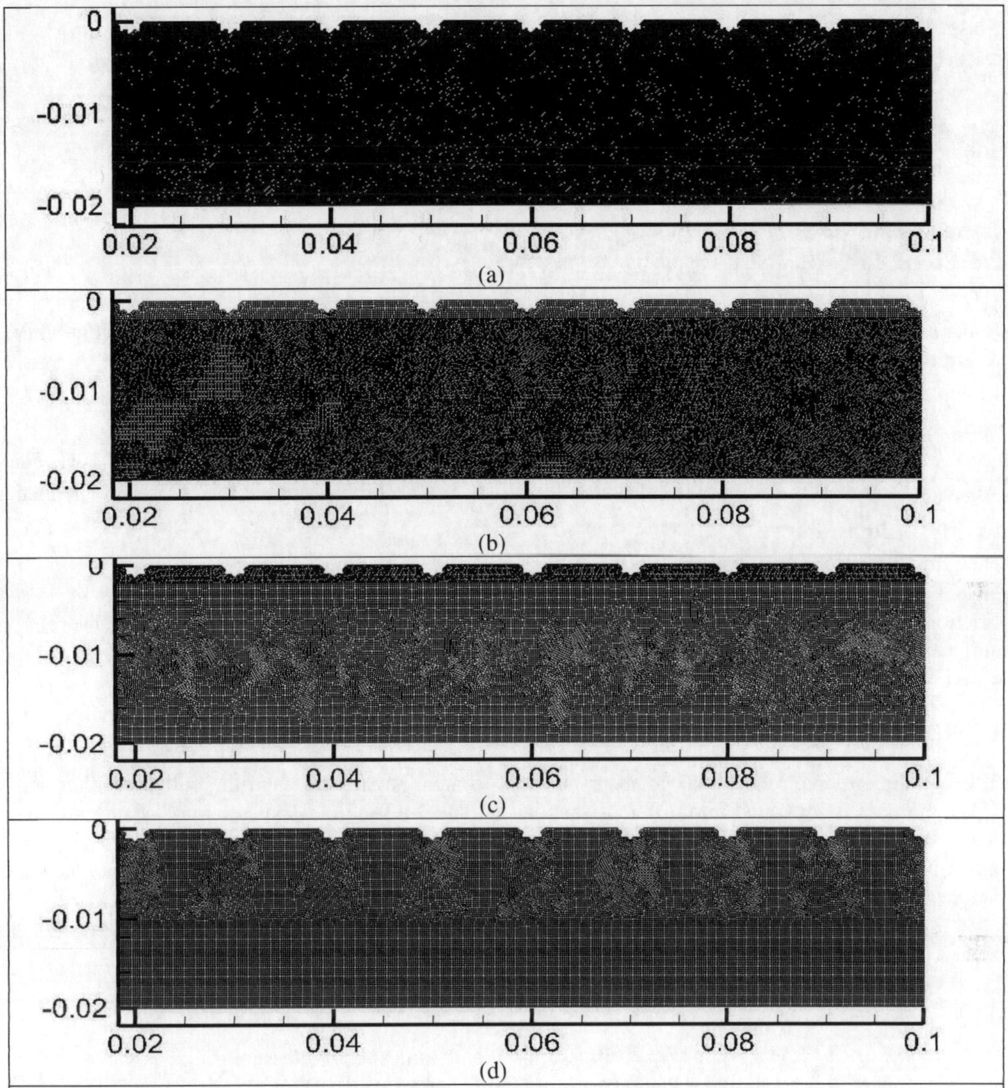

Figure 2. Different cell shape with element size 2.3 mm, (a) 464594 cells, (b) 372958 cells, (c) 276010 cells, (d) 175256 cells.

Generated meshes are tested with varying element size from 2.1 mm to 2.6 mm. The square mesh performance, mesh type (d) found better as compare to mesh types (a), (b) and (c). Uniform square mesh type (d) with 2.3 mm element size are found suitable in present study. The less than 1% error is considered for Nusselt (Nu) number and for (f) friction factor. All grid independence tests are done for rib height 1.4mm, pitch distance 10 mm with undulation (N) 8.

3.2 Governing Equation

Here the mass, momentum and the energy equations have been solved based on initial and boundary condition by using the commercial FVM-CFD coding. The governing equations are presented in the form of finite difference mode as below.

Mass

$$\frac{d\rho}{dt} + \nabla.(\rho\vec{v}) = 0 \quad (1)$$

Momentum

$$\frac{d(\rho\vec{v})}{dt} + \nabla\cdot(\rho\vec{v}\vec{v}) = \rho\vec{g} - \nabla p + \nabla(\vec{\tau}) + \vec{F} \quad (2)$$

Energy

$$\frac{d(\rho E)}{dt} + \nabla\cdot(v^{\rightarrow}(\rho E + p)) = \nabla\cdot(k\nabla T + (\vec{\tau}\cdot\vec{v})) \quad (3)$$

Some non-dimensional numbers are involved in our present study: Reynolds number, Nusselt number & Friction factor which can obtain by:

Reynolds Number: $Re = \frac{\rho v D_h}{\mu}$ (4)

Here, D_h is a hydraulic diameter.

Average Nusselt (Nur) number is calculated for roughened absorbing surface which can be obtained by: $Nu_r = h\frac{D_h}{k}$ (5)

Here, h is convective heat transfer coefficient.

Friction (fr) factor are depended upon the pressure (ΔP) drop due to given roughened absorbing surface across the testing section. It can be calculated by:

$$f_r = \frac{(\Delta P/l)D_h}{2\rho v^2} \quad (6)$$

Here, ΔP is pressure drop at testing section.

The two dimensional rectangular geometry at the x-y plane having inlet and the outlet at either end sides, the duct is separated into different section; entrance, testing and exit section. Testing section upper surface is taken as absorber plate (aluminum material). The working temperature of fluid and material are taken at ambient temperature. So the property of fluid and the materials are also taken at ambient temperature. The inlet fluid temperature is taken 300 degree Kelvin.

Table 3. Boundary conditions

	Boundary conditions
Duct inlet	velocity-inlet
Duct outlet	Pressure-outlet at atmospheric pressure
Absorbing Plate	wall (constant heat flux) $1000 W/m^2$
Bottom wall	Insulated (adiabatic)/ no slip

The wide ranges of models are present in Ansys (FLUENT) for capturing the turbulent effect. The selection of model doesn't depend upon only the fluid flow; the accuracy and the computational time satisfaction also needed. Figure 3 (a) describes the results of various turbulent models and compared their individual results with the predicted fluid flow for smooth (s) duct which was calculated from Dittus-Boelter correlation eq. (8). The renormalization (RNG) group k- ε model found less error of 2.5% from predicted values as compare to the others turbulent models.

3.3 Model selection and Validation

Numerically approach to fluid dynamic and thermal analysis problem are solved with the help of CFD (FLUENT v14.5) code are common nowadays. Present, investigation involves wide range of turbulent models which are SST model, k- ω model, standard k- ω, standard k- ε, k- ε RNG model, and k- ε realizable models for the present analysis and the best model is chosen for analysis. The predicted results of smooth duct are also validated. After comparing, the turbulent model renormalization (RNG) group k-ε is found to be best for the investigation. And also the existing results of semi-

circular rib SAH [8] have been taken to compare with present roughened geometry for validation and the similarity solution obtained and it have been presented in figure 3. (b).
Dittus –Boelter equation: $Nu_r = 0.023Re^{0.8}Pr^{0.4}$ (7)

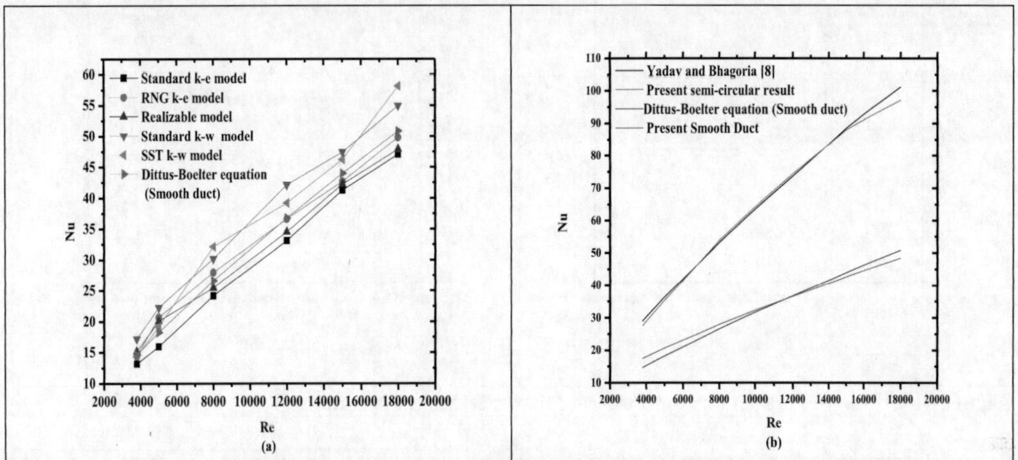

Figure 3. (a) Selection of turbulent model, (b) Validation.

The validation result describes and also comparisons of the prediction fluid flow result of smooth duct. Friction (f_s) factor of smooth duct can be predicted by using Blasius equation (8).

Blasius Equation: $f_s = 0.0791\ \mathrm{Re}^{-0.25}$ (8)
The turbulent kinetic energy (K.E) with the RNG ($k - \varepsilon$) transport model [10] as defined as

$$\frac{\partial}{\partial x_i}(\rho k u_i) = \frac{\partial}{\partial x_j}\left(\alpha_k \mu_{eff} \frac{\partial k}{\partial x_j}\right) + G_k - \rho\varepsilon \quad (9)$$

$$\frac{\partial}{\partial x_i}(\rho\varepsilon u_j) = \frac{\partial}{\partial x_j}\left(\alpha_\varepsilon \mu_{eff} \frac{\partial \varepsilon}{\partial x_j}\right) + C_{1_\varepsilon}\frac{\varepsilon}{k}(G_k) - C_{2_\varepsilon}\rho\frac{\varepsilon^2}{k} - R_\varepsilon \quad (10)$$

Where, $G_k$is the (K.E generation) energy due to mean velocity gradient, G_k energy developed by average velocity gradient as defined as

$$G_k = -\rho\overline{u'_i u'_j}\frac{\partial u_j}{\partial x_i} \quad (11)$$

μ_{eff} is the effective turbulent (τ) viscosity, μ_{eff} is define as

$$\mu_{eff} = \mu + \mu_t$$

Where μ_t is the turbulent (τ) viscosity, μ_t is define as

$$\mu_t = \rho C_\mu \frac{k^2}{\varepsilon}, \quad (12)$$

Where C_μ = Constant,
α_k, & α_ε, represents Prandtl number for kinetic energy k & dissipation, ε.

The present constant values are having default values form Launder and Spalding 1972 [11]:
$C_{1_\varepsilon} = 1.42$, $C_{2_\varepsilon} = 1.68$, $C_\mu = 0.0845$, $\alpha_k = 1.39$ and $\alpha_\varepsilon = 1.39$.

The turbulent model RNG K-e find more appropriate to capture the turbulent effect for this investigation. The conversing residual for continuity equation are taken as 10-3 and 10-6 for momentum and energy. The SIMPLE scheme is selected for pressure- velocity coupling and second orders upwind are selected for velocity and energy [Paulraj et al. 12 and 13]. The constant velocity inlet is taken at entrance zone, Atmospheric pressure is selected at exit of the duct and plate is maintained at constant heat (q) flux of 1000W/m2.

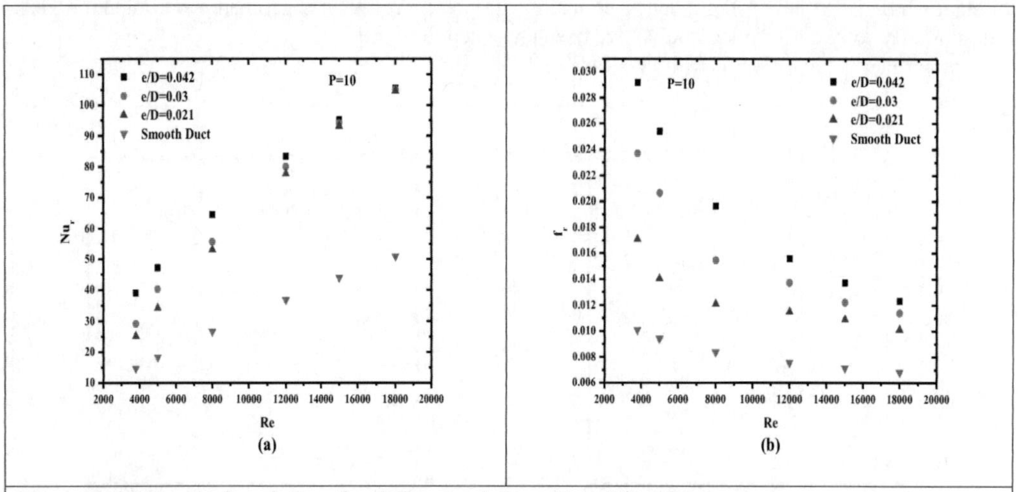

Figure 4. (a) Nu_r (b) f_r variations for different relative e/D and P= 10mm.

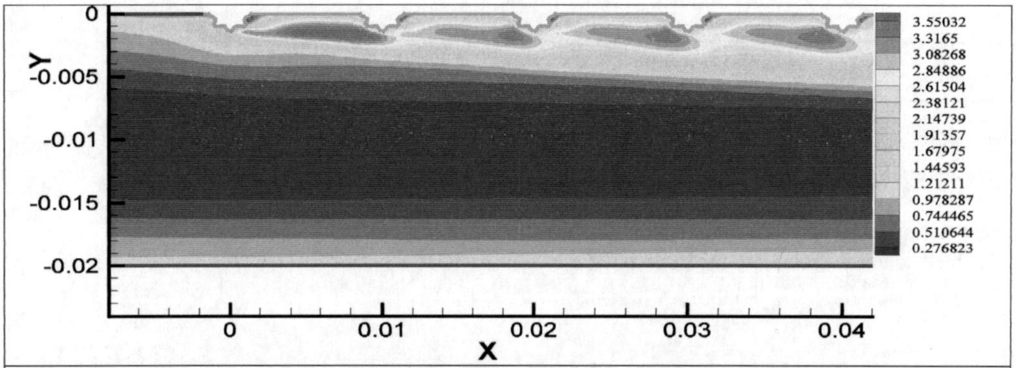

Figure 5. Turbulent Kinectic Energy of N=8 at Re 18000

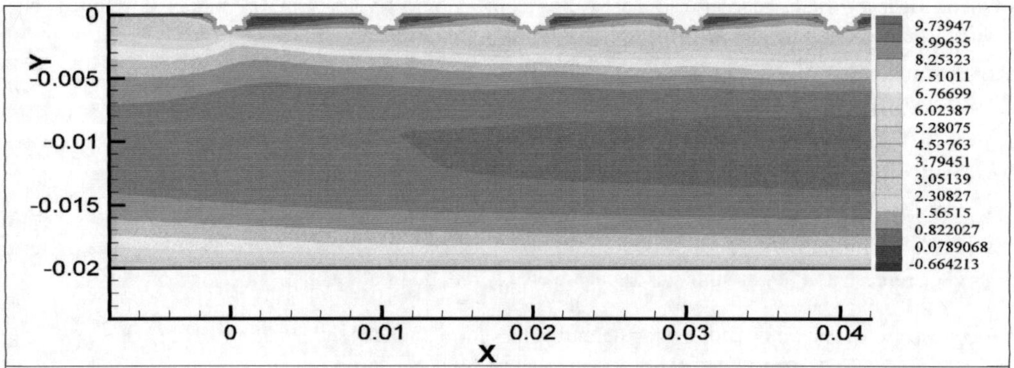

Figure 6. Velocity Magnitude of N=8 at Re 18000.

IOP Conf. Series: Earth and Environmental Science **312** (2019) 012002 doi:10.1088/1755-1315/312/1/012002

4. Result and Discussion

Average Nusselt (Nu) number and friction (f) factor are presented and comparing with various parameters for showing the optimum effects of thermal characteristics. The comparisons are done with existing results of the semi circular rib and the smooth (s) SAH duct. From the figure 4 (a), describe the deviations in Nu value and pressure drop (∇P) by various e/D in different Reynolds (Re) number. The Reynolds (Re) number 18000 gives the maximum Nu value with minimum friction in all the set of parameters taken. It is also noticed that the Nu value increases when e/D, roughness height increased. However, when we seen the higher Reynolds (Re) number, there is no huge variation in (Nu) Nusselt number noticed. From figure 4 (b) describes the influence of e/D on friction factor and the range from 0.042 to 0.021 relative roughened surface at different (Re) Reynolds number and compared the results with predicting smooth (s) duct results. The friction factor gives higher values when the relative roughened surface get increase and also lower Reynolds (Re) number values give increased pressure drop values because viscous layer get suppressed. Figure 5 & figure 6 describing the contour of (τ) turbulent K.E and velocity magnitude. The maximum turbulent (K.E) kinetic energy noticed at wavy ribs leading edge (except first rib) and which indicates that the higher turbulence occurs at near wavy ribs tailing edge. Also the K.E decreases along the flow direction. From figure 6, the higher velocity (υ) magnitude is noticed at wavy ribs leading side and lower velocity magnitude observed at tailing side. This velocity gets affected because of wavy surface and the low pressure regions take place at tailing edge. Turbulent kinetic energy (K.E) & dissipation increased due to increases of intensity of the turbulent. Figure 7 (a) describe the Nu and undulation (N) in different Reynolds (Re) number. The undulation (N) of value 4 gives lower Nusselt (Nu) number as compare to higher value for undulation (N) of 8 and 12 for Reynolds (Re) number range 3800-5000 because of undulation effect of wavy rib. It is also seen at figure 7 (b) the friction (f) factors for undulation (N) of value 4 are lower for Re 3800 and 5000 as compare undulation (N) value 8 and 12. The lower friction (f) factor gives the lower effect of heat transfer. The higher Reynolds (Re) number of range 8000- 18000 gives higher Nusselt (Nu) number and lower friction as compare to other values of N. Figure 8 shows the fluid behaviors in presence of circular wavy rib. The turbulence is taking place at the tailing side of the wavy rib.

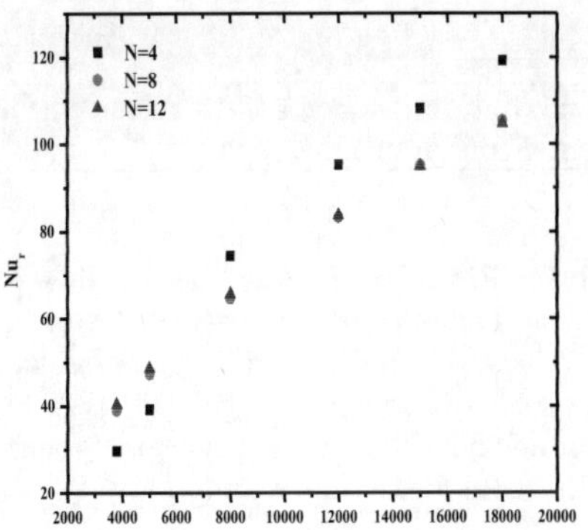

(a)

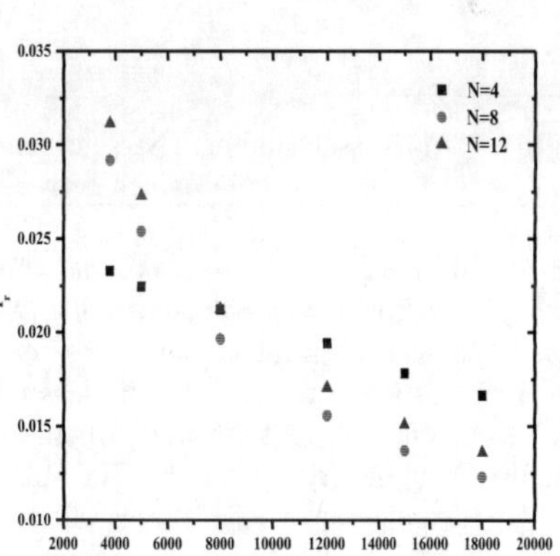

(b)

Figure 7. (a) Average Nu
(b) f_f variation for different undulation (N)

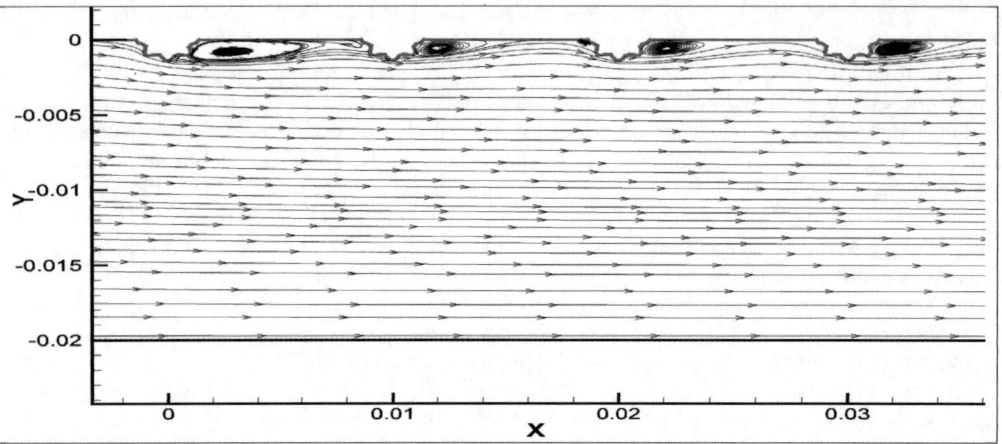

Figure 8. Stream line for N=8 at Re-18000.

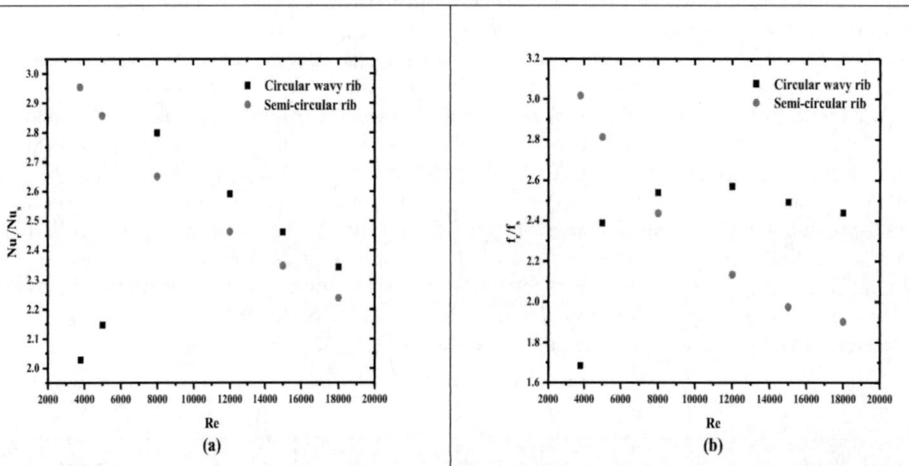

Figure 9. (a) Nusselt number (Nu_r/Nu_s) enhancement and (b) Friction factor (f_r/f_s) enhancement comparison between wavy rib and Semi-circular.

Figure 9(a) describes the Nusselt number (Nu_r/Nu_s) enhancement of wavy rib and semi-circular rib. The wavy ribs find fewer enhancements variation for lower Reynolds (Re) number of 3800-5000 because of effect of undulation (N). It is also noticed at Figure 9(b) the friction factor (f_r/f_s) enhancement of wavy ribs are having less enhancement values as compare to the semi-circular rib. The enhancement values of the wavy rib are higher for Reynolds (Re) number range 8000-18000 with the penalty of higher friction factor. The maximum Nusselt number (Nu_r/Nu_s) enhancement found for Reynolds (Re) number is 1.055 times of semi-circular ribs at Re 8000 with the friction factor (f_r/f_s) enhancement of 1.04 times of semi-circular ribs.

5. Conclusion

Circular wavy ribs are placed near to the absorbing surface for evaluating the friction factor and Nusselt number effect in two-dimensional rectangular channel flow. Reynolds (Re) number, relative roughened (e/D) height, relative roughened (P/e) pitch, and undulation (N) parameters are taken and compared. The height of ribs disturbed the fluid flow pattern, hence gives the higher heat transfer characters but with the penalty of pressure drop. The relative roughened height 0.042 at Reynolds (Re) number 18000 for N=4 gives maximum average Nusselt (Nu) number value 119.19 with the penalty of 0.0166 friction factor. It is found, the circular wavy roughened surface placed near to the

IOP Conf. Series: Earth and Environmental Science **312** (2019) 012002 doi:10.1088/1755-1315/312/1/012002

absorbing plate with N=4 gives higher heat transfer performance for higher Reynolds number as compare to N=8 and N=12.In present investigation the e/D=0.042, N=4 and the P/e=7.14 gives higher Nusselt number (Nu_r/Nu_s) enhancement Reynolds (Re) number range 8000-18000 as compare to the semi-circular roughness. At Re-8000 the wavy surface gives 1.055 times enrichment in the Nu as compare to semi-circular roughened surface and 2.791times of the smooth channel SAH.

Nomenclature

A_s	Surface area of absorbing plate, m^2	L_3	Length of exit zone, mm
D	Hydraulic diameter, mm	m	Mass, Kg/s
C_p	Specific heat of operational fluid, J/kg K	P	Pitch in mm
e	Height of the rib , mm	ΔP	Pressure drop, Pa (Pascal)
h	Heat transfer coefficient , W/m^2K	T_0	Outlet Temperature, K
H	Duct height, mm	T_i	Temperature at inlet, K
I	Intensity, W/m^2	T_a	Atmospheric temperature, K
k	Thermal conductivity, W/mK	T_a	Ambient temperature, K
L	Channel length, mm	T_w	Wall temperature, K
L_1	Entry zone length, mm	v	Velocity of fluid, m/s^2
L_2	Length of test zone, mm	W	Channel width, m
Dimensionless Parameter		Greek Symbol	
A	Amplitude	μ	Dynamic viscosity, Ns/m^2
e/D	Relative roughness height	ζ	Rotational angle , degree
f	Friction factor	ε	Dissipation rate
f_r	Friction of roughened surface	ω	Specific dissipation rate
f_s	Friction of smooth surface	δ	Transition sub-layer thickness, m
N	Number of undulation	k	Turbulent kinetic energy, m^2/s^2
Nu	Nusselt number	τ	Turbulent
Nu_r	Nusselt number for roughened surface	*Subscripts*	
Nu_s	Nusselt number of smooth surface	i	Inlet
P/e	The relative roughness pitch	o	Outlet
Pr	The Prandtl number	r	Roughness
Re	The Reynolds number	s	Smooth
W/H	Channel aspect ratio		

IOP Conf. Series: Earth and Environmental Science **312** (2019) 012002 doi:10.1088/1755-1315/312/1/012002

Reference

[1] Yadav S, Paulraj MP, Sahu SK. *Proceedings of the ASME 2018 Power Conference (POWER2018); June 24-28, Florida, USA*, pp. V002T12A006; 12 pages, DOI: 10.1115/POWER2018-7231.
[2] Prasad BN, Saini JS. *Solar Energy 1988*; **41(6)**:555-60.
[3] Aharwal KR , Bhupendra K. Gandhi , Saini JS, *Int. J. Heat and Mass Trans* **52** (2009) 5970–5977.
[4] Chaube A, Sahoo PK, Solanki SC. *Renew. Energy 2006*; **31**:317-31.
[5] Yadav AS, Bhagoria JL. *Energy* **55** (2013) 1127-1142
[6] Kumar S, Saini RP. *Rene. Energy* 2009; **34**: 1285-91.
[7] Hatami M, Song D, Jing D. *Int. J. Heat and Mass Trans* **98** (2016) 758–767.
[8] Yadav AS and Bhagoria JL. *Numer. Heat Trans, Part A*, **65**: 679–698, 2014.
[9] ASHRAE Standard 93. *Method of test to determine the thermal performance of solar collector.* 30329. Atlanta, GA: American Society of Heating, Refrigeration and Air Conditioning Engineers; 2003.
[10] ANSYS FLUENT 14.5. *Documentation. Ansys*, Inc; 2003-04.
[11] Launder BE and Spalding DB, *Lectures in mathematical models of turbulence*, London, New York, Academic Press, 1972.
[12] Paulraj MP, Parthasarathy RK, Taler J, Taler D, Oclon P, Vallati A, *Heat Trans. Research* 2018; **49(12)**: 1151–1170.
[13] Paulraj MP, Byon C, Vallati A, Parthasarathy RK, *Heat Trans. Eng* 2018; DOI: 10.1080/01457632.2018.1513629.

IOP Conf. Series: Earth and Environmental Science **312** (2019) 012003 doi:10.1088/1755-1315/312/1/012003

Experimental Analysis of Thermoelectric Device to Produce Cooling Effect Using Solar Energy and Phase Changing Material

P. Lakshmi Prasanna*, **Dr. K. Dharma Reddy**

Department of Mechanical Engineering,
Sri Venkateswara University college of Engineering, Tirupati-517502

Email: plprasanna24@gmail.com

Abstract. This technical paper presents the design and experimental analysis of Thermoelectric Device to produce cooling effect using solar energy as power input. Solar energy is a co_2 free renewable energy source. Solar power is an unlimited source of energy. Solar energy is stored in the battery through solar panel. Now the Dc current obtained from solar panel is stored in the battery and is converted into AC current using inverter. Thermoelectric device works on "Thermoelectric effect" which is based on the principles of Peltier effect and Seebeck effect. A model is created to understand the working of thermoelectric effect. Efforts are made in this paper to attain the lower temperature than the surroundings with and without application of phase changing materials. A Phase Changing Material (PCM) is a substance with a high heat of fusion, which is capable of storing and releasing high amount of energy at the time of melting and solidifying at a particular range of temperature. In present work, Formic acid is used as a PCM and results are compared to obtain the optimum process.

Keywords: Peltier effect; Seebeck effect; thermoelectric effect; solar energy; PCM; Formic Acid.

1. Introduction

In today's world there is an increasing demand for low temperature applications in various fields which led to the increase in utilization of electricity and that devices in turn release more amounts of harmful gases such as CFC's,CO_2. This increase in release of green house gases causes global warming. To minimize these effects, focus is shifted towards the utilization of renewable energy sources. Abundant source of renewable energy on the earth's surface is solar energy [1]. The new alternative method to produce cooling effect is to design and develop a thermoelectric device powered by solar energy. The thermoelectric modules consists materials made up of semiconductors, which are electrically connected in series configuration and thermally in parallel configuration for creating cool and hot temperatures on either side of the module [2]. Despite the fact that these devices are less efficient compared to conventional refrigerating systems, but they are weightless, cost efficient, less noisy, and are environmentally safe. In addition to this an attempt is made to obtain the lower temperature with the use of phase changing materials (PCM) [3]. A PCM will have lower thermal resistance because of that it will release more amount of heat during liquid to solid phase change and absorb more heat during solid to liquid phase change at the beginning of melting and solidifying process [4].

The main objective of this study is designing and fabricating a working model of thermoelectric device for cooling which employ peltier effect [5] [6], i.e., lowering the temperature at one junction and increasing the temperature at the other junction when electric current is passed in a circuit which consists of two dissimilar conductors; the effect in circuits which contain dissimilar semiconductors is more effective. The model is designed in such a way that the volume inside the cabin should be cooled and maintained within the temperature ranging 7 ^{0}C to 22 ^{0}C in less period of time and the chillness obtained should remain for at least half an hour. Formic acid, an organic phase changing material is used additionally to achieve the advantages of its charging and discharging processes. During the charging process of the formic acid it absorbs chillness and reaches to its freezing point. When the power input to the cabin of the thermo electric device is off then the heat enters into the cabin through the TEC module. To avoid this heat absorption into the cabin the formic acid undergoes discharging process so that the temperature inside the cabin can be retained for some time [7]. In this work organic PCM is preferred than inorganic PCMs because organic PCMs are chemically stable, environmentally safe, non-toxic and are non-corrosive in nature [8]. This device can be used in remote locations where generation of electric power is less and uncertain because it is powered from battery which can be charged through solar panel.

To achieve the cooling effect in the model a thermoelectric cooling module (TEC) or peltier cooling module is selected as per calculated cooling load [9] [10]. The internal structure of the peltier element comprises semiconductor pellets fabricated from N-type and P-type Bismuth Telluride materials. The array of pellets is electrically connected in series, but thermally arranged in parallel to maximize thermal transfer between the hot and cold ceramic surfaces of the module. A thermoelectric module comprising a Peltier element sandwiched between two ceramic plates of high thermal conductivity, with a power source, is effectively able to pump heat across the device from one ceramic plate to the other. Moreover, the direction of heat flow can be changed simply by reversing the direction of current flow. Applying a DC voltage causes the positive and negative charge carriers to absorb heat from one substrate surface and transfer and release it to the substrate on the opposite side. Therefore, the surface where energy is absorbed becomes cold and the opposite surface, where the energy is released, becomes hot. The thermoelectric modules must be used with a heat sink.

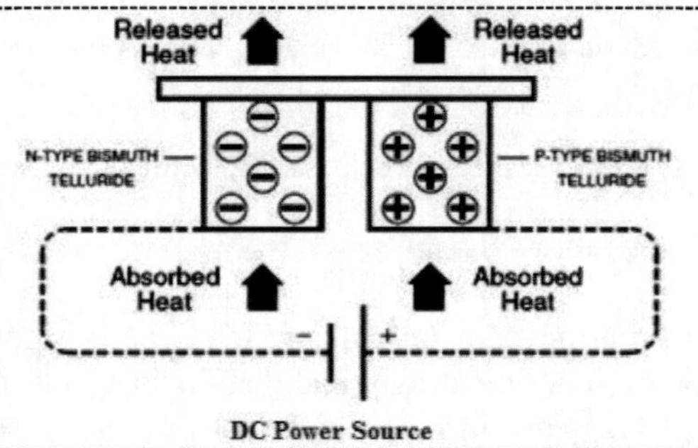

Figure1. TEC Module Working

1.1. *Literature review*

After studying a number of patents on models of thermoelectric devices, solar powered thermoelectric cooling systems are characterised in this section. In 1978 Beitner proposed a simple design which consists of thermoelectric modules powered by an external DC power source and the heat obtained on the hot side is dissipated to the ambient through natural convection. In 1982 Reed and Hatcher indicated an efficient manner to dissipate heat from hot side of TEC to the ambient through forced convection by employing cooling fans. In 1996 Park et al. proposed a new design of thermoelectric refrigerator working with low electrical power supply and also by bringing together the advantages of integrating super insulation materials and PCM's in thermoelectric system to achieve an environmental friendly device which is capable of saving energy and can retain the obtained temperature uniformly for prolonged periods of time.

2. Objective of the design

Present work aims at designing a model of thermoelectric device for cooling with cabin having a volume of 16L. The device is designed in such a way that temperature range inside the cabin should be in between 7 ^{0}C to 15 ^{0}C for a specified period of time. The proposed system is solar powered and the device works on DC power stored in the battery, calculation of DC power from voltage and current that is obtained from solar panel at a particular period of time in a day is required. Additionally PCM is used to avoid heat absorption from the TEC module into the cabin when the power is off. As the

IOP Conf. Series: Earth and Environmental Science **312** (2019) 012003 doi:10.1088/1755-1315/312/1/012003

system is used in outdoor applications proper care should be taken in providing insulation and radiation control.

3. Steps involved in designing thermoelectric cooling device

3.1. *Dimensions of the model.* In view of the conditions laid down in the objectives a cubic square cabin is selected with the given dimensions.
Outer dimensions of the cabin = 0.30 x 0.30 × 0.30m
Inner dimensions of the cabin = 0.25 x 0.25 × 0.25 m
Thickness of aluminium plate = 2mm

3.2. *Materials.* Aluminium sheets which have thermal conductivity of 205W/m-K are used as the inner walls of the vertical side panels. Expanded polystyrene sheets (EPS) which are 50mm thick, 30kg / m^3 density and 0.033W/m-K thermal conductivity are used to provide shielding for temperature losses from the cabin. Formic acid with properties as shown in the table 1 is poured into two aluminium boxes of dimensions 16cm×16cm×2cm. These aluminium boxes are attached on the cold side of the TEC module.

Table1. Properties of formic acid

Formula	Density	Boiling point	Melting point	Heat of fusion
CH_2O_2	1.220g/m-L	101 ^{0}C	7.8 ^{0}C	247 (kJ/kg)

3.3. P*rocedure for designing a thermoelectric device.* To design a model of thermoelectric cooling system, the crucial step involves understanding the thermal load concept. Thermal load is the amount of heat energy to be removed from an inner environment. Based on the calculated thermal load, a suitable TEC module is selected for the model. Each TEC module has unique capacity in heat removal process .To achieve required objectives of the design the amount of heat to be removed from the cabin is obtained from the thermal load calculations. After selection of TEC module, the difference in performance of the device without and with the application of organic PCM formic acid is observed.

3.4. *Calculation of heat load.* Thermal load can be classified as (i) active load and (ii) passive load.
In many thermoelectric applications active load can be neglected and therefore there is no need for calculating the active load. Whereas the passive load is defined as a small amount of energy that must be continuously moving in and out of the load to retain a temperature gradient between system and surroundings. In present work the main aim is to maintain the temperature inside the cabin lower than the ambient temperature. Even though utmost care is taken to provide insulation there will be possibility of some leakages in the system. Passive load through the vertical walls is calculated by considering heat transfer components of conduction and convection, from the following equation:

$$Q_v = \frac{\Delta T.A}{\frac{1}{h}+\frac{L}{k}}$$

Required cabin temperature = 10 ^{0}C
Room temperature = 35 ^{0}C
Temperature gradient = 35-10= 25 ^{0}C
K_{Al}= 205 W/m-K;
K_{EPS}= 0.033 W/m-K;
h_{air} = 10 W/m^2K
Area, A_1=A_2=A_3=0.30 x 0.30 = 0.09 m^2
Passive load through the vertical wall $Q_{v1} = \frac{\Delta T.A}{\frac{2}{h}+\frac{L1}{KEPS}+\frac{2\times L2}{KAl}} = \frac{1.562}{\frac{2}{10}+\frac{0.05}{0.033}+\frac{0.004}{205}} = 0.911$W

IOP Conf. Series: Earth and Environmental Science **312** (2019) 012003 doi:10.1088/1755-1315/312/1/012003

The passive load through the vertical walls is $Q_v = Q_{v1} + Q_{v2} = 0.911 + 0.911 = 1.822W$
For the remaining sides of the cabin the walls contain only EPS and the passive load will be only due to conductive heat transfer.

$$Q_{s1} = \frac{\Delta T.A}{\frac{l}{KEPS}} = \frac{0.3125}{\frac{0.05}{0.033}} = 0.206W$$

$Q_s = Q_{s1} + Q_{s2} + Q_{s3} + Q_{s4} = 0.206 + 0.206 + 0.206 + 0.206 = 0.824$ W
Penetration of heat load due to opening and closing of the cabin, Q_P is approximately 15W
Total passive load $Q_{TP} = Q_V + Q_S + Q_P = 1.822 + 0.824 + 15 = 17.646$ W; for safe design, $Q_{TP} = 25W$

3.5. *Design of the device*. The values that are obtained from the calculations needed for the designing of the system are: Heat load $Q_{TP} = 25$ W; Ambient air temperature $T_A = 35\ ^0C$; required temperature inside the cabin $T_C = 10\ ^0C$. Now with the help of these known values, we have to identify the following:

i. Hot side temperature of the thermoelectric module (T_H)
ii. Temperature gradient across the thermoelectric module (ΔT)

Hot side temperature of the thermoelectric module (T_H). The hot side temperature of the thermoelectric module is given from the equation:

$$T_H = T_A + (V \times I + Q)\, R_T$$

Where, T_A = Ambient temperature
Q = Total passive load
$(V \times I)$ = Thermoelectric module power
R_T = Thermal resistance ($^0C/W$)

While designing the thermoelectric device, an assumption is made for the heat sink temperature i.e., the heat sink temperature should not go beyond 15 ^0C more than the ambient temperature. Therefore, the hot side temperature of the thermoelectric module $(T_H) = 35\ ^0C + 15\ ^0C = 50\ ^0C$.

*Temperature gradient across the thermoelectric module (ΔT).*The difference of the temperature over the thermoelectric module can be obtained by subtracting the required temperature to be maintained inside the cabin (T_C) from the hot side temperature of the thermoelectric module (T_H).
Therefore, $\Delta T = T_H - T_C = 50\ ^0C - 10\ ^0C = 40\ ^0C$.

3.6. *Procedure for selection of thermoelectric cooling module (TEC).* A TEC module should have the capability to cool and maintain a proper temperature inside the cabin and also the model of the TEC module should cope up with the dimensional constraints of the design. A suitable thermoelectric cooling module is selected, based on the calculated values such as heat load Q, heat losses, temperature gradient (ΔT), power input etc. The thermoelectric modules are able to operate in two limit modes:

i. Q_{max} maximum cooling capacity mode, at $\Delta T = 0$;
ii. ΔT_{max} maximum object cooling mode, at $Q_C = 0$

More over in practically the combined operation mode is applied. A module's maximum cooling capacity (Q cmax) is determined from the formula:

$$Q_{cmax} = \frac{Qc \times \Delta T\ max}{\Delta T\ max - \Delta T}$$

Where, Q_c = cooling capacity
ΔT_{max} = max temperature difference across the module (72 ^0C for single stage modules)

Then a TEC module with maximum cooling capacity equal to or greater than the calculated maximum cooling capacity is selected from the TEC module performance data tables. It is observed that module TEC1-12706 meets the design requirements of the device. In present work only 2 TEC modules are

IOP Conf. Series: Earth and Environmental Science **312** (2019) 012003 doi:10.1088/1755-1315/312/1/012003

employed to obtain the nominal power rating or the cooling capacity. We can apply more TEC modules for the design to reduce the time taken for acquiring the desired temperature in the cabin.

Table2. Specifications of TEC

Model number	Operating voltage	Maximum Current	Maximum Voltage	Nominal power rating	Couples	Dimensions
TEC1-12706	12V DC	6 Amp	15.4 V	92.4	127	40 x 40 x 3.5 mm

3.7. *Selection of heat sink.* Based on the assumption made for the heat sink temperature i.e., the heat sink temperature should not go beyond 15 ^{0}C more than the ambient temperature, an effective heat removal system should be employed. Following are the different types of heat sinks: (i) Natural convection heat sink, (ii) forced type convection heat sink and (iii) liquid cooled heat sink. In this work forced convection system i.e., fans are used to drive away the heat from the hot side of the thermoelectric module to the surroundings. Generally the thermal resistance (R_T) of the forced convention type heat sinks ranges from 0.10 ^{0}C/W to 0.5 ^{0}C/W.
Substituting the values of T_A, V, I, and Q in the equation $T_H = T_A + (V \times I + Q) R_T$ and the value of R_T is determined and should be verified if it is in range.
$R_T = (T_H-T_A) / (V \times I + Q) = (50-35) / (12 \times 5 + 17.646) = 0.1869\ ^0C/W$
Therefore in present design using a TEC1-12706 module employed with a forced convection cooling system meets the required specifications.

4. Fabrication of thermoelectric device.

The model with the required dimensions is assembled as shown in the figure. The device is powered from a battery which is charged from the solar energy. A digital thermometer is fixed to check the temperature inside the cabin.

Figure 2. Model of thermoelectric cooling device

5. Results and discussions

5.1. S*olar power.* When panel is fixed position voltages values are taken by using multimeter and ammeter using rheostat and power is calculated from P=VI for different time intervals

IOP Conf. Series: Earth and Environmental Science **312** (2019) 012003 doi:10.1088/1755-1315/312/1/012003

Table 3. Calculating DC power from obtained voltage and current

TIME	09:30	10:00	10:30	11:00	11:30	12:00	12:30	13:00	13:30	14:00	14:30	15:00
Voltage (V)	19.93	19.95	19.90	19.85	19.94	19.92	19.93	19.81	19.87	19.77	19.80	19.63
Current (A)	2.47	2.74	2.50	2.51	2.50	2.50	2.47	2.36	2.15	1.98	1.93	1.65
DC Power (W)	49.23	49.28	49.75	49.82	49.85	49.80	49.23	46.75	42.72	39.14	38.21	32.39

Total number of cells used in solar panel 54;
Voltage of each cell 0.5 V
Maximum voltage obtained from panel=54 x 0.5=27 V
Capacity of battery used is 12 V- 26Ah
Average current available =2.3A
Number of hours required to charge the battery =26/2.3=11.3 hrs
Thermoelectric device capacity is 12V -92W
Current consumed by battery is 6A
Time taken for discharge of battery is 26/6= 4.3 hrs

5.2. *Comparision of time and temperature curves without and with application of PCM Formic acid*
The ambient temperature is observed as 35 ^{0}C which remains constant. The temperature of the cabin initially ie., at 0minutes is observed as 35 ^{0}C without and with the application of PCM. The variation of time and temperature is tabulated for every 10 minutes separately. The temperature of the cabin is observed with the help of digital thermometer fixed in the cabin. A graph is drawn between time and temperature with and without application of PCM. It is observed that the time taken for the cabin to cool down with the application of PCM is less when compared to the time taken to cool down without PCM.

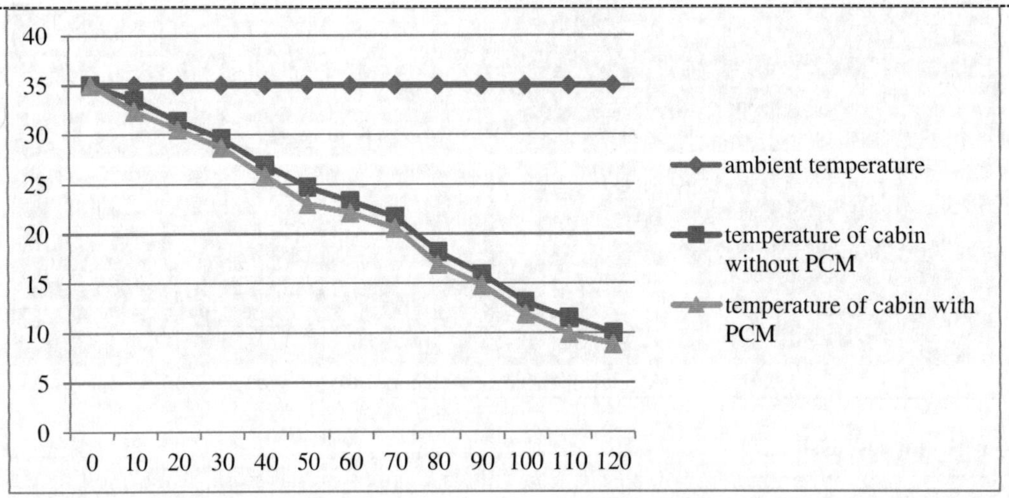

Figure 3. Variation of temperature (y-axis) with time (x-axis)

IOP Conf. Series: Earth and Environmental Science **312** (2019) 012003 doi:10.1088/1755-1315/312/1/012003

6. Conclusion

Therefore a thermoelectric device is designed successfully for producing cooling effect which fulfills the mentioned objectives. However there are some limitations in this system but it has the advantage of utilising the renewable energy i.e., solar power and the cooling inside the cabin will retain for more time with the help of PCM which made the system energy efficient. In the present work, the cooling effect is produced in the cabin of thermoelectric device with and without the application of PCM. This has a wide range of applications in optoelectronic devices, for storing medicines under controlled temperatures etc. Further improvements can be done in this work like adding metallic foams like copper, aluminium etc to the PCM to enhance its thermal conductivity so that the desired temperature can be obtained in less time and to retain the cooling effect for more time. As renewable energy is used as power input, this device can be used in remote and rural places where there is no electric supply.

References

[1] Swapnil B Patond, Feb 2015 *Experimental Analysis of Solar Operated Thermo-Electric Heating and Cooling System* (IJETT ISSN: 2231-5381) **20** 03

[2] Sujith G, Antony Varghese, Ashish Achankunju,Rejo Mathew, Renchi George and VishnuV, 2016 *Design and fabrication of thermoelectric refrigerator with thermosiphon system* (IJSEAS ISSN: 2395-3470) **2**

[3] Mehling and H Cabeza L.F. *Heat and cold storage with PCM. An upto date Introduction into basics and applications* (http:/www.springer.com)

[4] Kondakkagari Dharma Reddy, Pathi Venkataramaiah and P Praveen kumar, 2015 *A Study on Phase Change Material Based Thermal Energy Storage System Using Fuzzy Logic* (International Journal of Applied Engineering Research © Research India Publications ISSN 0973 4562) **10** pp 18089-18103

[5] Manoj S Raut and Dr P. V. Walke , May 2012 *Thermoelectric Air Cooling For Cars* (IJEST ISSN : 0975-5462) **4** 05

[6] Siva Kumar N, Agathiyan V, Dhanush S, Gokulnath V and Jagadeshan R, 2018 *Fabrication of solar refrigeration system by peltier effect* (IJARIIE-ISSN(O)-2395-4396) **4**

[7] Juan P Trelles and John J Duffy *Numerical Simulation of a Porous Latent Heat Thermal Energy Storage for Thermoelectric Cooling* (Energy Engg Dept, University of Massachusetts Lowell MA 01854, U. S. A)

[8] Sathiyaraj R, Rakesh R, Mithran N and Venkatesan M *Enhancement of heat transfer in phase change material using graphite- paraffin composites* (http://creativecommons.org/licenses/by/4.0/).

[9] Sabah A Abdul-Wahab, Ali Elkamel b, Ali M Al-Damkhi, Hilal S Al-Rubai'ey, Abdulaziz K Al-Battashi and Muhammad U Chutani, 2009 *Design and Experimental Investigation of portable solar thermoelectric refrigerator* (An International Journal for Renewable Energy).

[10] Mayank Awasthi and K V Mali, 2012 *Design and development of thermoelectric refrigerator* (Int. J. Mech. Eng. & Rob. Res. ISSN 2278 – 0149) **1**.

Optimisation of Parameters in Thermal Energy Storage System by Enhancing Heat Transfer in Phase Change Material

V. Sravani*, Dr. K. Dharma Reddy

Department of Mechanical Engineering,
Sri Venkateswara University, Tirupati, Chittoor – 517502
Andhra Pradesh, INDIA.

Email: sravaniveligaram@gmail.com

Abstract. Phase change material (PCM) is used to store the heat in the form of latent heat and sensible heat. Thermal energy storage by latent heat using PCM was recognized as attractive alternative to sensible heat storage, because large quantity of thermal energy is stored in small volume. In the present experimental analysis, Beeswax is used as PCM and it is enclosed in stainless steel capsules. Beeswax exhibits slow thermal response due to low thermal conductivity value, so an effort is made to dope PCM with copper, aluminium and graphite powders separately to enhance its heat transfer. The capsules are stored in insulated tank and hot water is supplied into it. The experimental design is prepared by considering parameters as Heat transfer fluid inlet temperatures, Flow rate and Number of Capsules. Experiments are conducted based on Taguchi analysis and the responses are recorded. The effect of considered parameters on Thermal Energy Storage (TES) system is studied by analyzing experimental data to find the optimal parameters for designing the effective TES system.

Keywords: Phase change material; Taguchi analysis; Thermal Energy Storage system

1. Introduction

In recent years, studies of phase change material have attracted much attention in view of their application in thermal energy storage system such as Green house heating, Building applications, Automobiles, Cooling of electronic devices, Solar power plants [1]. Thermal energy storage framework is the most popular method utilized for storing and releasing thermal energy in the form of sensible heat and latent heat [2-3]. Thermal energy storage in Latent heat form using phase change material is better method because it gives benefits like reduction in temperature variability and high thermal energy storage density [4-5]. Melting and solidifying phase of a phase change material will absorb and release large amount of heat during its phase change because of its low thermal resistance [6].

Various PCM's are classified mainly based on compositions (like organic PCM's and inorganic PCM's) and melting points (like high temperature PCM's and low temperature PCM's). The organic PCM's are normally preferred to use in Latent heat type thermal energy storage system because they

IOP Conf. Series: Earth and Environmental Science **312** (2019) 012004 doi:10.1088/1755-1315/312/1/012004

are chemically stable, biodegradable, non toxic and non-corrosive in nature [7]. Low Thermal Conductivity of PCM reduced the heat transfer performance of charging and discharging cycle during the phase change [8]. Much effort has been devoted to improve the thermal storage performance of PCM by adding different metal powders [9]. Composite PCM's exhibits larger thermal conductivity and the charging time were decreased with enhanced heat transfer characteristics [10].

The investigation on the parameters optimization in thermal energy storage system is still limited and much experimental work needs to be done in this direction. This report is concerned with the optimization of parameters in thermal energy storage system by enhancing heat transfer in phase change material with metal powders. In the present study, Beeswax is used as PCM and compared its characteristics of heat transfer by doping with various metal powders such as copper (Cu), aluminium (Al) and graphite (GR). The charging and discharging characteristics of composite PCM were studied. The influencing parameters on charging and discharging times are Heat transfer fluid inlet temperatures, Flow rate and Number of Capsules.

2. Experimental Setup

The fig.2.1 given below depicts the experimental setup. The Beeswax with 62°C melting temperature and 141.49 kJ/kg latent heat of fusion is used as phase change material. The heat transfer characteristics of composite beeswax is studied using stainless steel tank of 10 litres capacity, heat source (electric heater), flow meter, composite PCM filled spherical capsules of 55mm diameter and water storage tank. The tanks are insulated with glass wool. The storage tank is evenly packed with composite PCM filled capsules. The water is taken as both HTF and SHS in this experimental setup.

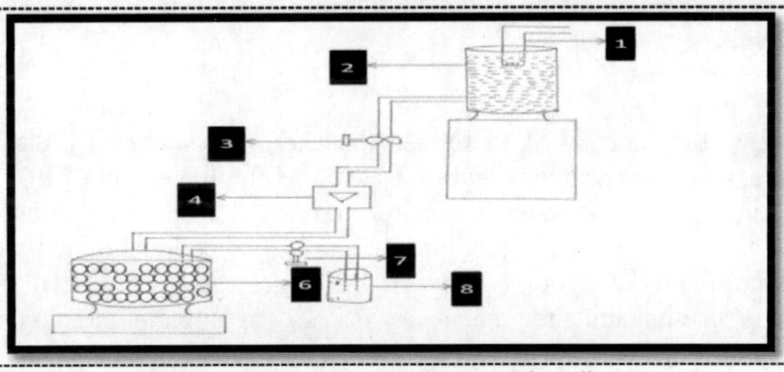

Figure 2.1. Experimental setup Line diagram
1. Heat source, 2. Heat transfer fluid storage tank, 3&7. Flow Regulators, 4. Flow meter, 5. PCM capsules, 6. Thermal energy storage Tank, 8.Outlet tank

Flow rate of Heat transfer fluid is measured by using flow meter. By using different tap openings flow rate is changed. The digital thermometers of ± 1° C accuracy are incorporated in TES tank to measure HTF and PCM temperatures. By using electric heat source HTF temperatures in storage tank is maintained constant. The properties of Bees wax is given in the table given below.

Table 2.1.PCM Properties

PCM	Melting point Temperature (°C)	Latent heat of fusion (kJ/kg)	Density, ρ (kg/m³)		Specific heat(J/kg°C)	
			Solid	Liquid	Solid	Liquid
Bees Wax ($C_{15}H_{31}COOC_{30}H_{61}$)	62	141.49	960	745	3400	2801

IOP Conf. Series: Earth and Environmental Science **312** (2019) 012004 doi:10.1088/1755-1315/312/1/012004

The influencing factors on thermal energy storage tank considered are HTF inlet temperature, HTF flow rate and number of PCM capsules. The experimental runs are obtained using Taguchi design in Minitab software. The factors and levels considered to create design are given in the table 2.2

Table2.2. Influencing parameters and their levels

S.No.	Parameters	Levels		
		L1	L2	L3
1	Heat transfer fluid inlet temperature(^{0}C)	65	70	75
2	Flow rate(liter/minute)	2	3	4
3	Number of capsules(No)	30	35	40

2.1. Charging Process.

In the charging process the heat energy is stored by circulating HTF continuously into the TES tank. Before starting charging process the PCM filled inside the capsules will be at room temperature of 34°C, which is lower than its melting point temperature. The PCM stores energy in the form of sensible heat until it reaches melting point, after that it stores large energy in the form of latent heat in liquid PCM at constant temperature phase change in small volumes. During this process the PCM and HTF temperatures are with every 2 min interval until they become equilibrium with each other. By this process the effect of given parameters on charging time is studied. Mostly less charging time is preferred for best results.

2.2. Discharging Process.

After charging process, the time taken to release the energy is studied by discharging process. Discharging process can be studied by taking out 2 liters of HTF liquid from TES tank for every 20 minutes and adding same quantity of water at room temperature of 34°C into TES tank. This process is continued until PCM temperature reaches room temperature. By this discharging time and quantity of water used to discharge are recorded for the study. High discharge time is preferred for best results. By separately adding copper, aluminum and graphite powders having different thermo physical properties to beeswax its heat transfer characteristics are studied comparatively.

2.3. Results and discussions.

The responses obtained from experimental runs for charging time(CT) in minutes, discharging time(DT) in minutes and discharge water quantity (QW) in litres for three combinations are detailed below in figure2.3.1

Minitab - Untitled - [Worksheet 1 ***]

↓	C1	C2	C3	C4	C5	C6	C7	C8	C9	C10	C11	C12	C13
	HTF TEMPERATURE	PCM CAPSULES(No)	FLOW RATE(lit/min)	CU-CT	CU-DT	CU-QW	Al-CT	Al-DT	Al-QW	GR-CT	GR-DT	GR-QW	
1	65	30	2	40	240	26	38	300	32	32	340	36	
2	65	35	3	38	260	28	42	300	32	30	340	36	
3	65	40	4	36	280	30	38	320	34	28	360	38	
4	70	30	3	34	300	32	36	340	36	26	380	40	
5	70	35	4	32	320	34	34	360	38	24	400	42	
6	70	40	2	36	280	30	38	320	34	28	360	38	
7	75	30	4	28	380	40	30	420	44	20	460	48	
8	75	35	2	30	320	34	32	380	40	22	400	42	
9	75	40	3	32	340	36	36	400	42	24	420	44	
10													

Figure 2.3.1 Experimental runs and responses

IOP Conf. Series: Earth and Environmental Science **312** (2019) 012004 doi:10.1088/1755-1315/312/1/012004

From the responses of all combinations beeswax and graphite powder combination with high thermal conductivity gave best results than copper and aluminium combinations. So the responses from graphite beeswax combination experimental runs are analyzed and predicted using Taguchi. The following results are obtained using Taguchi analysis. The obtained results are shown in tables 2.3.1 and 2.3.2. In this study response characteristics like signal to noise ratios and the means were selected to find out which factor is statistically determinant. To determine the relative importance of each factor in the model response characteristic coefficients are used. Relative strength of each factor is indicated by absolute value coefficient. If that absolute value coefficient is large then that factor will have large impact on response characteristics. The average of each response characteristics is shown in the response table for each level of each factor. The relative magnitudes of effects are compared using ranks based on delta statistics which is also given the tables. The difference in highest average to the lowest average in the table gives delta statistics. The predicted main effect plots for means and SN ratios are shown in figures 2.3.2 and 2.3.3.

Table2.3.1. Response Table for Means

Level	HTF TEMPERATURE(^{0}C)	PCM CAPSULES(No)	FLOW RATE(lit/min)
1	137.8	**153.6**	144.2
2	148.7	148.4	148.9
3	**164.4**	148.9	**157.8**
Delta	26.7	5.1	13.6
Rank	1	3	2
Optimal stages	3	1	3

Table2.3.2. Response Table for Signal/ Noise Ratios
Nominal is best (10×Log10(Ybar^2/s^2))

Level	HTF TEMPERATURE(^{0}C)	PCM CAPSULES(No)	FLOW RATE(lit/min)
1	-2.365	-2.597	-2.506
2	-2.593	**-2.606**	-2.561
3	**-2.810**	-2.566	**-2.701**
Delta	0.446	0.040	0.195
Rank	1	3	2
Optimal stages	3	2	3

IOP Conf. Series: Earth and Environmental Science **312** (2019) 012004 doi:10.1088/1755-1315/312/1/012004

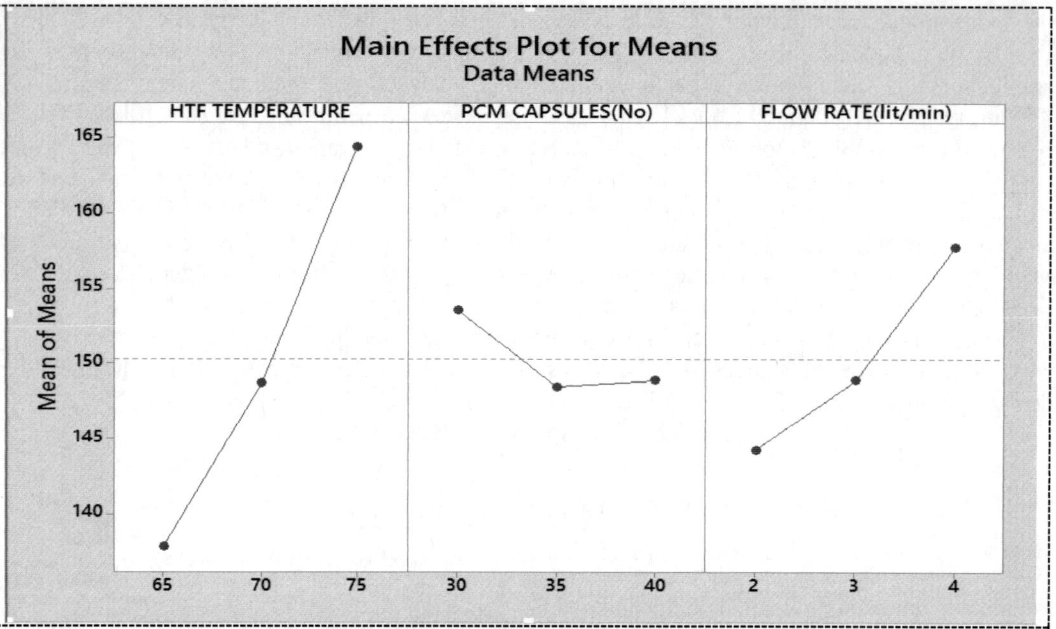

Figure 2.3.2. Means plot

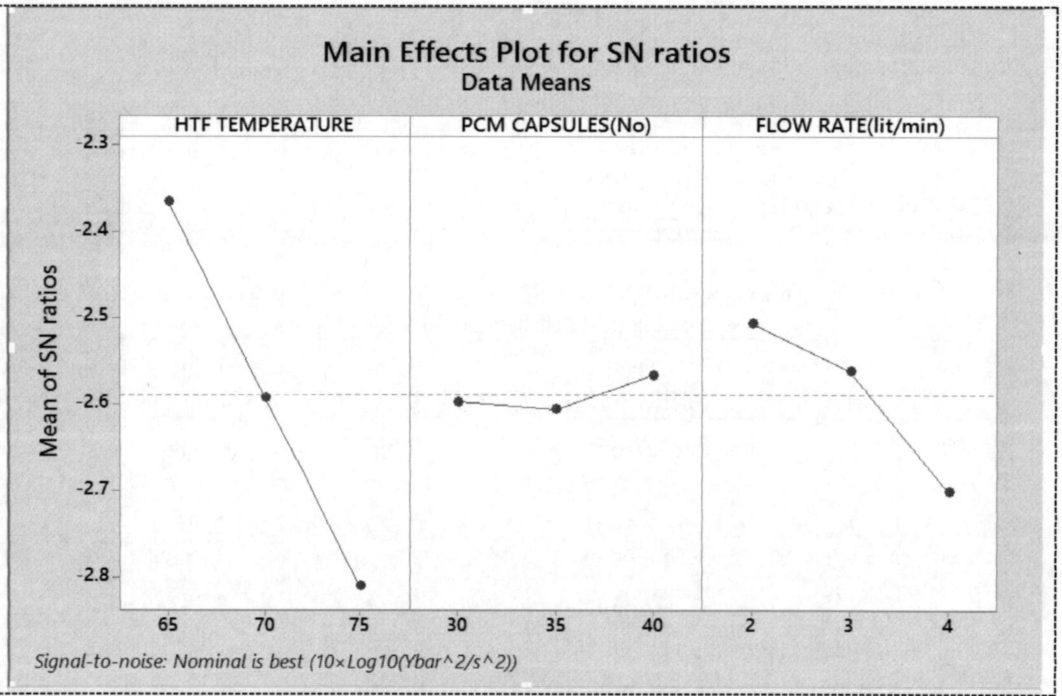

Figure 2.3.3. SN ratios plot

2.3.1. Regression. A statistical technique used to determine relation between two or more variables. The following equations are obtained from Taguchi method using Minitab software.

Regression equations

GR-CT	=	84.67 - 0.8000 HTF TEMPERATURE + 0.0667 PCM CAPSULES(No) - 1.667 FLOW RATE(lit/min)

GR-DT	=	-188.9 + 8.000 HTF TEMPERATURE - 1.333 PCM CAPSULES(No) + 20.00 FLOW RATE(lit/min)

GR-QW	=	-16.89 + 0.8000 HTF TEMPERATURE - 0.1333 PCM CAPSULES(No) + 2.000 FLOW RATE(lit/min)

3. Conclusion

From the results following conclusion can be drawn:

- The drawback of low thermal conductivity of beeswax is reduced by doping beeswax with metal powders. The graphite / beeswax composite showed better performance like less charging time and high discharging times when compared with copper and aluminium combinations.
- The effect of inlet temperature of heat transfer fluid at 75°C and flow rate of 4lit/min is more on charging time and discharging time
- From the result it is concluded that Taguchi design is able to select the optimum parameters and formulate regression equations to find relation between the variables.

References.

[1] Kondakkagari Dharma Reddy, Pathi Venkataramaiah, Tupakula Reddy Lokesh, *Energy and Power Engineering* 537-549, (2014)
[2] Kondakkagari Dharma Reddy, Pathi Venkataramaiah, P. Praveen kumar, *International Journal of Applied Engineering Research* ISSN 0973 4562 Volume **10**, pp. 18089-18103 © Research India Publications, (2015)
[3] Francis Agyenim, Neil Hewitt, Philip Eames, Mervyn Smyth, *Renewable and Sustainable Energy Reviews*. **14**, pp. 615–628, (2010)
[4] Xiao, X., Zhang, p., and Li, M., *Preparation and thermal characterization of paraffin/metal foam composite phase change material,* **112**, 1357-1366, (2013)
[5] Tian, Y., and Zhao, C.Y., *Natural convection investigation in porous, Phase change material,* **3**, 769-772, (2011)
[6] Murat Kenisarin, Khamid Mahkamov, *Renewable and Sustainable Energy Reviews.* **11**, 1913–1965, (2007)
[7] Thirugnanam. C, Marimuthu. P, *International Journal of Engineering and Innovative Technology,* **3(2)**, 372-376, (2013)
[8] Nihal S, Emel O, *Thermochim Acta.* **540**, pp. 7–60, (2012)
[9] P. Zhang a, Z.N. Meng a, H. Zhu a, Y.L. Wang b, S.P. Peng, *Applied Energy.* **185**, pp. 1971–1983, (2017)
[10] Mettawee ES, Assassa GMR, *Sol Energy*. **81**, pp. 839–45, (2007)

Production of methanol from biogas using methanotrophs for the application in a microbial fuel cell

Alok Kumar, Anand Ramanathan*

Department of Mechanical Engineering

National Institute of Technology Tiruchirappalli 620015, India

*Corresponding Author. Tel: 91-431-2503423 Fax: 91-431-2500133,
Email id: anandachu@nitt.edu

Abstract. Methane is available in abundance as natural and renewable gas. Methane-rich biogas has low power density so cannot be considered as a good substrate for the microbial fuel cell. The power density of methane-rich biogas can be increased by its conversion to methanol. Two-step strategy has been followed for the production of methanol from methane-rich biogas and its utilization in a microbial fuel cell. Methanol production using methanotrophs is an attractive way for the valorisation of waste-derived biogas. Besides this, methanotrophs have been used in converting the methane-rich biogas into a variety of bioproducts which include single cell protein, polyhydroxyalkanoate bioplastics, biodiesel, extracellular polysaccharides, propylene oxide, and human health supplements. This paper focusses on the biological conversion of methanol in which solid-state anaerobic digestate can be used for the isolation of methanotrophic bacterium. A fresh isolate possesses the characteristics which resembled to constrained methanotrophs from the methylocaldum genus. This freshly new isolated methanotroph evolved on purified methane or biogas and successful production of methanol is accomplished from the biogas. This study also deals with the effect of formate and various inhibitors of methanol dehydrogenase for the biological conversion of methanol and where the formate act as an electron donor. But isolate produced methanol without any electron donor with the help of phosphate. Moreover, isolate also produced methanol without methanol dehydrogenase inhibitor by the use of formate. The maximum quantity of methanol consists of an optimum quantity of formate and phosphate as a growth medium. This study reveals an efficient conversion of methanol from biogas using solid-state anaerobic digestate. The produced methanol will not go for the purification process and can be used directly in a microbial fuel cell for the power generation purpose.

Keywords: Methanotrophs; methylocaldum genus; methanol dehydrogenase; formate; phosphate; microbial fuel cell.

1. Introduction

Million tons of biomass wastes which include crop residues, food wastes, and animal manure are generated [1]. These generated biomass waste can be used for energy production and different types of valuable products. Anaerobic digestion technology can help in degradation of role in these biomass waste for the biogas production [2]. Biogas generally contains methane (CH_4), carbon dioxide (CO_2), and less quantity of contaminant gas which includes ammonia (NH_3) and hydrogen sulfide (H_2S). The major drawbacks of biogas are its low energy density and are costly as well difficult to store, distribute and transport. That's why biogas in small scale anaerobic digester (AD) is often gets flared and wasted [3]. Biogas can also be captured from landfills and can be used as renewable fuels. Biogas can be converted to compressed natural gas as well as liquid chemicals via thermochemical processes [4]. Biological conversions are generally more attractive than thermochemical conversions [5]. A diverse group of bacteria commonly known as methanotrophs uses monooxygenase (MMO) for the oxidation which converts methane to methanol. These bacteria consume two reducing equivalents for the oxidation of a single molecule of methane. Normally, formate dehydrogenase (FDH), methanol dehydrogenase (MDH), and formal dehydrogenase (FalDH) further result in oxidation of methanol to carbon dioxide (CO_2) which generates reducing equivalents of six per molecule of methanol. That's why, production of methanol with methanotrophs is accomplished by the use of MDH inhibitors such as phosphate and sodium chloride (NaCl), and usually provides external electron donors which are generally formatted to carry out the metabolic activity and survival of methanotrophs [5]. Biological upgradation of biogas using methanotrophs is an interesting approach to add the value to the waste-derived methane gas because methanotrophs grow at ambient pressure and moderate temperature, can utilize a low concentration of methane (CH_4) which is generally less than 20%, and capable of producing methanol at high efficiency [5-7].

Trickle bed bioreactors have a fascinating design for the cultivation of methanotrophs because of their low costs and limited power requirements as compared to that of membrane bioreactors [8, 9]. Trickle bed bioreactors (TBRs) are a type of cylindrical packed reactors which is packed with inert materials having high specific surface area [10]. A thin layer of liquid is provided on the packing surface by circulating the nutrient medium through the trickle bed bioreactor (TBR) while gases are blown up either co-current or counter-current to the liquid [8]. In this study, the two-step process has been discussed for the generation of power from methanol derived from biogas.

Methanotrophs isolation can improve the feasibility of methanol production from the biogas. Isolation of methanotrophs from various sources such as waste treatment facilities, natural gas fields, and soils [5]. Few studies also focus on the production of methanol by methanotrophs isolation from hydrogen-rich anaerobic digester using biogas as a source [11]. Methanotrophs can last long without oxygen and can uses biogas as their main source of carbon as well as carbon. This paper focusses at the role of formate and inhibitors of MDH for the biological conversion of methanol, Isolation and characterization of methanotrophs from the solid state anaerobic digester, comparison of purified methane (CH_4) and biogas as a substrate for isolation of methanotrophs. Finally, the produced methanol using methanotrophic isolation can be utilized in a microbial fuel cell (MFC). This paper also includes the details regarding the construction associated with a MFC.

2. Materials and methods

2.1. *Methanotrophs isolation*

Eight samples of digestate can be collected from solid-state AD reactors. The quantity of each sample should be 5g [12]. The switchgrass and anaerobically digested wastewater sludge in the ratio of 3:1 can be fed into one liter solid-state AD reactors. Total solids (TS) required for controlling the solid-state reactor is either 20% or 30%, and incubation period under mesophilic conditions (36±1) is 70 days. The variation in methane yield of solid-state AD reactors is varied from 88 l/kg VS_{add} (TS of 30%) to 113 l/kg VS_{add} (TS of 20%) [12]. The procedure of isolation is the modified version of protocols explained in Dedysh and Bowmen [13, 14].

The incubation of solid-state AD samples can be performed in 250 ml flasks which consisted of nitrate mineral salts medium (45 ml per flask). The nitrate mineral salts medium consisted of 0.2% (v/v) of chelated iron (Fe) solution along with 0.284 g/l of sodium diphosphate (Na_2HPO_4), 1.0 g/l of magnesium sulfate heptahydrate ($MgSO_4 \cdot 7H_2O$), 0.134 g/l of calcium chloride dihydrate ($CaCl_2 \cdot 2H_2O$),

IOP Conf. Series: Earth and Environmental Science **312** (2019) 012005 doi:10.1088/1755-1315/312/1/012005

0.272 g/l of Monopotassium phosphate (KH_2PO_4), 1.0 g/l of potassium nitrate (KNO_3), and 0.05% (v/v) of a trace element solution. The chelated iron (Fe) solution consisted of 2 g/l of Ethylenediaminetetraacetic acid (EDTA), 1 g/l of ferric (III) ammonium citrate, and 0.3% (v/v) of concentrated hydrochloric acid in deionized water. The composition of trace element in solution includes 200 mg/l of Iron(II) sulfate heptahydrate ($FeSO_4•7H_2O$), 30 mg/l of Boric acid (H_3BO_3), 10 mg/l of Zinc sulfate ($ZnSO_4•7H_2O$), 500 mg/l of EDTA, 20 mg/l of Cobalt(II) Chloride hexahydrate ($CoCl_2•6H_2O$), 3.0 mg/l of Sodium molybdate dihydrate ($Na_2MoO_4•2H_2O$), 1.0 mg/l of calcium chloride dihydrate ($CaCl_2•2H_2O$), 3.0 mg/l of Manganese(II) chloride tetrahydrate ($MnCl_2•4H_2O$), and 2.0 mg/l of Nickel(II) chloride hexahydrate ($NiCl_2•6H_2O$) in deionized water. The initial pH of nitrate mineral salts should lie in the range of 6-7. The rubber stopper can be used for sealing the flasks and 0.2 μm of a filtered mixture of pure methane (99% purity) should be filled in the headspace of the flasks and 1:4 (v/v) of a CH_4: air ratio can be selected. The incubation temperature of 37 °C and continuous shaking with 200 rpm can be followed. After the completion of three days, an enriched culture of 5–10 ml should be transferred into new flasks having 40–45 mL of freshly nitrate mineral salts (NMS) medium and similar ratio of methane to air can be followed. This process has to be repeated every three days. After the completion of 25 days, the enriched mixture has to be spread on the nitrate mineral salt agar plate. These plates will be incubated in an anaerobic jar at a temperature of 37 °C for 1-2 weeks having headspace of air and methane (CH_4) in the ratio of 1:4 (v/v). The freshly nitrate mineral salts plates could have been selected for streaking of selected individual colonies until single colonies based on size, shape, and color are visible. The obtained isolates should be shifted to 1 ml of nitrate mineral salts (NMS) medium in a sealed vial of 15 ml capacity with the reactor headspace containing 0.2 μm of filtered methane and air in the ratio of 1:4. The incubation temperature of 37 °C can be followed by shaking at 300rpm until the visible increase of turbidity in culture. Light microscopy can be used for determining cell morphology. If the culture doesn't grow in nitrate mineral salts (NMS) improved with glucose (0.05% (w/v)) [13, 14].

The analysis and classification of 16S rRNA gene sequences can be followed by the BLAST and RDP Classifier against the NCBI RefSeq database [15]. The Methylocaldum sp. 14B is one the fastest growing isolates that's why it can be selected for further experiment.

2.2. *Various physical and chemical inputs involved in cultivation.*

The strain 14B ability can be tested for the utilization in different sources of carbon by growing aerobically in 1 ml quantity of liquid nitrate mineral salts (NMS) medium which is added with 0.1% (w/v) of citrate, methanol, acetate, glucose, formate, or xylose. Cultivation can be performed in vials of 5 ml with shaking at 200 rpm and a temperature of 37 °C. The growth can be observed by the variation in optical density and light microscopy can be used for examining the cell morphology. A fixed quantity (20 ml) of nitrate mineral salts medium in the flask of 125 mL can be selected for culturing the strain 14B with air and methane (CH_4) ratio of 4:1 (v/v) in headspace in order to determine the effectiveness of sodium chloride (NaCl), copper (Cu^{2+}), temperature, nitrogen source, and pH on growth of cell using methane (CH_4) as the only source of carbon. The copper should be selected for testing as it had shown influence on the growth of the cell specifically for the methanotrophs which acquire particulate methane monooxygenase [16].

The assessed concentrations of Cu^{2+} in a fixed quantity of $CuCl_2$ that needs to be supplemented in nitrate mineral salts (NMS). The liquid nitrate mineral salts (NMS) containing a fixed concentration of NaCl can be used to find out the effect of various concentration of salts. The growth which will take place can be assessed by the replacement of potassium nitrate (KNO_3)in nitrate mineral salts (NMS) medium with ammonium sulfate ($(NH4)_2SO_4$). The cultures should be incubated at a temperature of 37 °C with continuous shaking of 200 rpm in order to determine the Cu^{2+}, pH, NaCl, and nitrogen source. The strain 14B should be cultured in nitrate mineral salts medium and incubation can be carried out at different temperatures with shaking at 200 rpm in order to determine the effect of temperature.

2.3. Formation of methanol

For the effective screening of methanol dehydrogenase (MDH) inhibitors, early growth phase should be selected for strain 14B cells which can be harvested by centrifugation process at 10000 rpm for 15 minutes, and resuspension can be done in 4.1 mL of nitrate mineral salts (NMS) medium containing a fixed quantity of sodium formate along with different concentration of methanol dehydrogenase inhibitors [17]. The most potent MDH inhibitors such as phosphate, magnesium chloride ($MgCl_2$), Ethylenediaminetetraacetic acid (EDTA), and sodium chloride (NaCl) provides a great impact on methanol accumulation. Before undergoing incubation, a fixed volume of cell suspension sample should go for removal and filtration process. Thus obtained filtrate can be analyzed using gas chromatography (GC) for determining the primal methanol presence. The leftover volume of the cell suspension should be shifted to glass vials which are closed with the help of rubber stopper. The purified methane (CH_4) having 99% purity should be added to vials to achieve a ratio of 5.15:1 for air: CH_4 in the headspace. The vials will be subjected to incubation at a temperature of 37 °C with continuous shaking at 200 rpm for 6 hours. After the completion of 6 hours, the cell suspension will be filtered and obtained filtrate will be analyzed using gas chromatography (GC) for the concentration of methanol.

Two replicates can be done for the vials. The volume of 6 ml of methane (CH_4) can be selected to provide an identical quantity of methane (0.25mmol) and 0.27mmol of oxygen (O_2) to the headspace of the reactor. Moreover, formate (80 mM) should be used to supply a quantity of formate (0.25 mmol) concordant to the levels of methane and oxygen in the reactor headspace. The impact of phosphate and formate on the production of methanol from biogas can be tested using 4 levels of sodium formate and 4 levels of phosphate (Na_2HPO_4-Disodium phosphate or KH_2PO_4- Monopotassium phosphate).

The reactors can be set up with above-described data and biogas have been considered as methane source (air: biogas of 3.1:1) in the headspace of the reactor with an optical density at 600nm. The time course of production of methanol can be tested by cultivating strain 14B at a temperature of 37 °C and shaking can be carried out at 200 rpm in nitrate mineral salts medium. This nitrate mineral salts medium will have phosphate with various sodium formate levels. The Tedlar gas bag will be connected to each of the flasks.

3. Construction of microbial fuel cell and its operation

Microbial fuel cell test can be performed in triplicate which consists of single-chambered, cubic shaped air-cathode, and an anode chamber of cylindrical shape (4 cm in length and 6 cm in diameter) [18]. The anode of MFC consists of fiber brush (both length and diameter are 2.5 cm) made up of graphite and is subjected to heat treatment at 450 °C in the presence of air for 30 minutes before the application and placement can be done horizontally in the middle of microbial fuel cell (MFC) chambers. The preparation of cathode can be accomplished using a hot-pressing method [19]. The layer of catalyst can be prepared by mixing 60% of polytetrafluoroethylene emulsion (PTFE) with activated carbon (AC) at a mass ratio of 1:6. A stainless steel mesh (say 42×42) can be selected as a cathode current collector. The selected membrane for microbial fuel cell should be hydrophobic in nature so that it can be used in the diffusion layer in order to minimize loss of water. That's why polyvinylidene fluoride membrane (0.45μm) is one the best choice for the microbial fuel cell. The pressure of 3×107 Pa at a temperature 60 °C for 15 s of should be selected for the pressing of the current collector, PTFE: AC and diffusion layer until the surface of the membrane becomes dry [19, 20]. The cathode which has been pressed can be taken out for drying so that it can be used later on.

The methanol produced from solid state anaerobic digestate (using methanotrophs isolation) can be filled directly in the microbial fuel cell. The potential of the anode can be obtained by placing a silver or silver chloride reference electrode in the central part of microbial fuel cell (MFC) chamber, while the cathode potential can be determined by using whole cell voltage and anode potential. Coulombic efficiency of the cell can be defined as the ratio of total charge displaced to the anode surface to the maximum theoretical charge extractable.

4. Results and discussion

4.1. Investigation of MDH inhibitors for production of methanol

It was interesting to note that without the addition of MDH inhibitors methanol gets accumulated with 80 Mm (Table 1). Few of the cases show less amount of methanol accumulated upon addition of MDH inhibitors (Table 1).

Table 1. Impact of MDH inhibitors for methanol accumulation [17].

MDH Inhibitors	Concentration of MDH inhibitors (mM)	Formate (Mm)	Productivity of methanol (g $l^{-1}d^{-1}$)
Phosphate	100	80	1.17 ± 0.04
	200	0	0.28 ± 0.00
	200	80	0.95 ± 0.13
Magenesium chloride ($MgCl_2$)	20	80	0.98 ± 0.13
	40	0	Not determined
	40	80	1.04 ± 0.01
Ethylenediaminetetraacetic acid (EDTA)	0.5	80	1.12 ± 0.03
	5.0	0	Not determined
	5.0	80	1.04 ± 0.14
sodium chloride (NaCl)	100	80	1.00 ± 0.04
	200	0	0.83
	200	80	0.85 ± 0.09
Without any addition	0	80	0.98 ± 0.05

4.2. Influence of phosphate and formate on the productivity of methanol

Production of methanol gets increased upon addition of phosphate in the absence of formate but maximum content of methanol was found out to be low which lies in between 0.04-0.05 g/l at the phosphate of 200 mM [17]. Moreover, interesting results were found which shows that higher productivity of methanol was obtained on addition of formate without phosphate which lies in between 0.5-1.0 g/l/d [17]. This results in hints that formate dehydrogenase provides limitation in methanol accumulation by strain 14B.

4.3. Methanol-powered microbial fuel cell

The methanol along with buffer solution of phosphate have been used for the operation of the microbial fuel cell. The maximum voltage and power density were found out to be 0.5 V and426 ± 17 mW/m^2 after the completion of the acclimation period of 45 days [21]. During the operation of the microbial fuel cell, fast consumption of methanol takes place and results in 51 ± 4 mg/L as final concentration after completion of 24 hrs of operation [21]. The acetate concentration during the operation reaches a maximum in 12 hrs and after completion of 24 hrs, acetate concentration reduces. The coulumbic efficiency of the microbial in a cycle was found out to be 22 ± 3% [21].

5. Conclusion

The two-step process can be followed to generate power from biogas obtained from the solid-state anaerobic digestion process. The methanotrophic isolate plays a crucial role in the production of methanol from biogas. Production of methanol can also be achieved with the addition of phosphate only in the cultivation medium but production can be further improved upon the addition of formate. Lastly, produced methanol from the biogas can be directly used in a microbial fuel cell without any purification process.

References

[1] *USDA, FY-2005 final manure and byproduct utilization, 2005*.https://www.arsa.usda.gov/SP2Userfiles/Programe/206/NP206FY2005AnnualReport.pdf

[2] L.De Bere, *Anaerobic digestion of solid waste;state-of-the art,water Sci,Technol*.**4**(2000), 283-290.

[3] World bank, flared gas utilisation strategy-opportunities for small-scale uses of gas, 2004.

[4] L yang, X. Ge, C. Wan, F.Yu.Li, *Progress and perspectives in converting biogas to transportation fuels, Renew. Sustain. Energy Rev*.**40**(2014).

[5] X. Ge, L.Yang, J.P.Sheets, Z. Yu, Y.Li, *Biological conversion of methane to liquid fuels:status and opportunities, Biotechnol.Adv*.**32**(2014)1460-1475.

[6] J.S han, C.M. Ahn, B.Mahanty, C.G Kim, *Partial oxidative conversion of methane to methanol through selective inhibition of methanol dehydrogenase in methanotropic consortium from landfill cover soil .Appl.Bichem.Biotech*.**171**(2013)1487-1499, isi:000326374700016.

[7] X.M. Wei, R. He, M. chem, Y.su, R.C Ma, *Conversion of methane-derived carbon and microbial community in enrichment cultures in response to O2 availability, Environ. Sci. Pollut. Res*.**23** (2016) 7517-7528.

[8] M.Devarapalli,H.K.Atiyeh, J.R. Philips, R.S. Lewis,R.L. Huhnke, *Ethanol production during semi-continuous syngas fermentation in a trickle bed reactor using clostridium ragsdalei, Bioresour. Technol.* **209**(2016)56-65.

[9] M.A. Deshusses, H.H.J, Cox, *A cost benefit approach to reactor sizing and nutrient supply for biotrickling filters for air pollution control ,Environ.Prog*.1**8**(1999) 188-196.

[10] V.V. Ranade, R.V.Chaudhari, P.R. *Gujal,Trickle Bed Reactors-Reactor Engineering and applications,Elsevier, Kidlington,Oxford,*2011.

[11] Wenxian Zhang,Xumeng Ge,Yueh-Fen Li,Zhongtang Yu,Yebo Li, *Isolation of a methanotroph from a hydrogen sulfide-rich anaerobic digester for methanol production from biogas*.

[12] Sheets, J.p,Ge,X,Li,Y. 2015. *Effects of limited air exposure and comparative performance between thermophilic and mesophilic solid-state anaerobic digestion of switchgrass.Bioresource.Technol*. **180**,296-303.

[13] Bowman J. *The methanotrophs—the families Methylococcaceae and Methylocystaceae. The Prokaryotes: Volume 5: Proteobacteria: Alpha and Beta Subclasses*. 2006:266-89.

[14] Dedysh SN, Dunfield PF. *Facultative and obligate methanotrophs: how to identify and differentiate them. InMethods in enzymology 2011* Jan 1 (Vol. **495**, pp. 31-44). Academic Press.

[15] Wang Q, Garrity GM, Tiedje JM, Cole JR. *Naive Bayesian classifier for rapid assignment of rRNA sequences into the new bacterial taxonomy. Appl. Environ. Microbiol.. 2007 Aug* 15;**73**(16):5261-7.

[16] Kalyuzhnaya MG, Puri AW, Lidstrom ME. *Metabolic engineering in methanotrophic bacteria. Metabolic engineering. 2015 May 1*;**29**:142-52.

[17] Johnathon P.Sheets, Xumeng Ge, Yueh-Fen Li, Zhongtang Yu, Yebo Li. *Biological conversion of biogas to methanol using methanotrophs isolated from solid-state anaerobic digestate.*

Bioresource Technology. Volume **201**, February 2016, Pages 50-57.

[18] H. Liu, R. Ramnarayanan, B.E. Logan, *Production of electricity during wastewater treatment using a single chamber microbial fuel cell, Environ. Sci. Technol*. **38** (2004) 2281–2285.

[19] W. Yang, B.E. Logan, *Engineering a membrane based air cathode for microbial fuel cells via hot pressing and using multi-catalyst layer stacking, Environ. Sci. Water Res. Technol*. **2** (2016) 858–863.

[20] H. Dong, H. Yu, X. Wang, Q. Zhou, J. Feng, *A novel structure of scalable air-cathode without Nafion and Pt by rolling activated carbon and PTFE as catalyst layer in microbial fuel cells, Water Res.* **46** (2012) 5777–5787.

[21] Jaewook Myung, Pascal E. Saikaly, Bruce E. Logan. *A two-staged system to generate electricity in microbial fuel cells using methane. Chemical Engineering Journa*l **352** (2018) 262–267.

Acknowledgments

The director of NIT, Tiruchirapalli provided the platform to learn the topic of my interest. The mechanical department specifically thermal engineering lab of this institute plays an important role in developing the interest in the study related to the fuel cell.

Development of computer aided modelling and optimization of microwave pyrolysis of biomass by using aspen plus

M Dineshkumar, Banagiri Shrikar, Anand Ramanathan*

Department of Mechanical Engineering,
National Institute of Technology, Tiruchirappalli-620 015, India.
*Corresponding Author. Tel: (0431) 2503423, Fax: (0431) 250 0133,
E-mail: anandachu@nitt.edu

Abstract. Biomass energy is nowadays recognized as a potential source which constitutes the main portion of expected renewable energy supplies in the future. Biomass resources such as municipal solid waste, agricultural residue food processing industry waste, energy crops are a common type of renewable energy source and it help to produce energy, chemicals, and fuels. The useful energy recovered from biomass can be effectively utilized by a bio-energy conversion technique known as pyrolysis. Pyrolysis is considered as a kind of thermo-chemical conversion technique of waste material in the inert atmosphere to harvest biochar, syn-gas and bio fuel. Bio fuel harvest from forestry biomass through the microwave pyrolysis technology has been significantly attracting attention in the renewable energy sector over recent years. This is because of fact that microwave pyrolysis can possibly reduce greenhouse gases and contribute to energy security. A simulation model of pyrolysis process is developed by using a software called advanced system for process engineering it is basically a simulation tool for computer-aided energy modelling. This software is used to optimize and analyse the efficiency of the process involved in pyrolysis and to increase the output product yield such as bio-oil, syn-gas, and biochar with respect to the function of input parameters such as pyrolysis temperature and physio-chemical character of the biomasses. The ASPEN PLUS simulation is carried out using four different variety of biomasses namely Acacia Nilotica, Calophyllum inophyllum seed, rice husk, and Bael shell. The simulation results demonstrate that Calophyllum inophyllum seed is best suited for bio-oil production through microwave pyrolysis as it contains low moisture content and higher cellulose. The obtained bio-oil yield is up to 48% can be found from this non-edible biomass.

Keywords: Microwave Pyrolysis; Biomass; Simulation; biofuel; Aspen plus.

IOP Conf. Series: Earth and Environmental Science **312** (2019) 012006 doi:10.1088/1755-1315/312/1/012006

1. Introduction

The global energy demand is increasing day by day. This is supported by the fact that the global primary energy consumption has grown by 2.2 % in 2017. This is the fastest growth recorded since 2013. Faced with a steep decline of fossil fuel consumption such as coal, the demand for renewable energy is on the rise. The Renewable energy generation rise by 17%, thus creating the largest increment at 69 million tons of oil equivalent (mtoe) [1]. Within the alternative energy scenario, bio mass energy is quickly become the major sources for energy. This is due to the fact that biomass is abundantly available, especially in developing countries like India which have large agriculture sectors. Biomass produces renewable solid, liquid and gaseous fuels. These alternative fuels are known as bio fuels [2]. A large number of studies were conducted to find alternate fuel sources.

Bioethanol and biodiesel are showing promising results in this regard. India proposed its biofuel policy which mandates the blending of 20% biofuel to conventional petroleum fuels. Thus, biomass is considered a very effective renewable energy source it very well utilized to meet the rising energy demand in the future [3]. India is an agricultural country, and it produces tons of organic waste which can be converted into biofuels by utilizing methods like pyrolysis which reduce their environmental side effects [4]. Pyrolysis process is a form of thermochemical conversion technique of biomass to maintain a reactor in an inert atmosphere. Biomass pyrolysis results in generation of a solid waste (alternatively known as bio-char), gases and a dark brown coloured liquid yield (commonly known as bio-oil) [5]. In this study, we have chosen microwave pyrolysis instead of the conventional pyrolysis. There are several reasons behind this choice the microwave irradiation produces hotspots which influence the yield of the products in a major way. The microwave pyrolysis has higher heating rate values and the surface area of the biochar is high. Other advantages of microwave pyrolysis include non-contact heating, quick starting and stopping mechanisms, material selective heating and volumetric heating. Microwave pyrolysis also possesses a unique feature where heating occurs from the interior of the material body [6]. Microwave pyrolysis, however is a very complex process. The decomposition reaction occurring in the pyrolysis of wood material comprises a complicated sequence of reactions it varying consequently with changing heat rate effects, particle physical properties, thermodynamic properties, sample preparation methods, temperature, moisture content and residence time. By take care all of these aspects, it is essential to pursue further study to enrich and improve simulation and modelling in this particular area.

Pyrolysis process modelling, especially for microwave pyrolysis has a widespread variety of approaches with various complexity [7]. A simulation model is developed in ASPEN PLUS for a medium temperature pyrolysis. The simulation model predicts that a modelled coal burning of residual volatiles and char compounds later to pyrolysis process not only provides sufficient energy essential for endothermic reaction, drying of moisture coal apart from this it also supplies additional energy for production of power. The ASPEN PLUS simulation model can efficiently compute the yield of pyrolysis and it used optimize the operating parameters through biomass chemical property values obtained from ultimate analysis results. Though, simulation, modelling are the very best alternatives for understanding the mechanism of pyrolysis reactions, and optimizing an efficient and economic pyrolysis process model instead of performing the time consuming and expensive experimental investigation [8].

The present study aims to develop more green and renewable technology trends which will be of use to the society. In this investigation, we have compared the bio-oil yields of four types of biomass namely Acacia Nilotica (Babul seeds), rice husk, Bael shell and Calophyllum inophyllum. We have used ASPEN PLUS to simulate microwave pyrolysis and as function of input parameters, we have optimized the process.

2. Principle of microwave-assisted pyrolysis process

The microwave assisted pyrolysis technique is a recently introduced method and was initially explored Tech-En limited in UK. This method is quickly developing as one of the greatest technologies among biomass pyrolysis processes. This is because of its effective heat transfer profile, very less residence time, opposing thermal gradients and volumetric heating, which ultimately leads to energy saving [9]. Microwave irradiation lies in the range of 0.3 and 300 GHz in terms of operating frequency and corresponding wave length of 0.01 to 1 meter on the electromagnetic spectrum, which means it lies between the radio wave region and the IR (Infra-red) regions [10]. The microwave irradiation produces microwave energy which penetrates the required material surface, and gets converted into heat energy in the particle core. This heat generated inside the material is carried towards the outside surface, which results in the generation of a temperature variation in between the surface of core in biomass and the outer layer of the particle, and volatile release in the same direction of the temperature gradient [11]. Therefore, microwave pyrolysis technology is considered a capable method to convert waste biomass into valuable green materials and high-quality biofuels.

3. Simulation and Modelling of Biomass

Advanced System for Process Engineering (ASPEN) is used to make computational modelling to simulate and optimize the process involved in microwave pyrolysis. ASPEN PLUS allows the user to develop a chain of sophisticated processes. Every block of a chain of processes can be individually tested with different parameters. In case, the block is complicated, it can be developed using FORTRAN subroutines. ASPEN PLUS is well equipped to calculate chemical, physical and biological parameters [12]. It is worth mentioning that this process of modelling can be performed by taking the biomasses as hypothetical solids in an approach quite different from the one described below. The process involves entering the various substances present in the biomass which are identified through Gas chromatography-mass spectrometry (GC-MS). This process is quite tedious and hence, was omitted in this study.

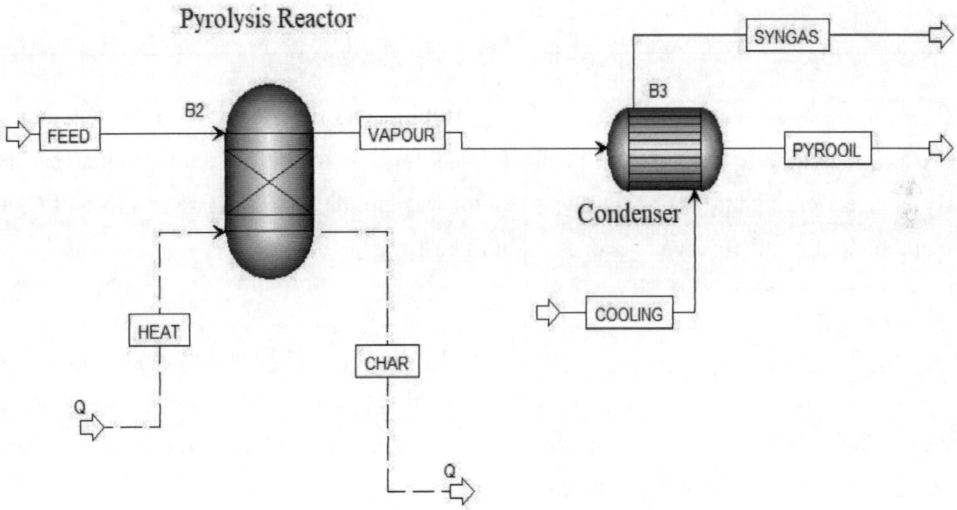

Figure 1. Flowsheet representing the microwave pyrolysis process.

This study uses four biomass wastes namely Acacia Nilotica, Calophyllum inophyllum seed, rice husk and Bael shell. Since, we are concerned with only the bio-oil yield of these seeds, the RYIELD reactor block is chosen. After the feed is passed through this block, the stream is passed through a condenser where the pyrolysis oil and syngas get separated. The flowsheet shown here illustrates the process.

IOP Conf. Series: Earth and Environmental Science **312** (2019) 012006 doi:10.1088/1755-1315/312/1/012006

Table 1. Biomass sample elemental analysis in terms of percentage weight.

Biomass samples	Carbon	Hydrogen	Nitrogen	Oxygen	Sulphur
Bael Shell [13]	40.67	6.23	1.34	51.76	0.235
Callophylum Innophylum [14]	43.82	6.35	3.15	45.98	0.695
Rice Husk [15]	45.28	5.51	0.67	41.6	0.29
Babul Seed [16]	54.1	6.12	5.23	34.53	0.362

Table 2. Proximate analysis of biomasses in terms of percentage weight in dry basis.

Biomass samples	Moisture (%)	Fixed Carbon	Volatile Matter	Ash
Bael Shell [13]	11.68	2.94	91.42	5.64
Callophylum Inophyllum [14]	3.56	21.1	72.61	2.73
Rice Husk [17]	1.3	16.8	69.3	13.9
Babul seed [16]	12.5	11	69.1	7.3

The entire system was simulated using Peng-Robinson (PENG-ROB) property methods. All of the above biomasses were defined as unconventional solids and their proximate analysis and ultimate analysis data were inputted. The pyrolysis reactor was simulated to operate at 450-600°C. The particle size at the input of microwave reactor is taken to be 1mm.

4. Results and Discussion

Callophylum inophyllum gave the highest oil yield with 48% as its maximum yield at a temperature of around 500°C. Babul seeds (Acacia Nilotica) was capable of a maximum yield of 42% in the temperature range of 500°C. Bael shell yield maximum bio oil of 35% at 500°C temperature and Rice husk gave an oil yield around 39% in the temperature of 500°C.

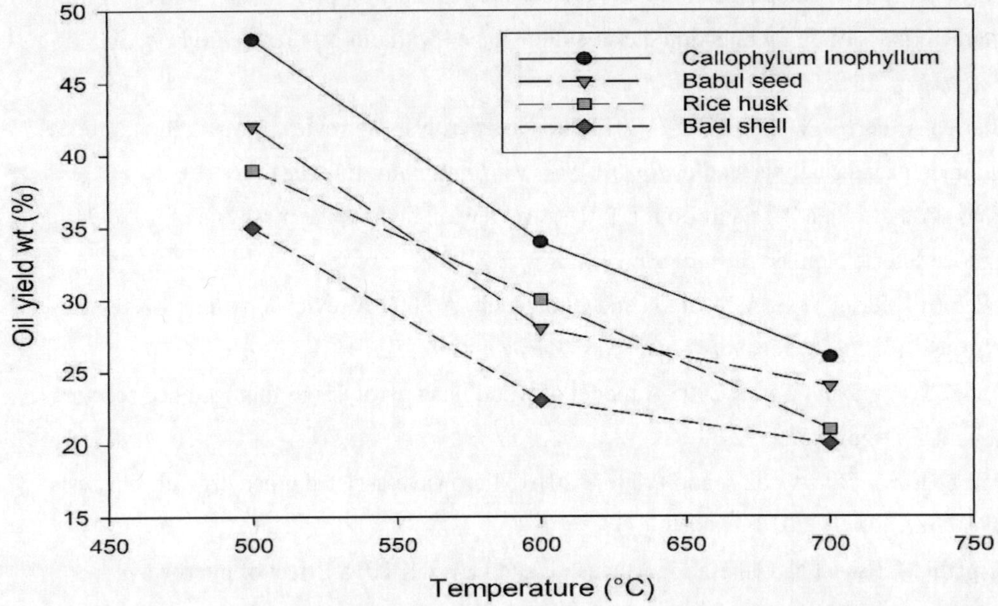

Figure 2. Percentage of bio-oil yield and corresponding temperature for each biomass sample.

There was an increase in percentage yield of syn-gas and bio-char as the temperature was increased. This was especially true for rice husk, whose bio-oil yield rapidly decreased to 25% at 600°C. This Aspen Plus model suggests that, though microwave pyrolysis is in many ways better than conventional pyrolysis it still produces lesser bio-yield and more bio-char with rise in temperature.

5. Conclusion

A computer based mathematical model was generated by utilizing ASPEN PLUS simulation software to investigate the efficiency and performance of the microwave assisted pyrolysis by selecting biomass feedstock with respect to the function of physical, chemical properties and operating conditions. Four different samples namely Acacia Nilotica, Calophyllum inophyllum seed, rice husk and Bael shell were used in the simulation separately to find out the yields of bio-oil, syn-gas and Char. The results obtained was showed that Callophylum inophyllum feedstock produced the highest output of bio-oil as 48% in the temperature range of around 500°C. and it is the best oil yielding seed amongst the four feedstocks for converting biomass into alternative fuel by using microwave pyrolysis technology.

References

[1] BP 2018 67th edition *Stat. Rev.* World Energy 1–56.

[2] D Vamvuka 2011 Bio-oil, solid and gaseous biofuels from biomass pyrolysis processes an overview *Int. J. energy Res.* **35**: 835–862

[3] Mohan D, Pittman C U and Steele P H 2006 Pyrolysis of wood/biomass for bio-oil: a critical review *Energy and Fuels* **20** 848–89

[4] Rajamohan S and Kasimani R 2018 Analytical characterization of products obtained from slow pyrolysis of calophyllum inophyllum seed cake: study on performance and emission characteristics of direct injection diesel engine fueled with bio-oil blends *Environ. Sci. Pollut. Res.* **25** 9523–38

[5] Sharma A, Pareek V and Zhang D 2015 Biomass pyrolysis - a review of modelling, process parameters and catalytic studies *Renew. Sustain. Energy Rev.* **50** 1081–96

[6] Huang Y F, Chiueh P Te and Lo S L 2016 A review on microwave pyrolysis of lignocellulosic biomass *Sustain. Environ. Res.* **26** 103–9

[7] Kabir M J, Rasul M G, Ashwath N and Chowdhury A 2003 A review on green wastes pyrolysis for energy recovery recent *Res. Environ. Geol. Sci.* 101–7

[8] Luo Z, Wang S and Cen K 2005 A model of wood flash pyrolysis in fluidized bed reactor *Renew. Energy* **30** 377–92

[9] De La Hoz A, Díaz-Ortiz A and Prieto P 2016 Microwave-assisted green organic synthesis *RSC Green Chem.* 2016–January 1–33

[10] Tripathi M, Sahu J N, Ganesan P, Monash P and Dey T K 2015 Effect of microwave frequency on dielectric properties of oil palm shell (ops) and ops char synthesized by microwave pyrolysis of ops *J. Anal. Appl. Pyrolysis* **112** 306–12

[11] Haeldermans T, Claesen J, Maggen J, Carleer R, Yperman J, Adriaensens P, Samyn P, Vandamme D, Cuypers A, Vanreppelen K and Schreurs S 2019 Microwave assisted and conventional pyrolysis of mdf – characterization of the produced biochars *J. Anal. Appl. Pyrolysis* **138** 218–30

[12] Puig-Arnavat M, Bruno J C and Coronas A 2010 Review and analysis of biomass gasification models *Renew. Sustain. Energy Rev.* **14** 2841–51

[13] Bardalai M and Mahanta D K 2016 Characterisation of the pyrolysis oil derived from bael shell (aegle marmelos) *Environ. Eng. Res.* **21** 180–7

[14] R. S and K. R 2017 Influence of temperature on yield, composition and properties of the sub-fractions derived from slow pyrolysis of calophyllum inophyllum de-oiled cake *J. Anal. Appl. Pyrolysis* **127** 159–69

[15] Tsai W T, Lee M K and Chang Y M 2007 Fast pyrolysis of rice husk: product yields and compositions *Bioresour. Technol.* **98** 22–8

[16] Garg R, Anand N and Kumar D 2016 Pyrolysis of babool seeds (Acacia nilotica) in a fixed bed reactor and bio-oil characterization *Renew. Energy* **96** 167–71

[17] Pattiya A and Suttibak S 2012 Influence of a glass wool hot vapour filter on yields and properties of bio-oil derived from rapid pyrolysis of paddy residues *Bioresour. Technol.* **116** 107–13

Acknowledgement

The authors thank the Director, National Institute of Technology, Tiruchirappalli, Tamilnadu, India for extending the facilities to carry out the research work.

IOP Conf. Series: Earth and Environmental Science **312** (2019) 012007 doi:10.1088/1755-1315/312/1/012007

Performance analysis of solar photovoltaic panel at Karaikal weather conditions

Sendhil Kumar Natarajan*, Subbarama Kousik Suraparaju, Elavarasan. E and Arjun Singh. K

Department of Mechanical Engineering,
National Institute of Technology – Puducherry, Karaikal 609609
U.T of Puducherry, INDIA.

Email:drsendhil1980iitmuk@gmail.com

Abstract. Increase in need of non-renewable energy sources has directed to the installation of Photovoltaic (PV) plants in many areas. Performance analysis of these PV plants could support in designing, operating and maintenance of PV systems. A PV panel was installed at the terrace of Science Block of NIT-Puducherry, Karaikal, India located at 10° latitude. The performance of this installed PV panel was analyzed by during August 2017 to February 2018. In this study, the average power output and system efficiency are analyzed. The results show that PV performance depends on geographical location and climatic conditions such as solar radiation and ambient temperature.

Keywords: Solar; Photovoltaics; Performance Analysis; Climatic Conditions.

1. Introduction

The demand for electricity keeps on increasing exponentially over the decades and also global warming and allergic diseases due to pollutions created by fossil fuels. One of the sustainable energy solutions for the problem stated above is through electric power generation using solar photovoltaic systems or power plants. Hyuna Kang et. al. [1] performed a technical analysis for improving the solar irradiation over the PV panel. More accurate tracking model was developed by considering the local climatic condition and found that 24.8% of efficiency can be improved with solar tracking. S. Martin Martinez et. al. [2] conducted a performance analysis of photovoltaic plants with various mountings in Spain. The dual axis tracking system was underestimated in the conclusion for its complexity and wind velocity came out as a major parameter influencing the performance of the photovoltaic power generation. Sebastijan Seme et. al. [3] made an analysis of photovoltaic system performance in Slovenia by comparing the results with the world around countries. In the analysis, it was found that the performance mainly relies on the intensity of solar irradiation and the temperature of the module. Also, the performance got affected by the inclination, azimuth angle of PV panel and snow barrier. John Boland and Marjolijn Dik [4] discussed the complexities in data to be considered for the analysis of the solar power systems performance. As a conclusion, simplified construction of the synthetic climatic data sets was given. Umberto Desideri and Pietro Elia Campana [5] analysed and compared the concentrating solar power plant with the photovoltaic plant. Economic analysis was also performed to determine the power plant type which suits for Italy. Manish Kumar and Arun Kumar [6] performed a review on the performance and degradation of the solar photovoltaic module by presenting the various solar

IOP Conf. Series: Earth and Environmental Science **312** (2019) 012007 doi:10.1088/1755-1315/312/1/012007

photovoltaic technology and its mathematical models. The study concluded that the climatic condition is influencing the photovoltaic systems performance. Daniel W. McKenney et. al. [7] developed spatial models of global insolation over Canada for providing the easy accessibility of the photovoltaic power production potential. Maxime Mussard and Mohamed Amara [8] performed a review on the performance of the solar PV modules in arid and semi-arid climates. The paper described the use of photovoltaic power production in semi-arid climatic conditions and also the parameters which affect the performance of the photovoltaic power production. BhubaneswariParida et. al. [9] reviewed the solar photovoltaic technologies and its importance. The paper evaluated the different techniques used to harness power and performance evaluation from the solar photovoltaic power production system. Elias Roumpakias, Anastassios Stamatelos [10] performed a comparative analytical study on grid-connected photovoltaic system. Three existing models are analysed and compared. An energy fluctuation of 10% was observed per year during the period 2013 to 2015. M. ShravanthVasisht et. al. [11] performed a feasibility study on the solar photovoltaic power generation system over the rail coaches. The performance analysis of the photovoltaic panel mounted over the rail coach of the Indian railway was submitted. The study mainly concentrated on quantifying the amount of reduction in diesel usage if the photovoltaic power system is incorporated over the coaches. M. ShravanthVasisht et. al. [12] analysed the performance of the rooftop solar photovoltaic systems to find the effects of the seasonal variations and discussed the parameters influencing the performance of the photovoltaic power production system. Rhythm Singh and Rangan Banerjee [13] estimated the rooftop power production potential from solar photovoltaic power production of a city. The peak load demand for various months also estimated. Sunderasan Srinivasan [14] segmented the market for photovoltaic electricity production in India. Also discussed the way the PV market entered into the Indian market. Wei Tian et. al. [15] investigated the effect on the performance of the photovoltaic system integrated overbuilding due to urban climate. In this current research, a photovoltaic panel (commercially available) is tested for its performance analysis in particular months (August 2017 to February 2018) so as to estimate its working at the weather conditions of Karaikal, India.

2. Photovoltaic Panel Details

In this regard, a photovoltaic panel (as seen in fig1.) consisting of 36 cells arranged as 18 × 2 and connected in series and parallel as shown in fig 2. The dimensions of the panel are of 0.8m length and 0.63m width. The panel is placed at terrace of Science Block of NIT Puducherry, located at Karaikal, India with an angle of 10° (tilted with respect to the latitude of location) and tested for the performance analysis from August 2017 to February 2018.The specifications of the photovoltaic panel are as follows:

Maximum Power (P_{MAX}): 75W
Open Circuit Voltage (V_{OC}): 22.50V
Short Circuit Current (I_{SC}): 4.45A
Voltage at max power (V_{PM}): 18V
Current at max. power (I_{PM}): 1.22A.

2.1. Data Collection

The performance of the solar photovoltaic panel is analysed by taking the readings during Seventh, Fourteenth, Twenty First and Twenty Eighth days of every month starting from August 2017 to February 2018 regularly. During these days, the ambient temperature, global solar radiation, operating voltage and short circuit voltage were recorded and tabulated in table 1. for analysing the performance of PV panel.

IOP Conf. Series: Earth and Environmental Science **312** (2019) 012007 doi:10.1088/1755-1315/312/1/012007

Figure 1. Solar PV panel used for analysis

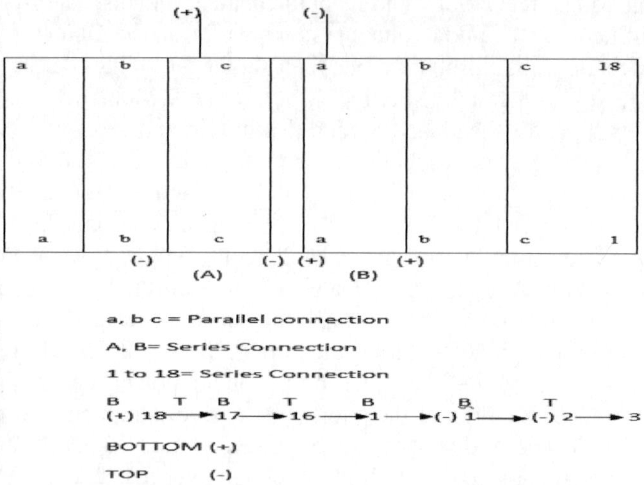

Figure 2. Line diagram of panel arrangement.

Table1. Recorded data

Date	Ambient Temperature(°C)	Global Radiation (W/m^2)	V_{OC}(V)	I_{SC}(A)
07.08.2017	28	407.5	20.1	0.41
14.08.2017	30	652.25	20.23	0.64
21.08.2017	30.8	1088.3	20.77	1.17
28.08.2017	31	1155	20.93	1.22
07.09.2017	30.8	703.3	20.4	0.7
14.09.2017	31	767.6	20.66	0.81
21.09.2017	33	940	20.75	0.99
28.09.2017	33.6	991	20.81	1.12
07.10.2017	26.8	556.2	20	0.59
14.10.2017	27	634.75	20.1	0.66
21.10.2017	28.4	682.5	20.54	0.67
28.10.2017	29	867.5	20.75	0.918
07.11.2017	26	459	20.4	0.49
14.11.2017	27.3	613.75	20.75	0.64
21.11.2017	28.3	861.25	20.9	0.885
28.11.2017	29	873.8	20.58	0.94
07.12.2017	27.6	659	20.2	0.98
14.12.2017	28	742	20.7	1.015
21.12.2017	29.9	978.5	20.8	1.15
28.12.2017	31	972.75	20.85	1.105
07.01.2018	27.1	838.75	20.5	0.65
14.01.2018	28.2	850	20.73	0.97
21.01.2018	29	900	20.8	0.875
28.01.2018	30.1	921.25	20.88	0.97
07.02.2018	27.3	854.75	20.53	0.88
14.02.2018	27.6	897.75	20.73	1.05
21.02.2018	28	930	20.83	1.05
28.02.2018	30	1001	20.9	0.935

From the above recorded data, power exerted is calculated from the following formula:

IOP Conf. Series: Earth and Environmental Science **312** (2019) 012007 doi:10.1088/1755-1315/312/1/012007

$$\text{Power (P)} = \frac{P_{MAX}}{V_{OC} \times I_{SC}} \times (V_{OC_{Actual}} \times I_{SC_{Actual}}) \quad (1)$$

Later, from the calculated power, efficiency is determined from the following formula:

$$\eta = \frac{V_{OC} \times I_{SC} \times F.F}{P_{in}} \quad (2)$$

Here, Fill Factor (FF) is calculated from:

$$\text{Fill Factor (FF)} = \frac{P_{MAX}}{V_{OC} \times I_{SC}} \quad (3)$$

Where
P_{MAX} is Maximum power;
V_{OC} is the open-circuit voltage;
I_{SC} is the short-circuit current;
P_{in} is the Input Power;
FF is the fill factor and
η is the efficiency.

The calculated values of power and efficiency were tabulated in table 2.

Table 2. Calculated values of Average Power and Average Efficiency

Month	Average Power (W)	Average Efficiency (%)
August	13.29	3.13
September	14.01	3.25
October	10.83	3.12
November	11.44	3.23
December	13.82	3.33
January	14.11	3.23
February	15.21	3.28

3. Results and Discussions.

The influence of solar radiation of the particular geometrical location (Karaikal, UT of Puducherry, India) on the efficiency of the installed photovoltaic panel is graphically depicted in the following figures with respect to each month starting from August 2017 to February 2018. From the figures 3,4,5,6,7,8,9; it is evident that the efficiency of the solar photovoltaic panel is highly influenced by the solar radiation. The variation of average global radiation and average efficiency of the installed photovoltaic panel is graphically depicted in the Figure 10, which shows that the efficiency is depending on the global radiation. During the analysis, the higher efficiency was observed in the months of December 2017 and February 2018. It is also noticed that the maximum average efficiency of the solar photovoltaic panel is about 3.33%.

The figure 11 shows that the variation of monthly average power and monthly average efficiency of the installed photovoltaic panel.

IOP Conf. Series: Earth and Environmental Science **312** (2019) 012007 doi:10.1088/1755-1315/312/1/012007

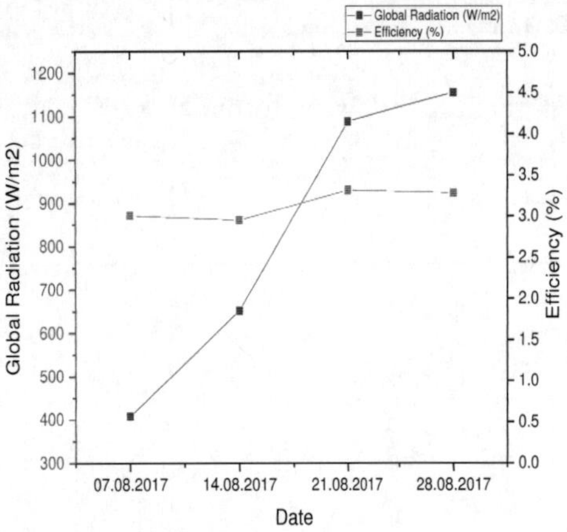

Figure 3. Variation of Global radiation and Efficiency during August 2017

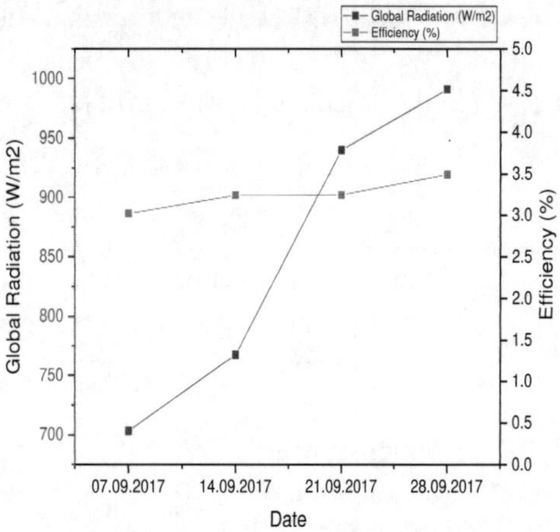

Figure 4. Variation of Global radiation and Efficiency during September 2017

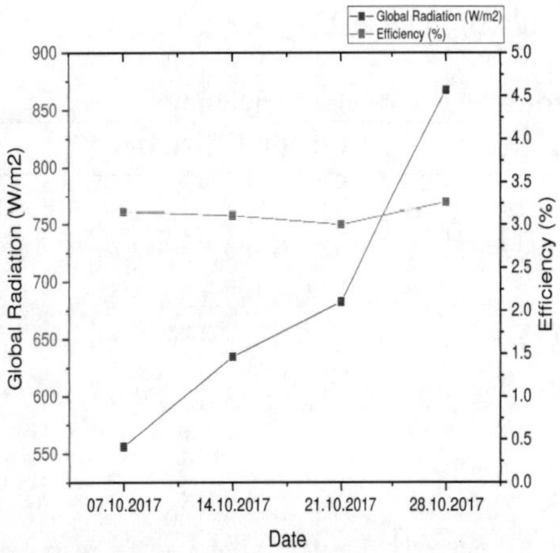

Figure 5. Variation of Global radiation and Efficiency during October 2017

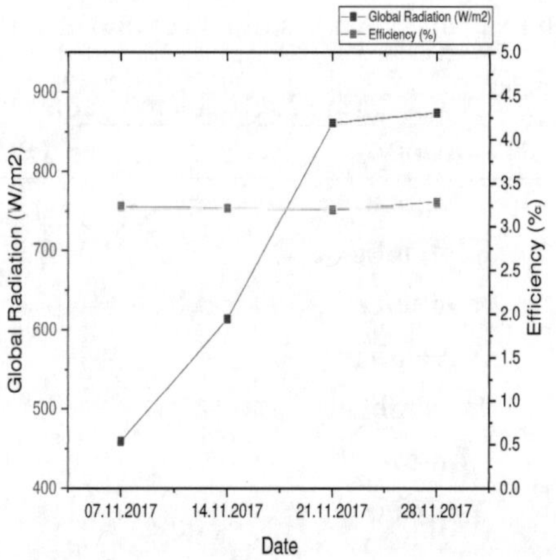

Figure 6. Variation of Global radiation and Efficiency during November 2017

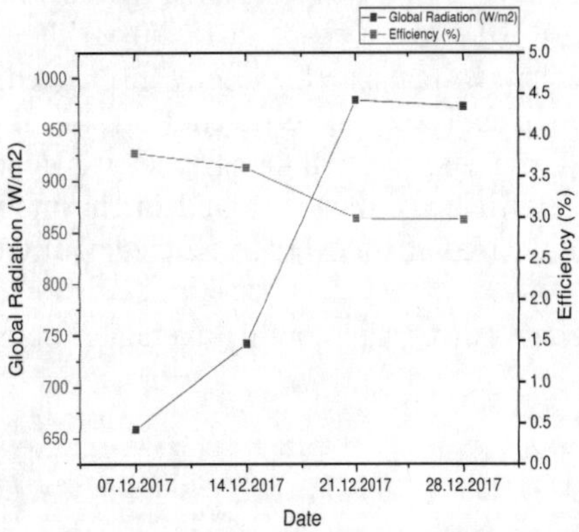

Figure 7. Variation of Global radiation and Efficiency during December 2017

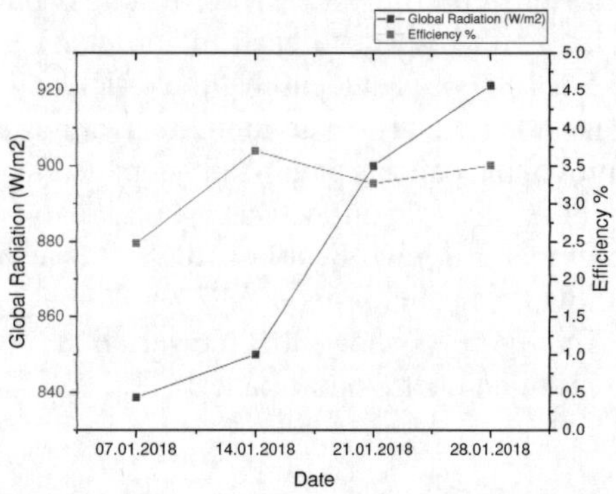

Figure 8. Variation of Global radiation and Efficiency during January 2018

IOP Conf. Series: Earth and Environmental Science **312** (2019) 012007 doi:10.1088/1755-1315/312/1/012007

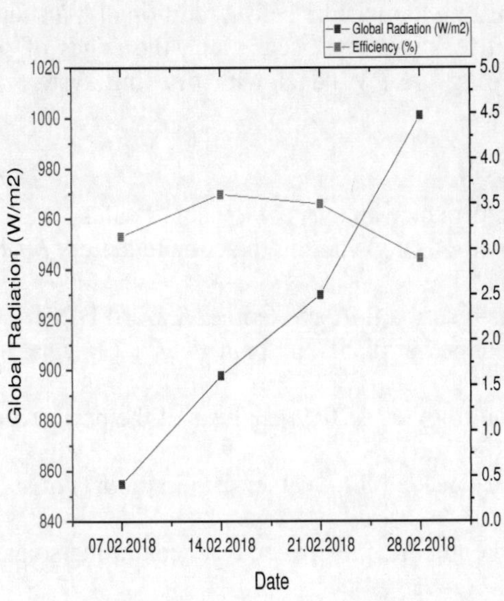

Figure 9. Variation of Global radiation and Efficiency during February 2018

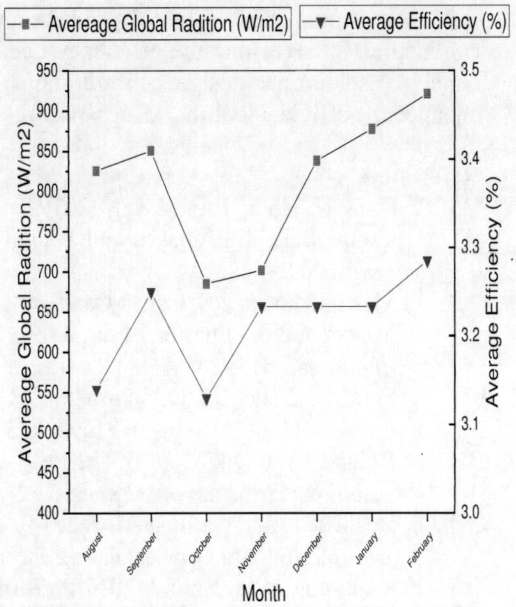

Figure 10. Monthly analysis of average radiation and average efficiency.

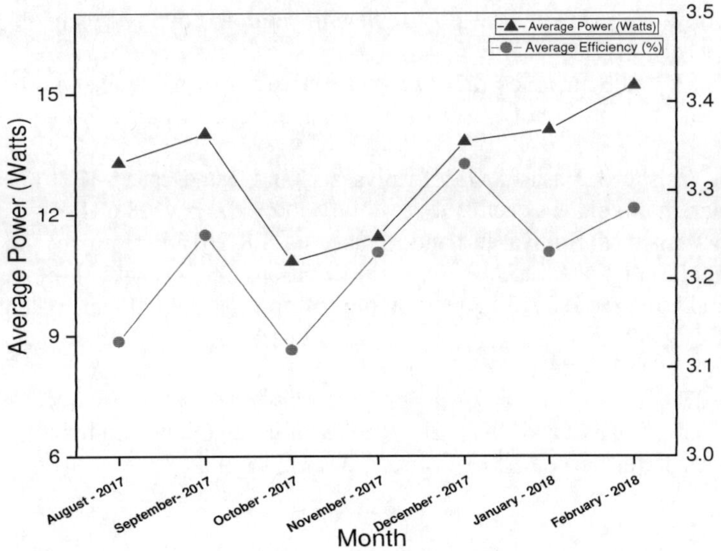

Figure 11. Graph of monthly average efficiency and average power.

4. Conclusions

In the current discussion, it is clearly seen that global radiation is key factor that influences the performance of the photovoltaic panel. It is observed that the average efficiency of the photovoltaic panel is in the range of 3.13% to 3.33% and average power of 10.83W to 15.21W during the investigation. The performance of PV panel can be increased by taking care of the several issues like shading, dust accumulation, scratches on the absorbing surface of PV panel which are the indirect influencing parameters of the solar radiation that transmitting to the solar cells. Additionally, insulation can also affect solar module efficiency because current leakages can occur along the edges of solar panels. Also, proper design configurations and equipping the PV panel with tracking system will enhance the efficiency of the solar panel.

References.

[1] Kang H, Hong T and Lee M 2019 Technical performance analysis of the smart solar photovoltaic blinds based on the solar tracking methods considering the climate factors *Energy Build.* **190** 34–48
[2] Martín-Martínez S, Cañas-Carretón M, Honrubia-Escribano A and Gómez-Lázaro E 2019 Performance evaluation of large solar photovoltaic power plants in Spain *Energy Convers. Manag.* **183** 515–28
[3] Seme S, Sredenšek K, Štumberger B and Hadžiselimović M 2019 Analysis of the performance of photovoltaic systems in Slovenia *Sol. Energy* **180** 550–8
[4] Boland J and Dik M 2001 The level of complexity needed for weather data in models of solar system performance *Sol. Energy* **71** 187–98
[5] Desideri U and Campana P E 2014 Analysis and comparison between a concentrating solar and a photovoltaic power plant *Appl. Energy* **113** 422–33
[6] Kumar M and Kumar A 2017 Performance assessment and degradation analysis of solar photovoltaic technologies: A review *Renew. Sustain. Energy Rev.* **78** 554–87
[7] McKenney D W, Pelland S, Poissant Y, Morris R, Hutchinson M, Papadopol P, Lawrence K and Campbell K 2008 Spatial insolation models for photovoltaic energy in Canada *Sol. Energy* **82** 1049–61
[8] Mussard M and Amara M 2018 Performance of solar photovoltaic modules under arid climatic conditions: A review *Sol. Energy* **174** 409–21
[9] Parida B, Iniyan S and Goic R 2011 A review of solar photovoltaic technologies *Renew. Sustain. Energy Rev.* **15** 1625–36
[10] Roumpakias E and Stamatelos A 2017 Comparative performance analysis of grid-connected photovoltaic system by use of existing performance models *Energy Convers. Manag.* **150** 14–25
[11] Shravanth Vasisht M, Vashista G A, Srinivasan J and Ramasesha S K 2017 Rail coaches with rooftop solar photovoltaic systems: A feasibility study *Energy* **118** 684–91
[12] Shravanth Vasisht M, Srinivasan J and Ramasesha S K 2016 Performance of solar photovoltaic installations: Effect of seasonal variations *Sol. Energy* **131** 39–46
[13] Singh R and Banerjee R 2015 Estimation of rooftop solar photovoltaic potential of a city *Sol. Energy* **115** 589–602
[14] Srinivasan S 2005 Segmentation of the Indian photovoltaic market *Renew. Sustain. Energy Rev.* **9** 215–27
[15] Tian W, Wang Y, Ren J and Zhu L 2007 Effect of urban climate on building integrated photovoltaics performance *Energy Convers. Manag.* **48** 1–8

IOP Conf. Series: Earth and Environmental Science **312** (2019) 012008 doi:10.1088/1755-1315/312/1/012008

Performance evaluation of compound parabolic concentrator with evacuated tube heat pipe

P.Vijayakumar[1] , G.Kumaresan[2] , S.Anish kumar[1] , U.Mohamed Faizal[1] , G.R.Vijay Chandran[1] , K.S.Vishnu Adharsh[1]
[1]Sri Shakthi Institute of Engineering and Technology, Coimbatore, Tamilnadu, India

[2]Bannari Amman Institute of Technology, Sathy, Erode, Tamilnadu, India

Corresponding author mail id: thermalvijay@gmail.com

Abstract. The keen interest of harvesting solar energy is increasing because of its clean, green and cheap nature. The pv modules collect the energy from the sun and produce electricity. The solar thermal collectors collect the solar energy in the form of heat. Compound parabolic concentrator is a type of solar thermal collector, which has a combined parabolic structure concentrating the solar radiation on a single line focus where the receiver tube is placed. The performance can be further increased by placing the evacuated tube, which traps the heat. The heat is efficiently conducted by using the heat pipe. A heat pipe is a heat-transfer device that transfer the heat between two solid interfaces. It combines the principles of both thermal conductivity and phase transition. The experimental study is carried out to find out the performance characteristics. The results according to the various parameters such as mass flow rate, tilting angles were plotted to find the better performance. The result shows that the maximum efficiency of the CPC is achieved between 12.00hrs. and 15.00hrs. The intensity of radiation during maximum efficiency is 1000 W/m^2 to 1050 W/m^2.

Keywords: Solar Energy, Compound Parabolic Concentrator, Evacuated Tube, Heat Pipe

1. Introduction

The sun's energy is available in most of the place, which is used for the advancement of technology. The solar energy is harvested using PV modules and solar thermal collectors for domestic and industrial applications at temperatures of 60°C to 300°C [1]. Solar thermal collectors as well as Solar concentrators are the devices used for concentrating solar energy for various purposes such as Solar water treatment technologies based on CPC [2], air heaters using solar concentrators [3], solar cookers based on CPC [4], Water heaters based on solar collectors [5], steam generators based on CPC [6].

Solar thermal collectors use heat-absorbing panels to absorb sunlight directly. The solar collectors commonly refer to solar hot water panels, but may refer to installations such as parabolic troughs [7] and solar towers. Here the solar energy is directly used for heating purpose [8] and in large it can be used to produce steam that can run the generator to produce electricity [6]. The types of solar collectors are imaging and non-imaging types. The compound parabolic concentrator is a non-imaging type concentrator which ha ability to concentrate rays to smaller absorber surface [10]. The sunray falling on the absorber is not focused, therefore concentration is achieved with a CPC design [11].

IOP Conf. Series: Earth and Environmental Science **312** (2019) 012008 doi:10.1088/1755-1315/312/1/012008

The compound parabolic collector is a combined parabolic structure, which concentrates the solar radiation on a single line focus [12]. The rays falling on the CPC is reflected to a single line focus based on the "edge ray principle" [10]. They are called "non-imaging" because they do not produce any optical image of the source. Compound parabolic concentrators (CPCs) is designed as stationary solar collectors to achieve relatively high temperature operations with high cost effectiveness [13].

The light rays coming from the edges of the sources are redirected to the edges of the receiver. This confirms that all light rays coming from the inner points in the source will make contact with the receiver as shown in figure 1.

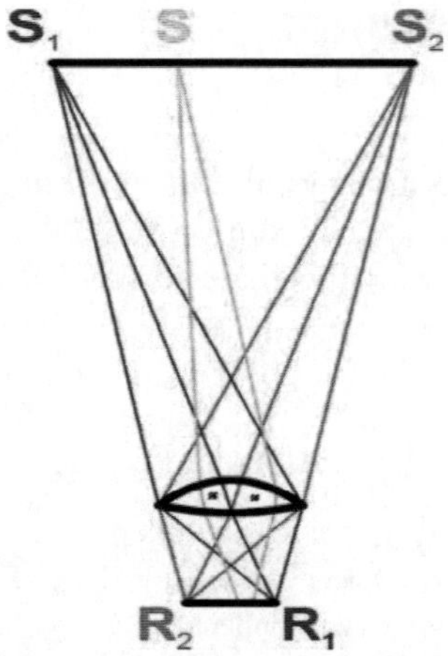

Figure 1. Edge ray principle

The rays falling on the CPC should be reflected to the receiver, so a reflective material is used which reflects the rays to the evacuated tubes. The reflective material should have high reflectivity.

Evacuated tube collectors are devices, which consists of cylindrical absorbing surface in which the vacuum is created between concentric glass tubes made up of borosilicate glass [16]. The outer layer is transparent allowing the light rays to pass through with minimal reflection. The inner layer is coated with a special selective coating (Al-N/Al) [17], which provides excellent solar radiation absorbing properties. The evacuated tube absorbs the solar energy reflected from the CPC and converts into the heat energy [18]. Thus, vacuum acts as an insulator, which does not lose the heat. To maintain vacuum between two glass layers a barium getter is used. This barium layer absorbs any CO, CO_2, N_2, O_2, H_2O and H_2 out-gassed from the evacuated tube during storage and operation, thus maintaining the vacuum.

A heat pipe is a heat-transfer device that transfer the heat between two solid interfaces. It works on both the principles thermal conductivity and phase transition [8]. Due to the very high heat transfer coefficients for boiling and condensation, heat pipes are highly effective thermal conductors. Therefore, when the heat pipe is introduced in the evacuated tube there would be more heat transfer resulting in a high efficiency solar thermal collector. Nano fluids can be used inside the heat pipe to increase the heat transfer [9]. The nana fluids such as CuO, graphene oxide nano fluids, etc. can be used inside the heat pipe [14].

2. Methodology

Description of various materials used and methodology used to make different parts detailed below.

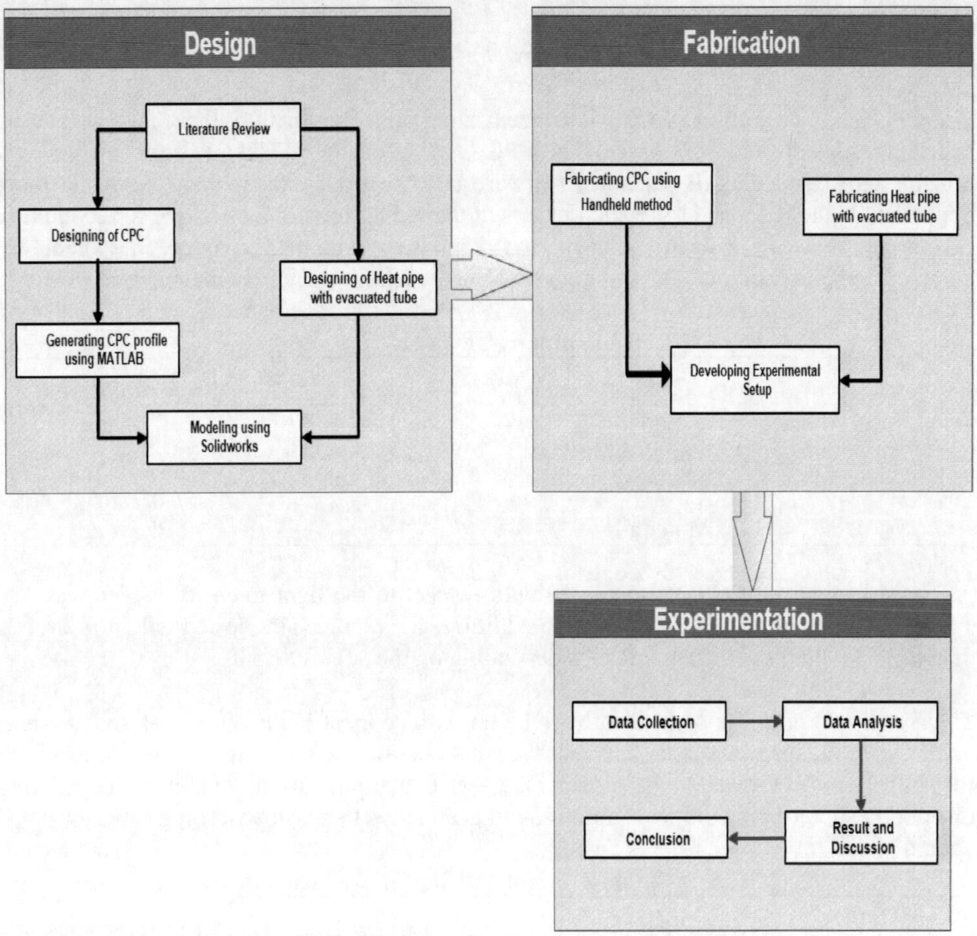

Figure 2. Methodology

2.1. Conceptual design

The conceptual design of CPC with evacuated tube was done with the SOLIDWORKS 3D modelling software shown in the figure 3.

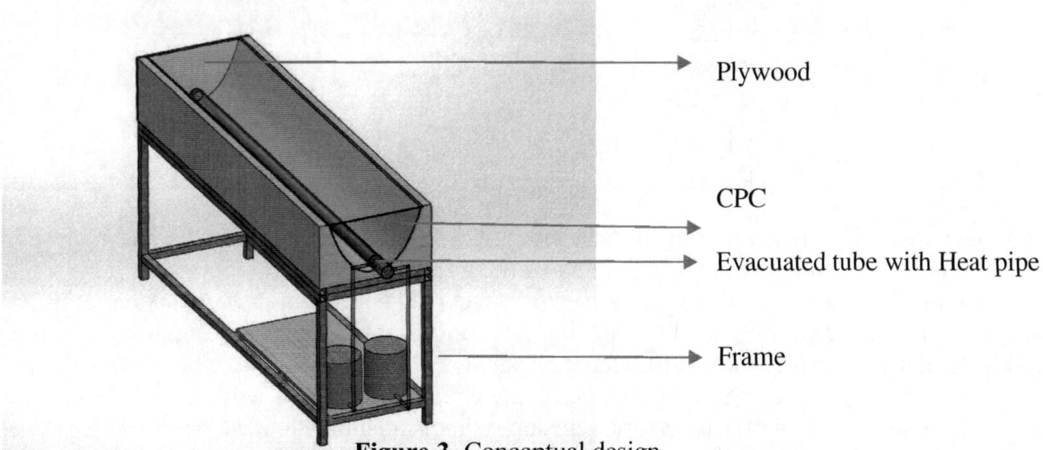

Figure 3. Conceptual design

IOP Conf. Series: Earth and Environmental Science **312** (2019) 012008 doi:10.1088/1755-1315/312/1/012008

2.2 Components

The components of the CPC are listed below

2.2.1 CPC The application of CPC is concentrating and illumination, the CPC is made by reflective material on its side wall. The base of the CPC is made of wood. The CPC dimensions were collected with the help of MATLAB statistical software and the truncation is done based on the experiment conducted by the M.M Isa [15], which results in the reduction of material size for economical purpose. The outline of the CPC is drawn on the plywood and is cut with the help of cutting wheel. The plywood acts as the supporter of the CPC and holds the Sheet metal, evacuated tube and heat pipe.

Table 1. CPC specifications

CPC	Specifications
Length	1500mm
Aperture width	341mm
Surface area	0.96m^2
Material	Plywood

The performance of CPC is increased with respect to the light intensity, which was proved by the experiment conducted by Ankur Geete. In addition, Giovanni casino tested CPC for larger acceptance angle (>30°) which shows better results depending on the solar intensity.

2.2.2 Evacuated tube The solar absorber tube has two concentric glass tubes closed at one end with an annular vacuum space and a selective surface absorber on the outer surface (vacuum side) of the inner tube. In heat pipe evacuated tube collectors, a sealed heat pipe, usually made of copper to increase the collector's efficiency in cold temperatures, and is attached to a heat absorbing reflector plate within the vacuum sealed tube.

Table 2. Characteristics of evacuated tube

Evacuated tube	Characteristics
Length	1500mm
Outer Diameter	47mm
Inner Diameter	37mm
Glass thickness	1.6-2.0mm
Material	Borosilicate Glass
Absorbance	Above 93%
Emittance	Above 8%
Heat loss	<0.8W/(m^{2}ö °C)
Startup temperature	<=25°C (77F)
No of coatings	3
Coatings	Al/N/Al

2.2.3 Heat Pipe. The manufacturing of heat pipe involves selection of materials, cleaning process, checking for leaks, creating vacuum, and filling working fluid, instrumentation works and the testing of the heat pipe. The material selected for the heat pipe is copper because of its high thermal conductivity and it does not corrode. Also it has high melting point and copper allows heat to quickly pass through it. Therefore, it is used in quick heat transfer applications.

The cleaning of metal is necessary to remove any grease and impurities associated with the metal. The metal is placed in pool of hydrochloric acid and sodium-di-chromate. It will remove the grease and impurities. The copper tube is checked by allowing the helium inside the copper pipe. The flow of helium

IOP Conf. Series: Earth and Environmental Science **312** (2019) 012008 doi:10.1088/1755-1315/312/1/012008

shows any cracks and holes if any present in the copper pipe. There may be chances for micro cracks in the copper pipe which can bedetected using helium.

Then the vacuum is created inside the copper tube with the help of vacuum pump. The vacuum pump sucks the air inside the copper pipe and maintains the pressure inside the copper pipe at 0.0004 bar. Then the acetone is filled inside the copper pipe to 40% of total volume. Then the pipe is sealed with the help of hydraulic press.

The thermocouple wires were fixed to the copper heat pipe to check the temperature at various points. The thermocouple wires were welded to the surface of copper heat pipe and sealed by silicon paste. Then the thermocouple wires are connected to the temperature indicator.

Table 3. Specifications of heat pipe

Heat pipe	Specifications
Material	Copper
Diameter	19mm
Thickness	1mm
Length	1400mm
Working fluid	Acetone

The working fluid used here is Acetone which is colorless, volatile, flammable liquid. Its chemical formula is CH_3-CH_3-CO. the boiling point of acetone is about 56°C. The ph value of acetone is 7 i.e. neither acidic nor basic it is neutral. The acetone has excellent compatibility with copper and it has lower boiling point which is suitable for heat pipes.

2.2.4 Aluminium foil. The Aluminium foil is the reflective material used to reflect the falling radiation to the receiver tube. The reflectivity of the material is important and it shouldn't absorb the falling radiation. The Foil is made of Al8011 type Aluminium material.

Table 4. Characteristics of Al foil

Aluminium foil	Characteristics
Reflectivity	Above 85%
Thickness	30 microns
Surface area	$0.96m^2$

2.2.5 Condenser. The condenser is the important component of the setup. Condenser is attached to the heat pipe in which heat is transferred from the working fluid to the water. The working fluid in the heat pipe condenses and gives away its heat to the surrounding water and returns back to the evaporator region. The length of the condenser portion is 100mm and is made up of plastic pipe with one inlet and outlet tubes. The size of the inlet and outlet tube is 6.35mm.

2.3 Design calculations

2.3.1 Concentration angle.

Concentration ratio = D/ (π*d)

Where, D = aperture area

d = receiver diameter

2.3.2 Half acceptance angle.

Half acceptance, θ = arc sin (1/C.R)

IOP Conf. Series: Earth and Environmental Science **312** (2019) 012008 doi:10.1088/1755-1315/312/1/012008

2.4 Fabrication process

The fabrication process listed is done with the help of conventional machines available. The components and methods used to fabricate the components were discussed in the table.

Table 5. Fabrication of components

S.No.	Component name	Method of fabrication	Function
1	Stand	Cutting Drilling Welding Grinding	Provides support for the base and tanks
2	Base	Cutting Drilling Screw joint	It holds the CPC and the evacuated tube
3	CPC	Hand bending	It reflects the solar radiation to the receiver tube
4	Heat pipe	Cutting Welding Vacuum pumping Hydraulic sealing	It transfer the heat concentrated on the receiver to the condenser
5	Condenser	Cutting Joining using silicon paste	In condenser the heat is transferred to the water flowing through it

Mild steel pipes of size 1inch were brought and cut using cutting wheel. Then the stand is made by welding the parts. Then the CPC structure were drawn on the plywood and cut using cut wheel. The sheet metal is placed over the plywood and screwed. The evacuated tube is placed on the CPC and hold with the help of clamp. The reflective material is fixed above the CPC. Then the heat pipe is placed inside the evacuated tube and the thermocouples were attached to the temperature indicator. The base of the CPC is placed above the stand and a screw rod is fixed to adjust the angle of the CPC.

2.5 Testing methodology

2.5.1 Mass flow rate. The various mass flow rates such as .016 kg/s, .025 kg/s, .033 kg/s, .041 kg/s, .05 kg/s were tested and the corresponding temperature were taken out. The water gets more heat when the mass flow rate is minimum. When the mass flow rate is maximum, the heat is transferred to more volume of water, which results in low temperature increase.

2.5.2 Tilting angle. The effect of a solar tilt angle on energy output may be up to 20% percent compared to flat plate collectors. Thus, various tilting angles like 10°, 15°, 20°, 30° and 45° and their corresponding results were plotted. The optimum tilting angle is based on the latitude and altitude of the location. The optimum angle for the Coimbatore is 11° facing south direction. This is the angle where the efficiency of the CPC is high.

The CPC is placed such that it facing south because that is generally where they would receive most sunlight. This is because India lies in northern hemisphere. Due to the tilt of the earth and elliptical orbit around the sun, the sun apparently moves from east to west from the south portion of the sky. Therefore, the roofs have high exposure to sun from the south direction.

IOP Conf. Series: Earth and Environmental Science **312** (2019) 012008 doi:10.1088/1755-1315/312/1/012008

3. Results and Discussions

The CPC was designed and fabricated as explained. The testing is done based on the testing methodology. Overall specifications with economic analysis, and test results found were detailed.

3.1 Experimental setup

Figure 4. Experimental Setup of Compound parabolic concentrator with evacuated tube heat pipe

3.2 Solar radiation intensity

The radiation intensity has a direct impact on the thermal efficiency of the CPC. If the radiation intensity is high, then the thermal efficiency of the CPC also high. The solar radiation intensity varies with the time. It increases steadily up to 14:00hrs and then decreases gradually. The performance of CPC is evaluated by exposing it to the indoor and outdoor conditions.

The CPC is placed to faces south. The CPC is kept at different tilting angles such as 0°, 11°, 30°, 45°. The temperature obtained at each tilting angle is noticed and the effective one is found. The solar radiation intensity varies with time. At first intensity is low which increases up to 14hrs and then it gradually decreases. The radiation intensity over the testing time period is explained below in figure 5. The efficiency of the CPC is increased greatly with the high intensity of radiation. The intensity of radiation is not constant. It differs each day due to atmospheric conditions. The maximum radiation intensity observed during the experiment was 1050W/m^2. The average intensity of radiation is about 933W/m^2.

IOP Conf. Series: Earth and Environmental Science **312** (2019) 012008 doi:10.1088/1755-1315/312/1/012008

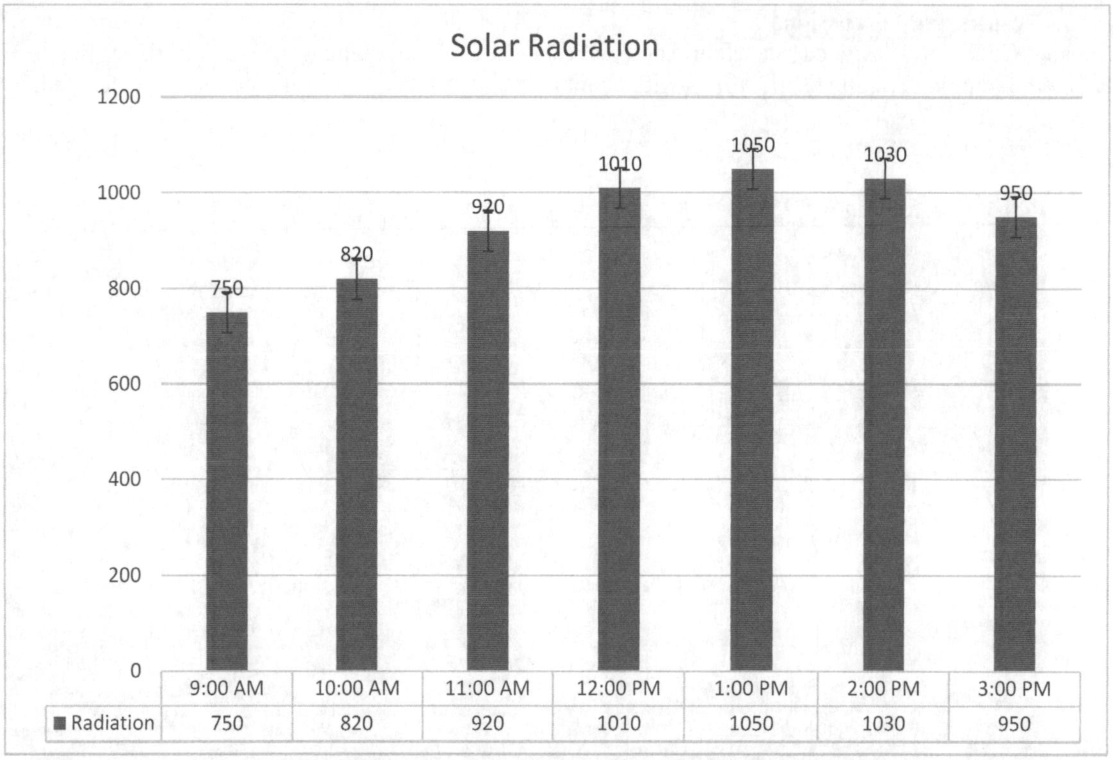

Figure 5. solar radiation intensity

The heat pipe requires some time to attain start up temperature. When heat pipe attains its start-up temperature, acetone evaporates and takes away the heat to the condenser portion. Acetone condenses and returns back to the evaporator portion. The phase change of liquid to vapor and vapor to liquid takes place in a cycle. The heat is transferred to the water flowing through the condenser portion. The heat absorbed by the heat pipe during the time is described in the figure 6. There are four thermocouples used which represents the temperature at various portions described in the table.

Table 6. Representation of thermocouple

Thermocouple No.	Indication
T1	Atmospheric temperature
T2	Temperature at end portion of heat pipe
T3	Temperature at middle portion of heat pipe
T4	Temperature of the condenser portion

The evaporator and condenser portion of the heat pipe are attached with thermocouple. Thus the temperature indicator shows the temperature at various portions. The temperature was noted at a time interval of every two minutes then the water flow is done. Again changing the tilting angle, the reading was taken. This repeated throughout the experiment.

Assumptions made in the experiment

- The heat loss due to the clamp to hold evacuated tube was negligible
- The flow is constant throughout the experiment
- The intensity of tungsten halogen light is constant throughout the area

IOP Conf. Series: Earth and Environmental Science **312** (2019) 012008 doi:10.1088/1755-1315/312/1/012008

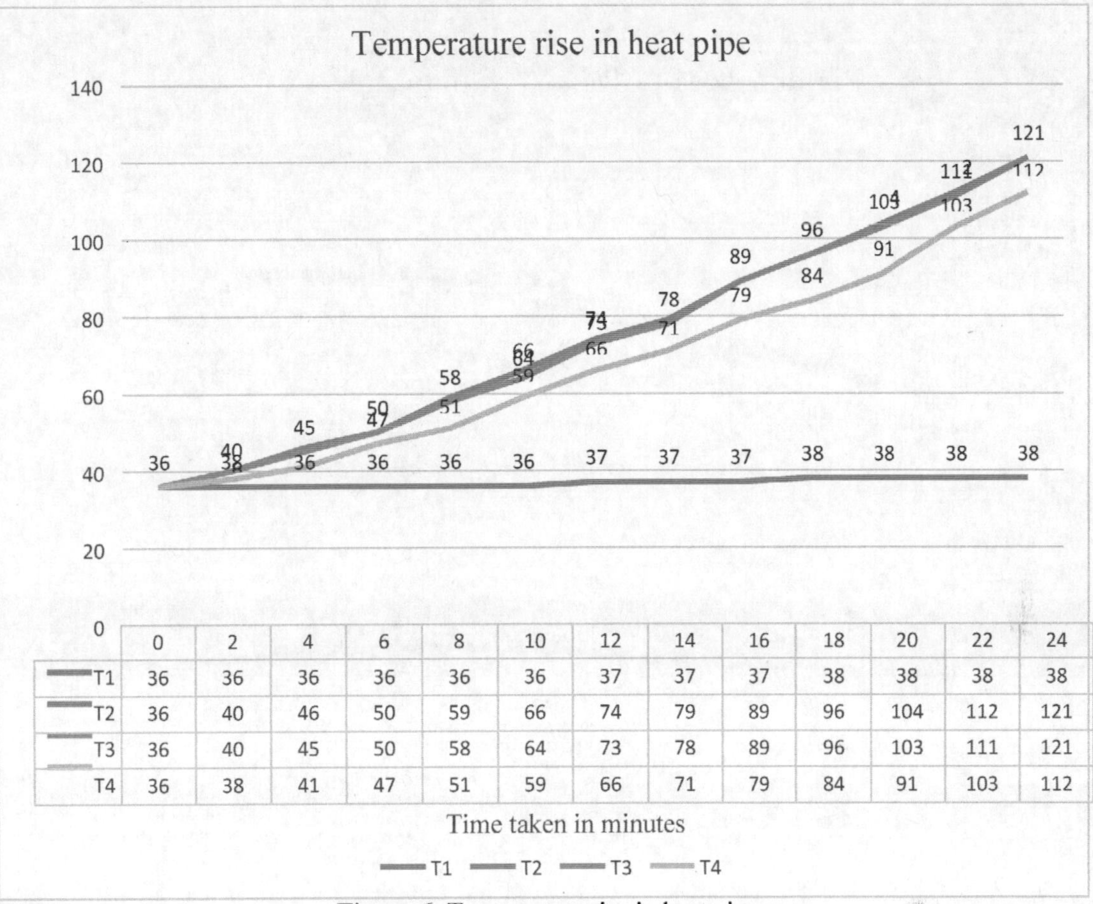

	0	2	4	6	8	10	12	14	16	18	20	22	24
T1	36	36	36	36	36	36	37	37	37	38	38	38	38
T2	36	40	46	50	59	66	74	79	89	96	104	112	121
T3	36	40	45	50	58	64	73	78	89	96	103	111	121
T4	36	38	41	47	51	59	66	71	79	84	91	103	112

Figure 6. Temperature rise in heat pipe

3.3 Performance of CPC at varying tilting angles

The heat absorbed by the CPC at varying tilting angles like 0°, 11°, 30° and 45° are compared. The results show that increasing the tilting angle increases the efficiency of the CPC [19]. At high tilting angles the heat absorbed by the CPC increases up to 13:00hrs uniformly and then the efficiency decreases. At some angle the efficiency is nearby constant throughout the time period. Thus angle with constant efficiency throughout the day is chosen as optimum tilting angle. The results show that the optimum tilting angle is nearly 11° which is shown in figure 8. The CPC is facing towards south because this direction receives more heat throughout the day. The optimum tilting angle varies with the altitude. The efficiency can be increased by using tracking mechanism. The figure 7 represents the performance of CPC at tilting angle 0°.

The temperature of CPC is increased constantly up to 14:30hrs and then decreases. The figure 9 represents the performance of CPC at tilting angle 30°. At 30° the temperature increases steadily and then decrease after 14:00hrs. The reason for this is the sun moving towards the west the intensity of radiation falling on the CPC decreases. At 45° the temperature attains maximum value up to 14:00hrs. Then temperature drops rapidly. Because the radiation falling on the sun is minimum after 14:00hrs. The figure 10 represents the thermal performance of CPC at 45°.

IOP Conf. Series: Earth and Environmental Science **312** (2019) 012008 doi:10.1088/1755-1315/312/1/012008

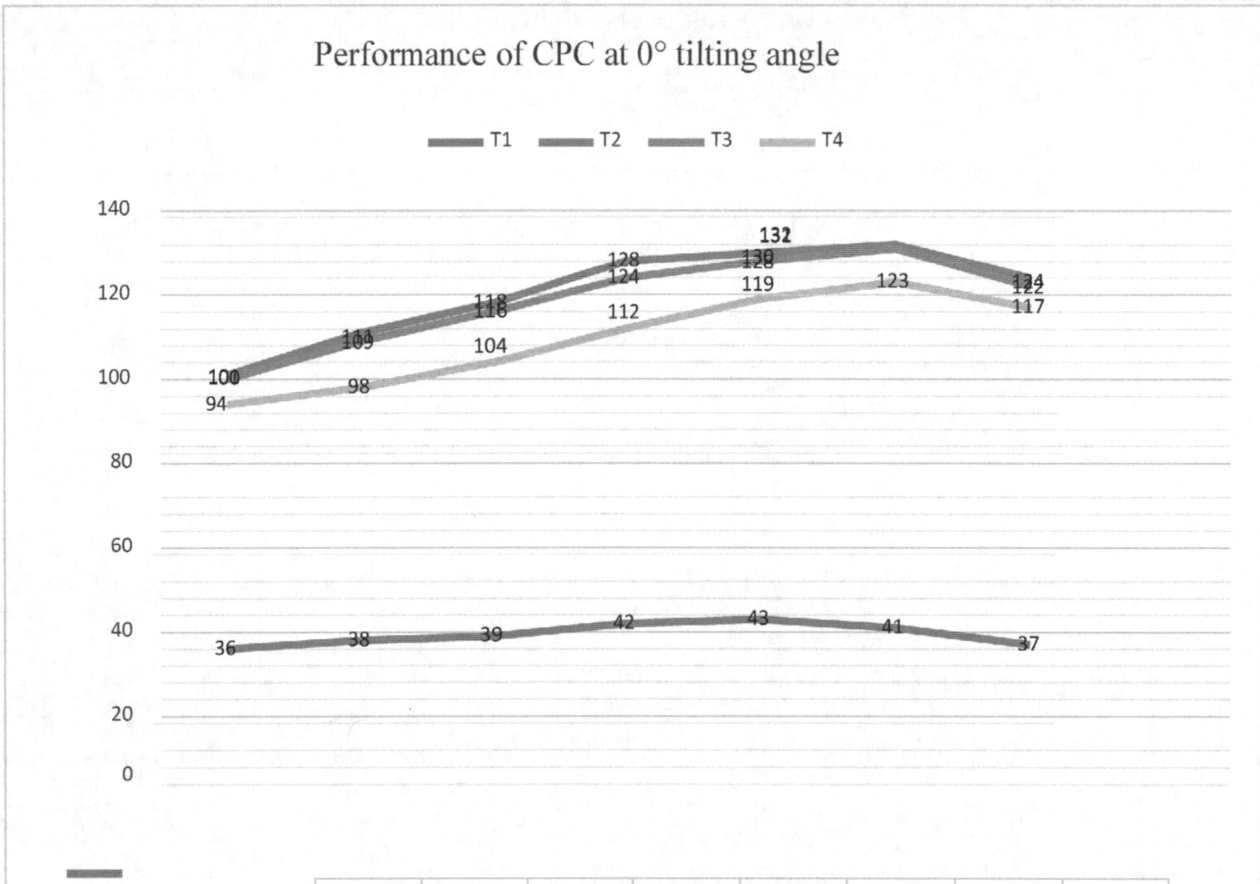

	9:30 AM	10:30	11:30	12:30 PM	1:30 PM	2:30 PM	3:30 PM
T1	36	38	39	42	43	41	37
T2	101	111	118	128	130	132	124
T3	100	109	116	124	128	131	122
T4	94	98	104	112	119	123	117

Figure 7. Thermal performance of CPC at 0° tilting angle

IOP Conf. Series: Earth and Environmental Science **312** (2019) 012008 doi:10.1088/1755-1315/312/1/012008

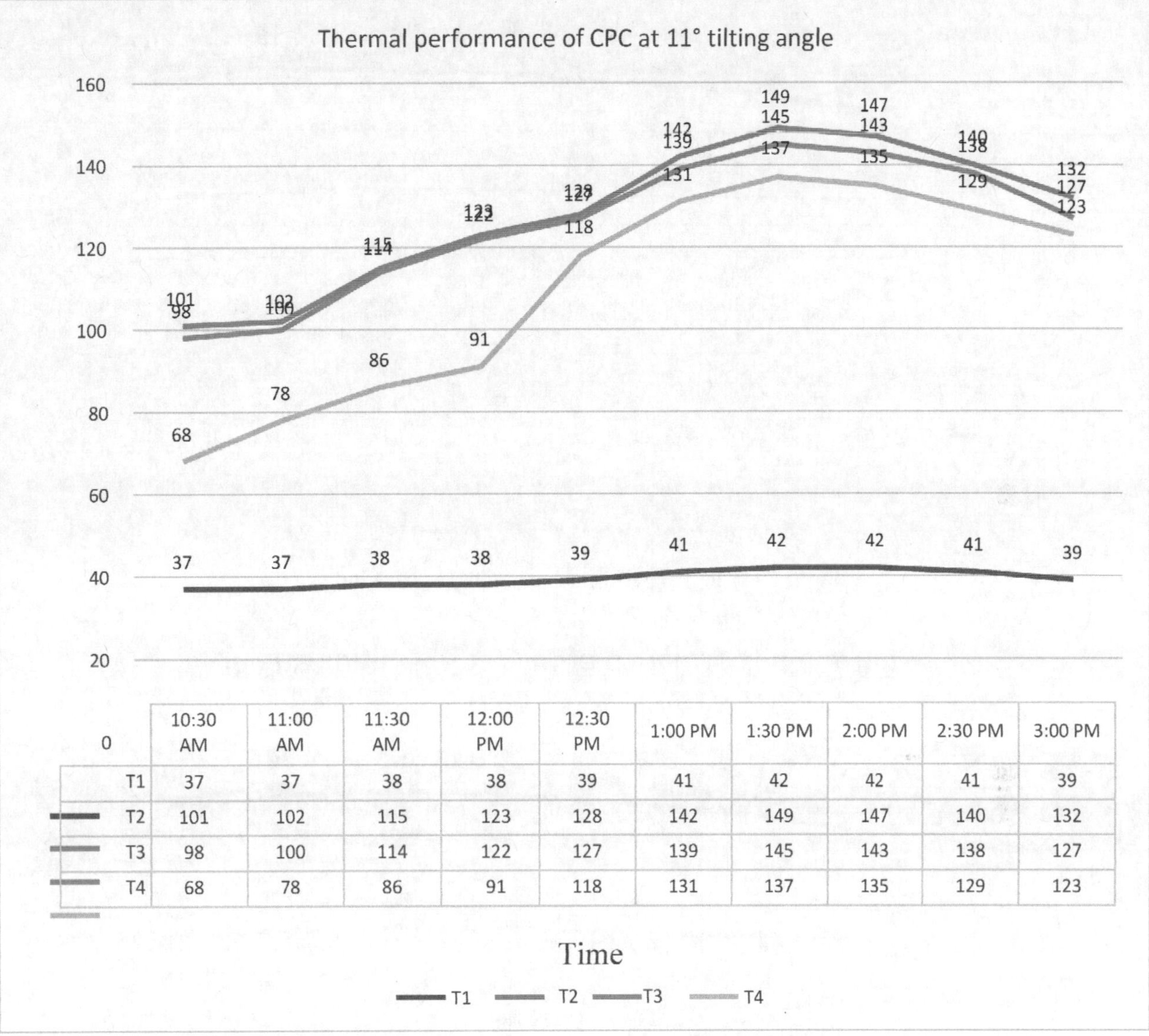

	10:30 AM	11:00 AM	11:30 AM	12:00 PM	12:30 PM	1:00 PM	1:30 PM	2:00 PM	2:30 PM	3:00 PM
T1	37	37	38	38	39	41	42	42	41	39
T2	101	102	115	123	128	142	149	147	140	132
T3	98	100	114	122	127	139	145	143	138	127
T4	68	78	86	91	118	131	137	135	129	123

Figure 8. Thermal performance of CPC at 11° tilting angle

In the above chart, the CPC receives less heat at first and then the temperature of CPC increases to maximum value at 13:30hr and starts to drop. During 9:00 to 13:00hrs the efficiency is similar to other tilting angles. But after 13:00hr the efficiency of CPC is higher for 11°. Thus, the overall efficiency of CPC is high at the tilting angle 11°.

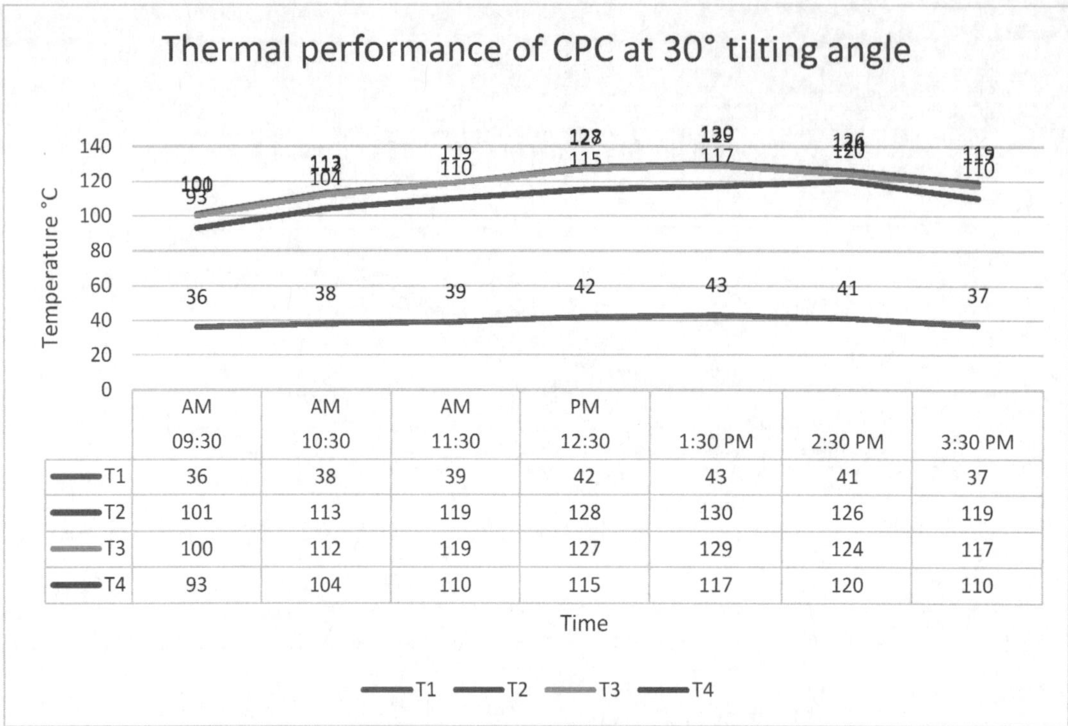

	AM 09:30	AM 10:30	AM 11:30	PM 12:30	1:30 PM	2:30 PM	3:30 PM
T1	36	38	39	42	43	41	37
T2	101	113	119	128	130	126	119
T3	100	112	119	127	129	124	117
T4	93	104	110	115	117	120	110

Figure 9. Thermal performance of CPC at 30° tilting angle

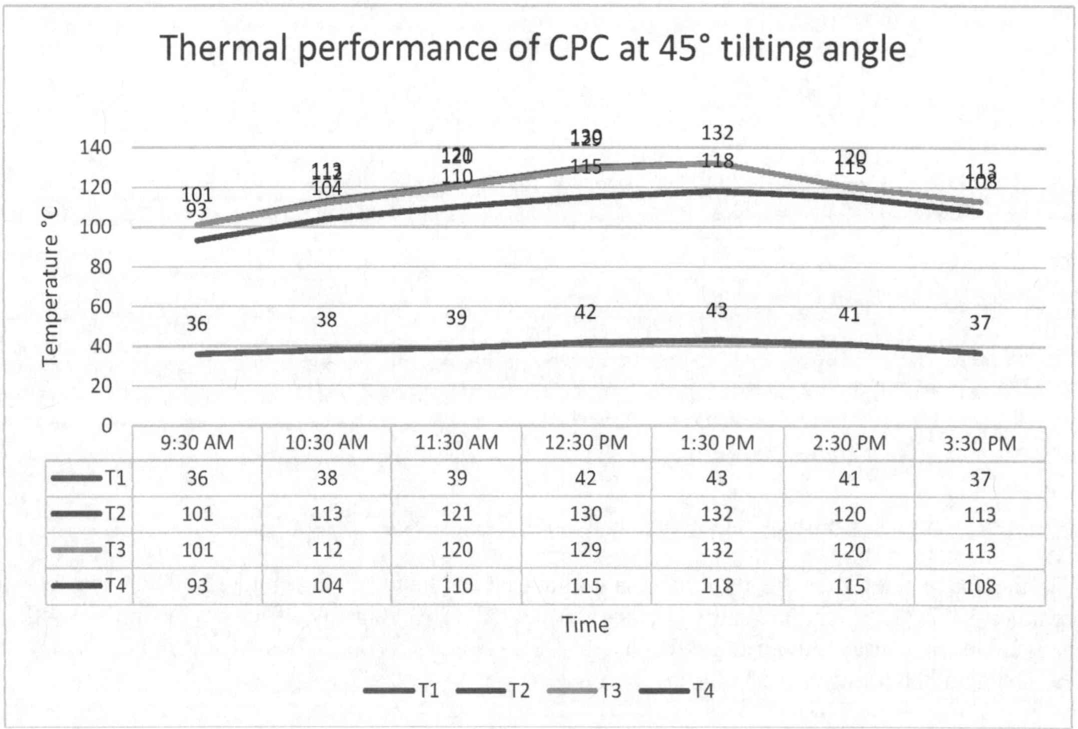

	9:30 AM	10:30 AM	11:30 AM	12:30 PM	1:30 PM	2:30 PM	3:30 PM
T1	36	38	39	42	43	41	37
T2	101	113	121	130	132	120	113
T3	101	112	120	129	132	120	113
T4	93	104	110	115	118	115	108

Figure 10. Thermal performance of CPC at 45° tilting angle

IOP Conf. Series: Earth and Environmental Science **312** (2019) 012008 doi:10.1088/1755-1315/312/1/012008

The experiment shows that at angle 11° the CPC receives heat throughout the experimental period and the overall efficiency is high for the whole day. The angles 30° and 45° shows the CPC receives more heat up to 14:00hrs and after 14:00hrs the heat absorbed by the CPC drops gradually. Thus optimum tilting angle for the compound parabolic concentrator thermal collector is 11°.

3.4 Thermal performance of CPC at indoor condition

The thermal performance of CPC at indoor condition was done which was represented in figure 11. The tungsten-halogen light acts as the light source for the CPC. The tungsten halogen light produces more heat and the radiation intensity of the tungsten halogen light is more than 2000W/m^2. The indoor testing does not produce the results obtained in the outdoor condition. The halogen light intensity can be varied by varying the supply voltage.

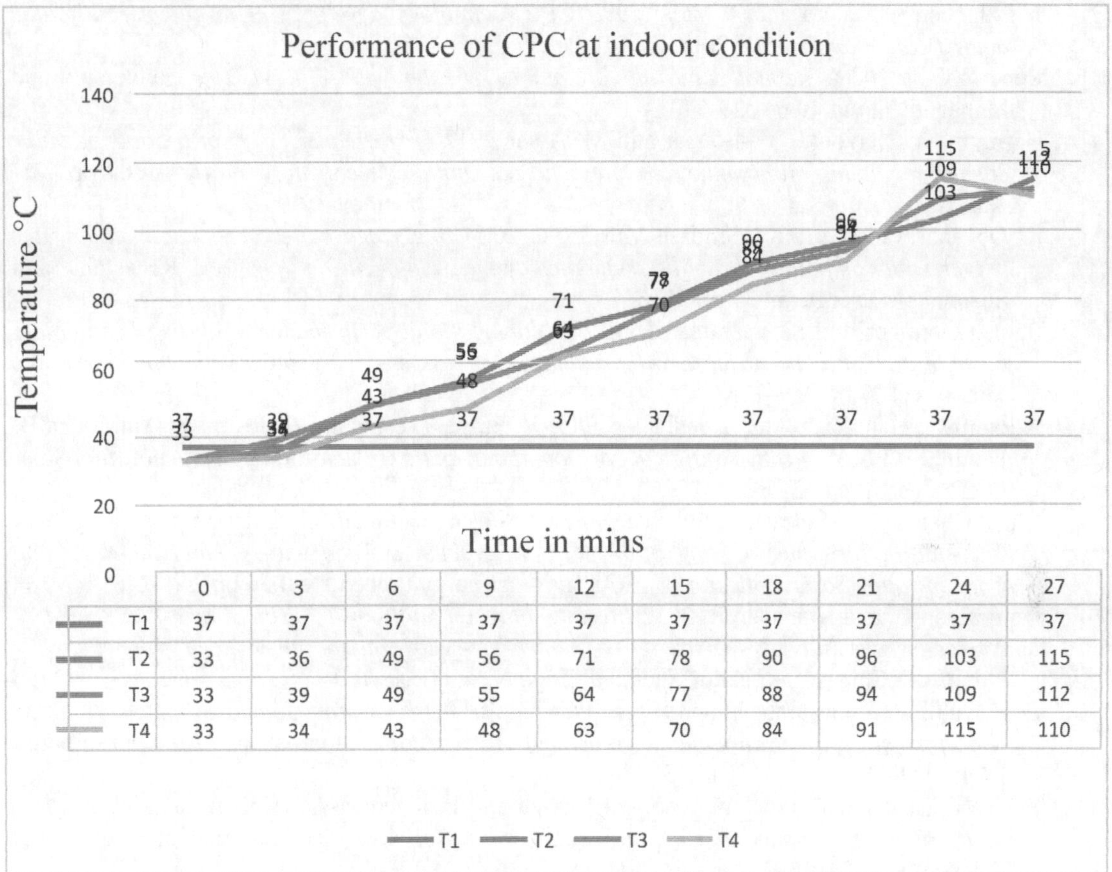

Time in mins	0	3	6	9	12	15	18	21	24	27
T1	37	37	37	37	37	37	37	37	37	37
T2	33	36	49	56	71	78	90	96	103	115
T3	33	39	49	55	64	77	88	94	109	112
T4	33	34	43	48	63	70	84	91	115	110

Figure 11. Performance of CPC at indoor condition

The experiment shows that the efficiency of the CPC can be increased with high reflectivity material. When the water flow is increased the temperature of the heat pipe drops suddenly. This is because the high volume of water takes away the heat, which results in low temperature rise. If the flow is very small, then the volume of water taking the heat is also small which results in high temperature of water. Thus, a very small flow of water is important to achieve the high temperature. In addition, the use of high reflective material increases the heat concentrated on the CPC, which further increases the efficiency of the CPC. The efficiency of the CPC can also be increased by attaching a solar tracking equipment. The solar tracking equipment tracks the solar radiation throughout the day and changes the direction of CPC to obtain maximum efficiency.

4. Conclusion

The testing of CPC results in improved performance, which resembles more energy, can be obtained using CPC. The additional setup of ray tracing technique further increases the performance of CPC. Because the ray tracing setup adjusts the CPC, such that the maximum rays can be utilized which results in higher concentration of rays. Thus, the efficiency is further increased. The results recommend using of CPC results in better efficiency, which can meet the energy demand increasing day to day. The efficiency of the CPC is also higher than other flat plate solar collectors.

References

[1] A. Ustaoglu, M. Alptekin, J. Okajima and S. Maruyama 2016 *Evaluation of uniformity of solar illumination on the receiver of compound parabolic concentrator (CPC)* Solar Energy **vol 132** pp 150-164

[2] J. Balanco, S. Malato and A Vidal 1999 *CPC technology development to commercial solar detoxification application* Solar Energy **vol 67** pp 317-330

[3] Rene tchinda 2008 *Thermal behaviour of solar air heater with CPC* Energy Conversion and Management **vol 49** pp 529-540

[4] Harmim, M. Merzouk , M. Boukar and M. Amar 2012 *Performance study of a box-type solar cooker employing an asymmetric compound parabolic concentrator* Energy **vol 47** pp 471-480

[5] K. Devanarayanan and K. Kalidasa Murugavel 2014 *Integrated collector storage solar water heater with compound parabolic concentrator - development and progress* Renewable and Sustainable Energy **vol 39** pp 51–64

[6] Rachel Oommen and S. Jayaraman 2002 *Development and performance analysis of compound parabolic solar concentrators with reduced gap losses - 'V' groove reflector* Renewable Energy **vol 27** pp 259–275

[7] Ajitkumar S. Gudekar, Atul S. Jadhav, Sudhir V. Panse, Jyeshtharaj B. Joshi and Aniruddha B. Pandit 2013 *Cost effective design of compound parabolic collector for steam generation* Solar Energy **vol 90** pp 43–50

[8] Wisut Chamsa-ard, Sukruedee Sukchai, Sorawit Sonsaree and Chatchai Sirisamphanwong 2014 *Thermal Performance Testing of Heat Pipe Evacuated Tube with Compound Parabolic Concentrating Solar Collector by ISO 9806 – 1* Energy Procedia **vol 56** pp 237–246

[9] Kumeresan G. and Kamatchi R. 2017 *Investigations on pool boiling critical heat flux, transient characteristics and bonding strength of heater wire with aqua based reduced graphene oxide nanofluids* Chinese Journal of Chemical Engineering (Elsevier)

[10] D. K. Patel, P. K. Brahmbhatt and Hitesh Panchal 2017 *A review on Compound parabolic solar concentrator for sustainable development* International Journal of Ambient Energy DOI:10.1080/01430750.2017.1318786

[11] Subhi S. Mahammed Tadahmun Ahmed Yassen and Hameed Jassim Khalaf 2012 *Theoretical Study of the Compound Parabolic Trough Solar Collector* Tikrit Journal of Engineering Sciences **vol 19** pp 1-9

[12] Ari Rabl 1976 *Optical and thermal properties of compound parabolic concentrators* Solar Energy **vol 18** pp 497-511

[13] Donghao Xu and Ming Qu *Compound parabolic concentrators in solar thermal applications: a review* Proceedings of the ASME 2013 7th International Conference on Energy Sustainability

[14] Kumaresan G., Venkatachalapathy S. and Lazarus Godson Asirvatham 2014 *Experimental investigation on enhancement in thermal characteristics of sintered wick heat pipe using CuO nanofluids* International Journal of Heat and Mass Transfer **vol 72** pp 507-516

[15] M. M. Isa, R. Abd-Rahman and H. H. Goh 2015 *Design Optimisation of Compound Parabolic Concentrator (CPC) for Improved Performance* International Journal of Electrical, Computer, Energetic, Electronic and Communication Engineering **vol 9**

[16] M. Gajic, N. Karwa, A. Mojiri, and G. Rosengarten *Modeling reflection loss from an evacuated*

tube inside a compound parabolic concentrator with a cylindrical receiver Solar energy **vol 23** DOI:10.1364/OE.23.00A493

[17] Atul A. Sagade, N.N. Shinde and P.S. Patil 2014 *Effect of receiver temperature on performance evaluation of silver coated selective surface compound parabolic reflector with top glass cover* Energy Procedia **vol 48** pp 212-222

[18] Qiang Wang, JinfuWang and Runsheng Tang 2016 *Design and Optical Performance of Compound Parabolic Solar Concentrators with Evacuated Tube as Receivers* Energies **vol 9** pp 795 DOI:10.3390/en9100795

[19] T. Arunkumar, R. Velraj, D.C. Denkenberger, Ravishankar Sathyamurthy, K. Vinoth Kumar and Amimul Ahsan 2016 *Productivity enhancements of compound parabolic concentrator tubular solar stills* Renewable Energy **vol 88** pp 391-400

IOP Conf. Series: Earth and Environmental Science **312** (2019) 012009 doi:10.1088/1755-1315/312/1/012009

Development of Solar Oven employed with Parabolic Concentrator

P.Vijayakumar[1] , G.Kumaresan[2] , S.Anish kumar[1] , S.Sudhagar [1] , G.R.Vijay Chandran[1] , K.S.Vishnu Adharsh[1]
[1]Sri Shakthi Institute of Engineering and Technology, Coimbatore, Tamilnadu, India

[2]Bannari Amman Institute of Technology, Sathy, Erode, Tamilnadu, India

Corresponding author mail id: vijaychandrangr@gmail.com

Abstract. The main purpose of our project is to replace old stoves with an eco-friendly solar oven. This parabolic solar oven is a form of outdoor cooking and uses the energy of sunlight to heat food or drink or to sterilize it. They use no fuels and cost, nothing to operate; they reduce air pollution and deforestation. This device works on the principle of converting sun's light energy into heat energy, which is retained for cooking and sterilizing the medical equipment. It reflects the sun light through a parabolic trough which surface is covered with a reflective film and concentrates it on the evacuated tube, which is fitted on the focus point of the parabolic trough. Parabola with a diameter of 600mm and depth 150mm is used to manufacture a parabolic trough. The evacuated tube's inner surface is coated with aluminium, which generates heat inside the hollow region. Aluminium foil is used as a reflecting film to reflect sun light, which has a reflectivity of 90%. So that the heat generation increases inside the evacuated tube. Whatever the subject (e.g. Food, water, medical equipment, etc.,) we need to heated are placed inside the evacuated tube through a tray for heating. The maximum temperature achieved inside evacuated tube is 302°C.

Keywords: Solar Energy, Parabolic Trough, Evacuated Tube, Solar Oven

1. Introduction

LPG and other fuels that we use are not infinitely available and also due to the consumption of wood and other sources we exploiting our natural resources. So the alternative way is to use the solar energy efficiently in every possible manner [15]. As the solar energy is easily available until the sun stays on the sky. So we use the solar reflecting films to reflect the sun rays in a parabolic trough to concentrate on its focus point to attain a maximum amount of heat that can possible. This solar oven works on a basic principle of converting sun's light energy into heat energy which is retained for cooking [12].

This solar oven helps in cooking foods, sterilizing the medical equipment in rural areas, etc., Day to day usage of LPG and other sources cost our future of mankind [13]. So utilizing the renewable resources such as sun light, wind, etc., will preserve the future of mankind and reduces deforestation, desertification, air pollution, etc., and lead mankind in a healthy way [11].

The solar collectors commonly refer to solar hot water panels, but may refer to installations such as parabolic troughs [1] and solar towers. Solar thermal collectors use heat-absorbing panels to absorb sunlight directly. Here the solar energy is directly used for heating purpose [2] and in large it can be used to produce steam that can run the generator to produce electricity [3]. The two types of solar collectors are imaging and non-imaging types. The parabolic trough is a non-imaging type concentrator that is able to concentrate rays to smaller absorber surface [4]. The sunray falling on the absorber is not focused, therefore concentration is achieved with a parabolic trough design [5].

The parabolic trough reflector is a collector which concentrate solar thermal energy in its focus point [10]. No matter what part of the collector the sunlight hits first but all incoming sunlight should be reflected from the surface and concentrate on its focus point [14]. The technical feasibility of using PTSC results in the thermal energy at temperatures up to 150°C for solar irradiance about 400-500 W/m^2.[6].

The light rays coming from the edges of the sources are redirected to the edges of the receiver. This confirms that all light rays coming from the inner points in the source will make contact with the receiver as shown in figure 1.

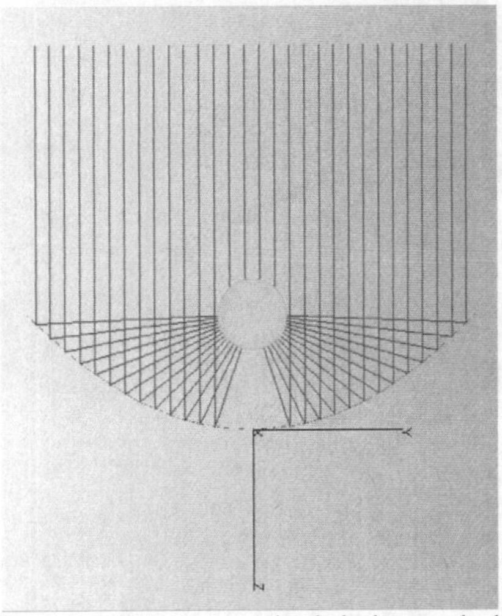

Figure 1. Edge ray principle in parabola

The rays falling on the CPC should be reflected to the receiver, so a reflective material is used which reflects the rays to the evacuated tubes. The reflective material should have high reflectivity.

Evacuated tube collectors are devices, which consists of cylindrical absorbing surface in which the vacuum is created between concentric glass tubes made up of borosilicate glass [7]. The outer layer is transparent allowing the light rays to pass through with minimal reflection. The inner layer is coated with a special selective coating (Al-N/Al), which provides excellent solar radiation absorbing properties. The evacuated tube absorbs the solar energy reflected from the CPC and converts into the heat energy. Thus, vacuum acts as an insulator, which does not lose the heat. To maintain vacuum between two glass layers a barium getter is used. This barium layer absorbs any CO, CO_2, N_2, O_2, H_2O and H_2 out-gassed from the evacuated tube during storage and operation, thus maintaining the vacuum [9].

2. Methodology

The components involved, their respective design, analysis and materials involved for the Parabolic Solar oven are briefed.

Methodology has three steps they are,

- Design
- Fabrication
- Experimentation

IOP Conf. Series: Earth and Environmental Science **312** (2019) 012009 doi:10.1088/1755-1315/312/1/012009

2.1. Conceptual design

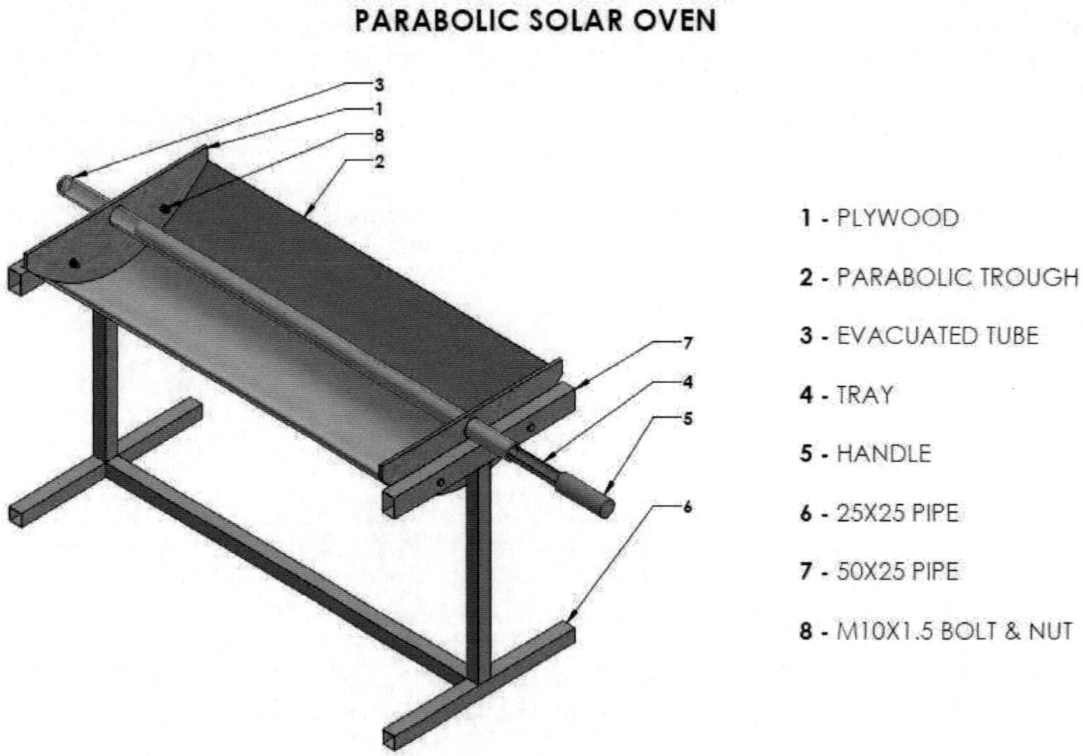

Figure 2. Conceptual design

2.2. Components

The components of the parabolic solar oven are listed below

2.2.1. Parabolic trough. Parabolic trough and other solar concentrators are increasing the attention by industry as well as engineering students which will provide a heat energy without help of any other external resources. Considering the application of parabolic trough for concentrating and illumination, the side wall of parabolic trough is made by reflective material. The base of the parabolic trough is made of wood. The outline of the parabolic trough is drawn on the plywood and is cut with the help of cutting wheel. The plywood acts as the supporter of the parabolic trough and holds the aluminium sheet metal and evacuated tube.

Table 1. Parabolic trough specifications

Parabolic trough	Specifications
Length	1200mm
Diameter	600mm
Height	150mm
Material	Plywood and Aluminium metal sheet

2.2.2. Evacuated tube. The solar absorber tube has two concentric glass tubes closed at one end with an annular vacuum space and a selective surface absorber on the outer surface (vacuum side) of the inner tube [8]. In heat pipe evacuated tube collectors, a sealed heat pipe, usually made of copper to increase the collector's efficiency in cold temperatures, and is attached to a heat absorbing reflector plate within the vacuum sealed tube.

Table 2. Characteristics of evacuated tube

Evacuated tube	Characteristics
Length	1500mm
Outer Diameter	47mm
Inner Diameter	37mm
Glass thickness	1.6-2.0mm
Material	Borosilicate Glass
Absorbance	Above 93%
Emittance	Above 8%
Heat loss	<0.8W/(m$^{2\dot{0}}$ °C)
Startup temperature	<=25°C (77F)
No of coatings	3
Coatings	Al/N/Al

2.2.3. Aluminium foil. The Aluminium foil is the reflective material used to reflect the falling radiation to the receiver tube. The reflectivity of the material is important and it shouldn't absorb the falling radiation. The Foil is made of Al8011 type Aluminium material.

Table 3. Characteristics of Al foil

Aluminium foil	Characteristics
Reflectivity	Above 85%
Thickness	30 microns
Surface area	0.96m^2

2.2.4. Frame. The frame is the base component which carries all other components of the experimental setup. It is made up of mild steel square pipe of side 25mm and thickness 1mm. The frame was welded using normal electrode arc welding.

2.2.5. Tray. It is a simple structure made of aluminium metal sheet inside the evacuated tube to hold the food and other materials which is to be subjected to high temperature.

2.3. Collection of materials

The following table shows the materials required to do the fabrication work and the other components readily available in the market.

Table 4. Collection of materials

Part No.	Description	Quantity No.	Material
1	Frame	1	Mild Steel
2	Parabolic trough	1	Plywood and Aluminium metal sheet
3	Reflective material	1	Al8011
4	Evacuated tube	1	Borosilicate glass
5	Tray	1	Aluminium metal sheet

IOP Conf. Series: Earth and Environmental Science **312** (2019) 012009 doi:10.1088/1755-1315/312/1/012009

2.4. Design calculations

2.4.1. Design of a parabolic trough

General equation for parabola

$$x^2 = 4 * f * y$$
$$r^2 = 4 * f * h$$

Where,

r = x = radius of the parabola
h = y = depth of the parabola
f = focal length of the parabola

Calculation of focal length for r = 300mm and h = 150mm

$$f = 150 \text{ mm}$$

2.4.2. Heat flow

Heat transferred from outer surface to the inner region of the evacuated tube is

$$Q = \frac{\Delta T}{R} = \frac{T_2 - T_1}{R}$$

Where,

T_2 = Temperature inside evacuated tube in °C

T_1 = Temperature outside evacuated tube in °C

$$R = \frac{1}{2\pi L}\left[\frac{1}{K}\ln\left(\frac{r_2}{r_1}\right)\right]$$

Where,

K = Thermal conductivity of glass = 1 W/(m.K)

r_2 = inner radius of the evacuated tube

r_1 = outer radius of the evacuated tube

Calculation of heat flow at different ambient temperature and time for r_2=23.5 mm, r_1=17.5 mm and L=1.5m

$$R = \frac{1}{2\pi(1.5)}\left[\frac{1}{(1)}\ln\left(\frac{23.5}{17.5}\right)\right] =$$

$$R = 0.0312 \text{ K/W}$$

IOP Conf. Series: Earth and Environmental Science **312** (2019) 012009 doi:10.1088/1755-1315/312/1/012009

2.5. Fabrication process

The fabrication process listed is done with the help of conventional machines available. The components and methods used to fabricate the components were discussed

- Drawn a parabola in a sheet by using eccentricity method.

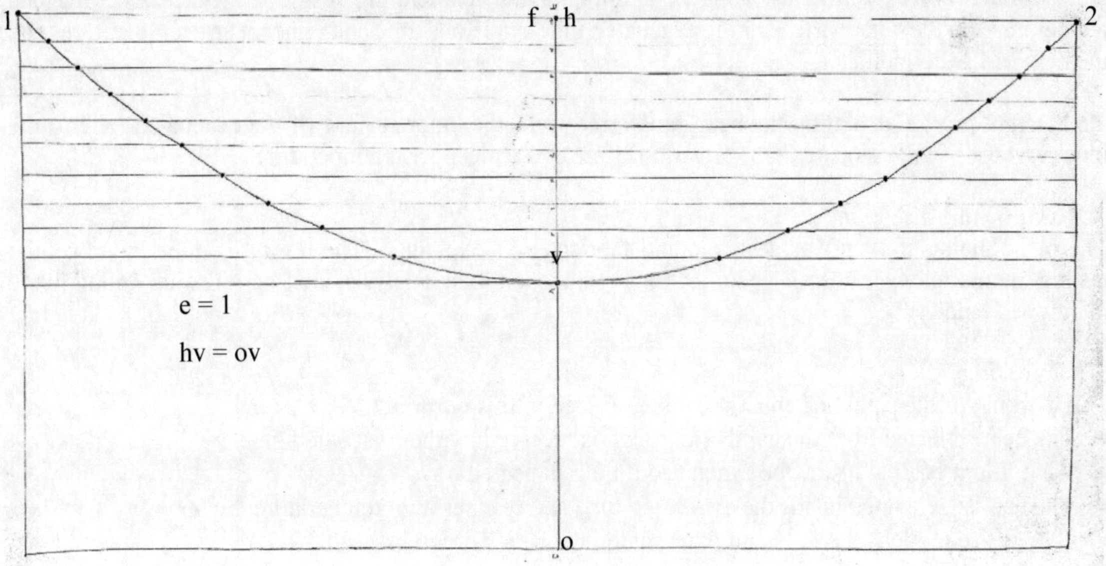

Figure 3. Eccentricity method

Where,

1f = f2 = radius of the parabola

hv = depth of the parabola

f = focus of the parabola

v = vertex of the parabola

- Pasting the sheet in the plywood.
- Cutting the plywood in a parabolic shape.
- Nailing the aluminium sheet in a parabolic shaped plywood.
- Through above process parabolic trough is manufactured.
- Frame is welded and bolted with a parabolic trough.
- Evacuated tube is fitted on the focus point of a parabolic trough.
- By rolling the aluminium sheet the tray is being manufactured.
- Handle is fitted to the tray.
- Reflecting film is pasted on inner surface of the parabolic trough to increase the reflectivity.

IOP Conf. Series: Earth and Environmental Science **312** (2019) 012009 doi:10.1088/1755-1315/312/1/012009

2.6. Testing methodology

2.6.1. Boiling of water. Time taken to reach 100°C of water at an atmospheric temperature of 36°C with the help of stop watch.

2.6.2. Temperature inside Evacuated tube. Finding the temperature inside the evacuated tube and temperature on the outer surface of the evacuated tube at various time and atmospheric temperature with the help of IR thermometer.

2.6.3. Heat flow. Calculating the heat transferred from the outer surface of the evacuated tube to the inner surface of the evacuated tube at various time and atmospheric temperature.

3. Results and discussions

The parabolic solar oven was designed and fabricated as explained. The testing is done based on the testing methodology. Overall specifications with economic analysis, and test results found were mentioned below.

3.1. Working

The working of the parabolic solar oven is described in this portion.

- The heat reflected from the parabolic trough is received by the Evacuated tube.
- Aluminium coating inside the evacuated tube gets heated.
- The heat is generated inside the evacuated tube and reaches high temperature.
- Tray is used to load the food and other materials, which are to be heated.
- IR thermometer is used to measure the temperatures at certain points.
- The readings at various time intervals were taken and compared.

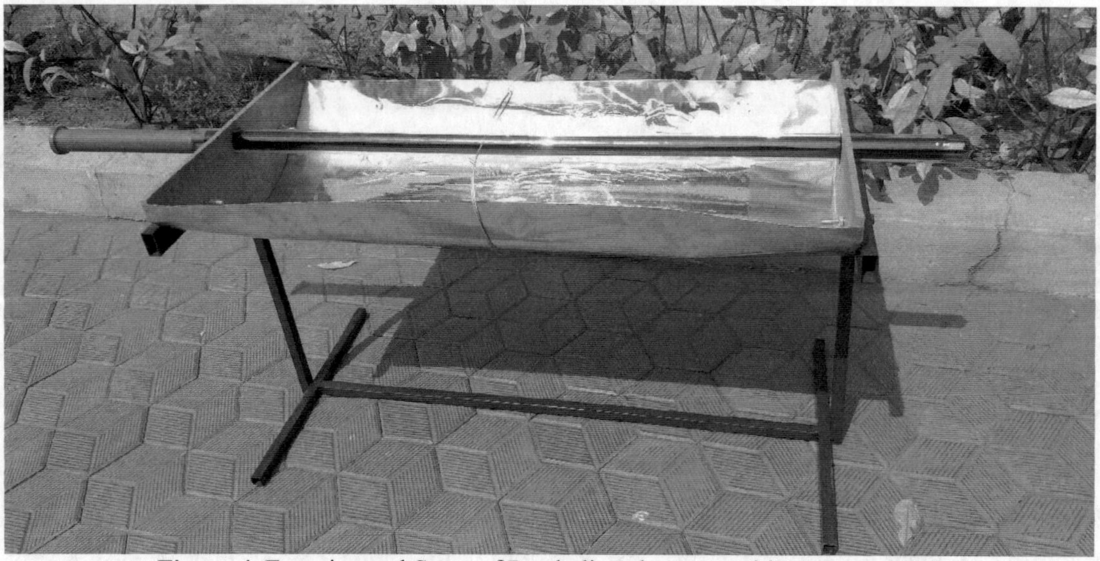

Figure 4. Experimental Setup of Parabolic solar oven with evacuated tube

IOP Conf. Series: Earth and Environmental Science **312** (2019) 012009 doi:10.1088/1755-1315/312/1/012009

3.2. Result

The fabricated prototype is made to work and we observed for its results.

Test is conducted on 36°C of atmospheric temperature.

Time taken to attain 100°C for 50ml water = 220 seconds

Maximum temperature achieved inside evacuated tube = 302°C

Heat flow at various ambient temperatures and certain time in day hours (in railways time)

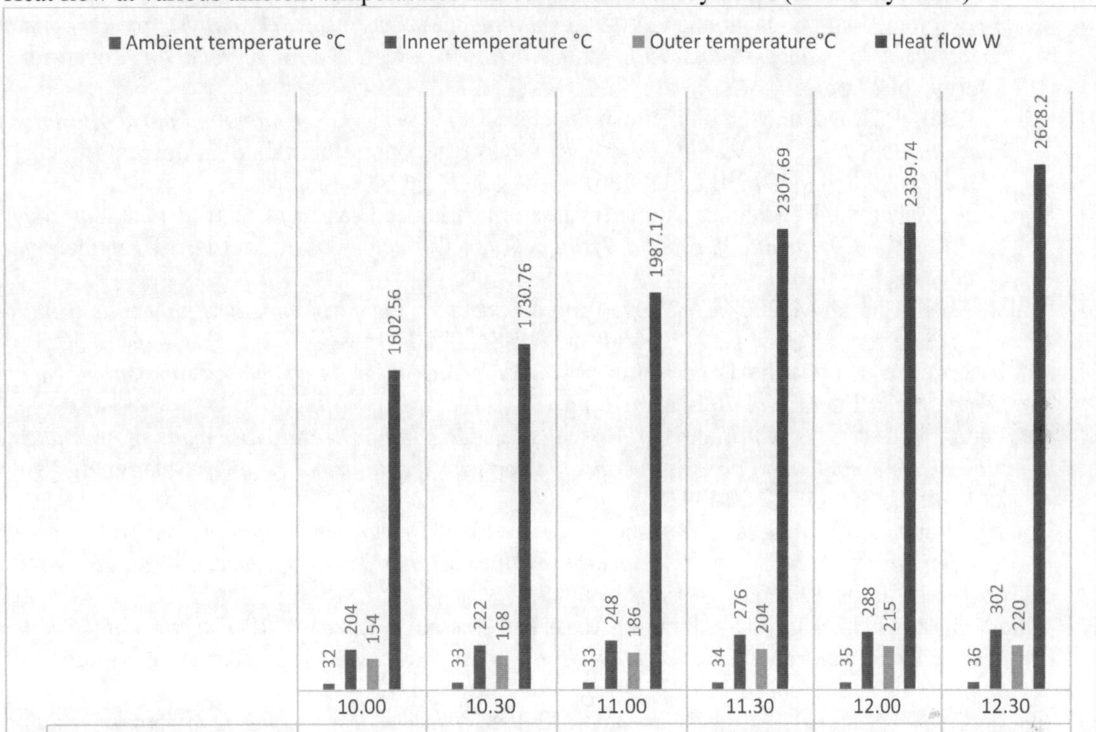

TIME	10.00	10.30	11.00	11.30	12.00	12.30
Ambient temperature °C	32	33	33	34	35	36
Inner temperature °C	204	222	248	276	288	302
Outer temperature°C	154	168	186	204	215	220
Heat flow W	1602.56	1730.76	1987.17	2307.69	2339.74	2628.2

Figure 5. Result chart

4. Conclusion

The testing of solar oven results in improved performance by using parabolic trough. The additional setup of evacuated tube further increases the temperature. The edge ray principle helps to achieve a parabolic trough, so that the maximum rays can be utilized which results in higher concentration of rays. Thus, the efficiency is further increased. This solar oven employed with parabolic concentrator has better efficiency, which can meet the energy demand increasing day to day. This provides higher temperature than box type solar ovens.

IOP Conf. Series: Earth and Environmental Science **312** (2019) 012009 doi:10.1088/1755-1315/312/1/012009

References

[1] Ajitkumar S. Gudekar, Atul S. Jadhav, Sudhir V. Panse, Jyeshtharaj B. Joshi and Aniruddha B. Pandit 2013 *Cost effective design of compound parabolic collector for steam generation* Solar Energy **vol 90** pp 43–50

[2] D. K. Patel, P. K. Brahmbhatt and Hitesh Panchal 2017 *A review on Compound parabolic solar concentrator for sustainable development* International Journal of Ambient Energy DOI:10.1080/01430750.2017.1318786

[3] Rachel Oommen and S. Jayaraman 2002 *Development and performance analysis of compound parabolic solar concentrators with reduced gap losses - 'V' groove reflector* Renewable Energy **vol 27** pp 259–275

[4] D. K. Patel, P. K. Brahmbhatt and Hitesh Panchal 2017 *A review on Compound parabolic solar concentrator for sustainable development* International Journal of Ambient Energy DOI:10.1080/01430750.2017.1318786

[5] Subhi S. Mahammed Tadahmun Ahmed Yassen and Hameed Jassim Khalaf 2012 *Theoretical Study of the Compound Parabolic Trough Solar Collector* Tikrit Journal of Engineering Sciences **vol 19** pp 1-9

[6] Falah Abed Alhasan M. 2002 *Challenges and prospect for the development of parabolic trough solar collectors (PTC's)* IJSRP vol 5 Issue 8 ISSN 2250-3153

[7] Morrison G. L, Budihardjo I and Behnia M. 2004 *Water in glass evacuated tube solar water heaters* Solar Energy **vol 76** Issues 1-3 pp 135-140

[8] M. Gajic, N. Karwa, A. Mojiri, and G. Rosengarten *Modeling reflection loss from an evacuated tube inside a compound parabolic concentrator with a cylindrical receiver* Solar energy **vol 23** DOI:10.1364/OE.23.00A493

[9] Qiang Wang, JinfuWang and Runsheng Tang 2016 *Design and Optical Performance of Compound Parabolic Solar Concentrators with Evacuated Tube as Receivers* Energies **vol 9** pp 795 DOI:10.3390/en9100795

[10] Miqdam Tariq Chaichan, Khalil Ibrahim Abass and Hussain A Kazem 2012 *The effect of variable designs of the central receiver to improve the solar tower efficiency* International journal of engineering and science ISSN:2278-4721 **vol 1** Issue 7 pp 56-61

[11] Mullick S. C., Kandpal T. C. and Saxena A. X 1987 *Thermal test procedure for box-type solar cookers* Solar Energy **vol 39** pp 353-360

[12] Nahar N. M. 1990 *Performance and testing of an improved hot box solar cooker* Energy Conversion and Management **vol 30** pp 9-16

[13] Nahar N. M. 2001 *Design, development and testing of a double reflector hot box solar cooker with a transparent insulation material* Renewable Energy **vol 23** pp 167-179

[14] Soteris A. Kalogirou 2004 *Solar thermal collectors and applications* Progress in Energy and Combustion Science **vol 30** Issue 3 pp 231-295

[15] Suhail Zaki Farooqui 2013 *A gravity based tracking system for box type solar cookers* Solar Energy **vol 92** pp 62-68

IOP Conf. Series: Earth and Environmental Science **312** (2019) 012010 doi:10.1088/1755-1315/312/1/012010

Performance analysis of solar assisted vapour Jet refrigeration system with regenerator (CRMC method)

Sathiya PrakashP[1],Kalaisselvane A[2]

[1]Research Scholar, Department of Mechanical Engineering, Pondicherry Engineering College, Puducherry, 605014, India.

[2]Associate professor, Department of Mechanical Engineering, Pondicherry Engineering College, Puducherry, 605014, India.

E-mail: sprakash292@gmail.com& kalaisselvane@pec.edu

Abstract. There are various forms of low-grade thermal energies that could all be useful in powering an ejector refrigeration system. Effective utilization of low-grade thermal energy is feasible from many available sources from industrial areas, solar and the automobile exhausts etc, . The normal Vapour Jet Refrigeration system is compared with the efficient Vapour Jet Refrigeration system fitted with a solar assisted regenerative heat exchanger unit. The ejectoris designed using Constant Rate Momentum change (CRMC) methodology to enhance the Coefficient of Performance by the usage of the refrigerant R134a as the working medium. The ejector cycle has been recognized as the promising technology for utilization of solar energy for cooling. The heat absorbed by the solar flat plate collector is partially used to heat refrigerant in the ejector cycle. Study on the effect of refrigerant temperature due to heat provided by boiler and solar flat plate in the Vapour Jet Refrigeration system on Coefficient of Performance, pumping power, entrainment ratio, diffuser outlet temperature & heat rejected at condenser is carried out. Simulation of ejector designed through Constant Rate Momentum change (CRMC) methodology is done using FLUENT software.

Keywords: Ejector, Refrigerant R134a, Entrainment ratio, Mach number, Constant Rate Momentum change (CRMC), regenerator, COP.

1. Introduction

The low-grade thermal energies which are available in various forms such as vehicle exhaust heat, industrial processes, waste heat, and the hot water which is generated by solar collector could all be useful in powering the system ejector refrigeration system. The technology investigated herein in addition to energy savings could result in the reduction of harmful greenhouse emissions. It is understood that fossil fuels burnt in power plants to generate electricity is associated to be linked to environmental climate shifts. Over several decades, the application of low-grade thermal energy has gained interest for many researchers. The availability of low grade energy is more in industrial processes, solar collectors and automobile exhausts. Since a huge part of industrial waste heat is deserted to the atmosphere, many reported researchers work carried out in the area of low grade energy for economic significance[1]. Refrigeration is found to be an emerging area to use the low-grade energy, since it is economically reasonable and ecological pleasant applications to utilize the thermal energy, which is otherwise go as waste. Refrigeration is carried out in producing necessary cooling effects in many processing industries, such as pharmaceutical, food, petrochemical and other related industries. Among the available refrigeration system mechanical vapour compression and vapour absorption refrigeration are very popular. Ejector refrigeration is a also emerging as a promising technology that uses waste heat energy as main energy source for refrigeration with different types of refrigerants such as R134a, R717 etc[4] [5].

2. Ejector refrigeration

IOP Conf. Series: Earth and Environmental Science **312** (2019) 012010 doi:10.1088/1755-1315/312/1/012010

The essential components of ejector refrigeration system are ejector, a generator and a pump and they perform the function of thermal compression. Ejector system, replaces the mechanical Vapour-compressor by utilizing low-grade energy. This permits the system to be driven by thermal energy. The key component of the system is ejector, which is used to entrain, mix and recompress both the fluids by transfer of momentum and energy using a primary jet with high velocity. Traditionally in ejector refrigeration, water is used as refrigerant. In early 1900 steam is used in first cooling and refrigeration jet systems. In order to recognize and increase the performance of steam jet systems extensive research has been carried out. The major drawback of this systems is the low COP values and not able to produce the cooling effect below zero degree Celsius. Even though refrigeration system using ejector is a valuable alternative to traditional refrigeration systems of vapour compression, the refrigeration system of vapour compression remains dominant in the market.

The performance of the refrigeration system-using ejector is improved by the use of refrigerants like halocarbon compound. To increase the ejector efficiency and reduce the unit cost production more research and enhancement are required. In the recent years vapour absorption refrigeration system are commercially available. To make the ejector refrigeration system commercially available product the benefits of the ejector refrigeration system should be compared to that of absorption refrigeration system.

3. Ejector

The essential components of jet ejector are a supersonic nozzle, chamber for secondary stream suction, a mixing section and a diffuser. To achieve very high kinetic energy of motive stream, a typical converging diverging nozzle is used The kinetic energy of the motive stream is increased by interior energy and related flow work due to pressure.

3.1 Working Principle

A supersonic flow is created at the outlet of the motive nozzle. The condition of the fluid changes from liquid phase to liquid vapour phase when it passes through the nozzle, due to the drop in pressure in the nozzle, which is even lower than the saturation pressure. The gleaming of the motive fluid is delayed in the interior of the nozzle by the effects of hydrodynamic non equilibrium and thermodynamics, Due to lower pressure created by the primary nozzle, the secondary fluid is drawn in to the jet ejector through the secondary nozzle. In the suction chamber the high-speed motive fluid begins to interact with the entrained fluid. The secondary fluid is accelerated due to the momentum transferred from the primary flow. To pre accelerate the flow of secondary an extra suction nozzle can be used. This helps to reduce the difference in velocity between the motive and entrained fluid streams and thereby excessive shearing losses is reduced.

The mixing of the motive and entrained fluid takes place at constant pressure. The process is repeatedly affected by the shock wave phenomena, which results in a significant rise in pressure and drop in velocity. The mixed flow then passes through the diffuser. The role of the diffuser is to recover the available kinetic energy at the mixing exit and transform it into the pressure energy.

The total outlet pressure at the exit of the diffuser is always in between the inlet flow pressures of the primary and secondary fluid. The two main characteristics, which regulates the ejector working performance, are the compression ratio and the entrainment ratio, where the compression ratio is the ratio between the discharge and entrained fluid pressure.

3.2 Advantages of ejector system

- Since the installation and the operating costs of the Ejector systems are low they are considered the most reliable technology in refrigeration
- Because of its high reliability and easiness, they are extensively used in power plants, aerospace, propulsion and cooling applications
- The system is with low noise and virtually no vibration since no moving parts are involved.

IOP Conf. Series: Earth and Environmental Science **312** (2019) 012010 doi:10.1088/1755-1315/312/1/012010

- The ejector systems have a great potential application in saving energy and give security to the environmental effects by utilizing the waste and thermal energy of low grade such as in air-conditioning and refrigeration.
- By removing compressor, related power loss disappears.
- Problems such as vibration, noise of compressor and driving belt have been eliminated.

3.3 Disadvantages of ejector system

- The limiting factor is that the efficiency of the conventional ejectors are typically low.
- It has very low performance values in comparison with the systems of vapor absorption and compression.

4. Vapour jet refrigeration system

The system consists of a boiler, an evaporator, a condenser, a liquid pump, an expansion device and an ejector. There are two loops in this refrigeration system. One is power loop, which consist of liquid pump, boiler, condenser and jet ejector, where primary fluid is pumped to higher pressure by liquid pump and heated in the boiler. The high-pressure and high temperature refrigerant is fed to the jet ejector which draws the refrigerant of low pressure and temperature, mixes with primary refrigerant and compresses to high pressure and high temperature gas. The refrigerant then passes through the condenser and it gets converted to liquid refrigerant. A portion of liquid refrigerant flows back to the pump and the remaining refrigerant is sent to the refrigeration cycle.

Other loop is refrigeration cycle where the refrigerant is expanded in the expansion device and the low temperature refrigerant is allowed to pass through the evaporator and refrigeration occurs in the evaporator coil.

5. Design of ejector

The various assumptions for the design of jet ejector are

- The flow is one dimensional and steady.
- Velocities are negligible at all inlet and exit conditions of ejector such as in motive and secondary inlet and at the diffuser exit.
- The losses due to the friction and mixing in the nozzle and the diffuser are distinct in terms of isentropic efficiency.
- The walls of the nozzle and the diffuser are adiabatic.
- The motive and the entrained fluids are identical and they should have the same molecular weight and specific heat values.
- The two streams mixing is assumed to be finished before the occurrence of normal shock wave and carried out at constant static pressure.
- Normal shock due to mixing takes place at end of mixing chamber at constant area.
- The motive and the entrained fluids are dry saturated vapours and obey the ideal gas behaviour.

The primary fluid from the boiler enters the ejector with a high pressure and lowvelocity which is called as the initial stagnation condition. As the motive fluid flow enters the ejectors converging part, its pressure gets decreased due to the reducing area and its velocity is increased.

5.1 Mass flow rate

$$QE = m(h_{g,e} - h_{f,c}) \quad (1)$$

Where QE = cooling load in kW and it is the heat absorbed at the evaporator
$h_{g,e}$= vapour's specific enthalpy at the evaporator outlet temperature
$h_{f,c}$= liquids specific enthalpy at the condenser saturation temperature

5.2 Entrainment ratio

$$ER = \left(\frac{m_2}{m_1}\right) \quad (2)$$

Where ER = Entrainment ratio
m_1 = mass flow of primary fluid
m_2 = mass flow of secondary fluid

5.3 Static pressure and temperature

The ratio between the fluid to sound velocities is called Mach number

$$M = \frac{V}{a} \quad \textbf{(3)}$$

$$a = \sqrt{\gamma \text{x R x T}} \quad (4)$$

Where M= Mach number
R= gas constant of the refrigerant in J/kgK
γ= specific heat ratio
a=velocity of sound in m/s
V=velocity of the fluid in m/s

Total temperature and total pressure can be calculated by using Mach number and static conditions at different section in the nozzle.

$$T_0 = T \times [1 + ((\gamma - 1) \times M^2] \quad (5)$$

$$P_0 = P \times [1 + ((\gamma - 1/2) \times M^2]^{(\gamma-1)/\gamma} \quad (6)$$

$$P_0/P = [T_0/T]^{(\gamma-1)/\gamma} \quad (7)$$

Where To = Total temperature of the fluid
Po = Total pressure of the fluid
P= Static pressure of the fluid
T= Static temperature of the fluid

5.4 Mixing section

The motive and entrained flows entrainment process is governed by momentum equation

$$\oint Adp = (1 + R_m)V_m - m_1V_1 - m_2V_2 \quad (8)$$

$$V_m = \frac{(m_{1\times}V_{1)}+(m_2V_2)}{m} \quad (9)$$

$$T_{0,m} = \frac{(m_{1\times}T_{0,1)}+(m_2T_{0,2})}{m} \quad (10)$$

$$D = 2 \times \sqrt{\frac{mXRXT}{\pi XPXV}} \quad (11)$$

Where Vm, Tm is velocity & temperature at the mixing region of ejector

D is the diameter of the diffuser.

m_1& m_2 is the mass flow rate of primary and secondary fluid.

5.5 Losses in ejector

The major losses in a jet ejector are frictional loss, mixing and kinetic energy loss and loss due to shock waves. The loss due to friction of the ejector occurs throughout the length, whereas the mixing occurs in the mixing section where the both streams interact. Kinetic energy losses are those, which occur within the convergent part of the nozzle and within the diffuser.

These result from frictional effects caused by flow separation. Flow separation occurs mostly on the divergent passages when flow area increases rapidly. This leads to strong turbulence near the wall. Shock wave loss is an irreversible process that occurs at the end of the mixing section. This loss leads to decrease in kinetic energy of the fluid flow and drastically the ejectors pressure lift ratio gets reduced.

6. Constant Rate of Momentum Change method

This method is designed with the diffuser geometry which eliminates the thermodynamic shock process that occur in the mixing chamber. At constant rate the fluid flow momentum is changed, which resulted in the elimination of shock wave and, which results in the gradual increase of static pressure from inlet to that of exit[3]. The momentum change at constant rate improves the entrainment ratio. It increases the pressure lift ratio and improves the performance of ejector[7].
CRMC design is based on the equation:

$$\frac{dM}{dX} = \gamma \quad (12)$$

Where $\frac{dM}{dX}$= rate of change of momentum constant

γ = specific heat ratio

7. Vapour jet refrigeration system with regenerator

The schematic of Vapour jet refrigeration cycle is shown in figure 1. The system consists of a boiler, solar flat plate collector, an evaporator, a condenser, a liquid pump, an expansion device, an ejector and a regenerator. Heat addition to the primary fluid is carried out in three stages. In the first stage waste heat available in the refrigerant that flows from the ejector outlet to the condenser is conserved by transferring the heat to low temperature primary fluid using a counter flow heat exchanger called regenerator. The super-heated refrigerant from the diffuser section of the ejector is condensed in the

IOP Conf. Series: Earth and Environmental Science **312** (2019) 012010 doi:10.1088/1755-1315/312/1/012010

condenser. A large amount of coolant is required to condense super-heated vapour to saturated liquid. In order to avoid this, a heat exchanger is incorporated between diffuser and the condenser. The liquid refrigerant from the pump can be used to cool this super-heated vapour wherein it preheats the fluid entering the boiler. Consequently, the heat supplied at the boiler can be reduced. The temperature of the primary refrigerant increases from T2 to T3'. In the second stage heat addition occurs in the solar flat plate collector. The temperature of primary refrigerant further increases from T3'to T3'.The third stage of heat addition occurs in the boiler. Temperature is increased from T3'to T4, ie till the refrigerant is heated to the required temperature.

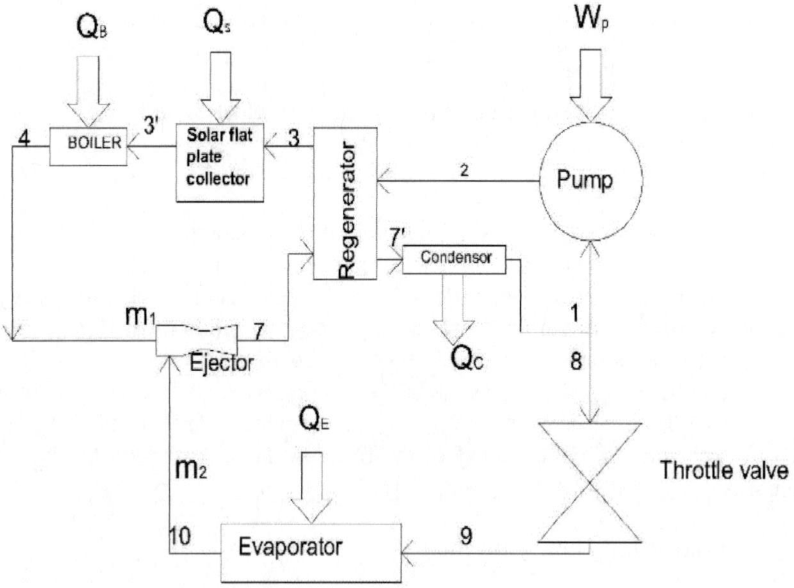

Figure1. Schematic diagram of Vapour Jet refrigeration system with regenerator

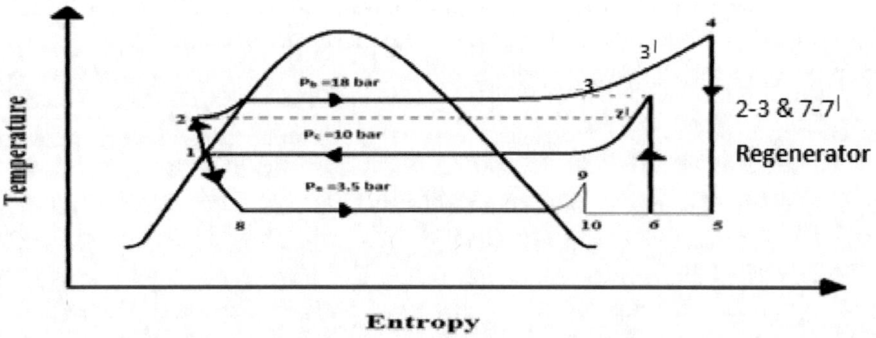

Figure2.T-S diagram of Vapour Jet refrigeration system with regenerator

7.1 Heating loads

IOP Conf. Series: Earth and Environmental Science **312** (2019) 012010 doi:10.1088/1755-1315/312/1/012010

The equations of energy balance under the steady state conditions are:

$$Q_S = m_1 \times (h_{g,b} - h_{f,p}) \quad (13)$$

Where Q_S = Heat supplied at the boiler in and flat plate collector kW

$h_{g,b}$= specific enthalpy of vapor at the boiler outlet

$h_{f,p}$ = specific enthalpy of liquid at the pump outlet

$$Q_C = m_1 \times (h_{g,d} - h_{f,c}) \quad (14)$$

Where Q_C= Heat rejected at the condenser in kW

$h_{g,d}$= vapour's specific enthalpy at the ejector outlet

$h_{f,c}$= liquids specific enthalpy at the condenser outlet

$$P_p = m_1 \times W_P \quad (15)$$

$$W_P = V_C \times (P_b - P_c) \quad (16)$$

Where P_p- pump power required in kW

W_P – specific work done by the pump

m_1- mass flow rate of primary fluid(kg/s)

V_C - specific volume (m^3/kg)

P_b - boiler pressure in bar

P_c - condenser pressure in bar.

7.2 Performance Evaluation

Coefficient of performance

COP is defined as the ratio of refrigeration effect obtained to the work done by the refrigerant.

$$(COP)th = \frac{CL}{Q_{S+W_P}} \text{ (Without regenerator)} \quad (17)$$

Where CL – Cooling load

Q_S -Heat supplied to the boiler

W_P–Pump work

$$(COP)th = \frac{CL}{Q_{R+W_P}} \text{ (With regenerator)} \quad (18)$$

Q_R- Heat supplied to the preheated fluid coming from the regenerator

$$Q_R = Q_S - Q_{reg} \quad (19)$$

Q_{reg} – heat gained by the refrigerant in the regenerator

The refrigerant properties of R134a are taken from standard refrigerant property tables.

IOP Conf. Series: Earth and Environmental Science **312** (2019) 012010 doi:10.1088/1755-1315/312/1/012010

Assuming the nozzles, diffuser, pump are cent percent efficient and neglecting the pressure drop in the condenser, evaporator and boiler calculations are made.

7.3 Inference

It is observed that the incorporation of regenerator in a system with Conventional ejector increases the COP of the cycle. It is evident that with regenerator having 10% effectiveness, the COP of the system is increased by 6.25%. For every 10% increase in effectiveness of regenerator, it is seen that COP is increased by 11.7%.

It is observed that the incorporation of regenerator in the system with CRMC ejector increases the COP of the cycle. It is evident that with regenerator having 10% effectiveness, the COP of the system is further increased by 3.23%. For every 10% increase in effectiveness of regenerator, it is seen that COP is increased by 13.5%. Graph.1 shows the Variation of COP with regenerator effectiveness.

Table1.COP Comparison of Vapor Jet Refrigeration System with &Without Regenerator for Conventional Ejector System & CRMC Ejector System

	(COP)TH(WITHOUT REGENERATOR)		**(COP)TH(WITH REGENERATOR)**	
REGENERATOR EFFECTIVENESS(%)	**CONVENTIONALEJECTOR SYSTEM**	**CRMC EJECTOR SYSTEM**	**CONVENTIONAL EJECTOR SYSTEM**	**CRMC EJECTOR SYSTEM**
100	0.16	0.3	1.86	3.24
90	0.16	0.3	0.91	1.54
80	0.16	0.3	0.59	1.06
70	0.16	0.3	0.44	0.82
60	0.16	0.3	0.35	0.66
50	0.16	0.3	0.29	0.55
40	0.16	0.3	0.25	0.48
30	0.16	0.3	0.22	0.42
20	0.16	0.3	0.19	0.37
10	0.16	0.31	0.17	0.32

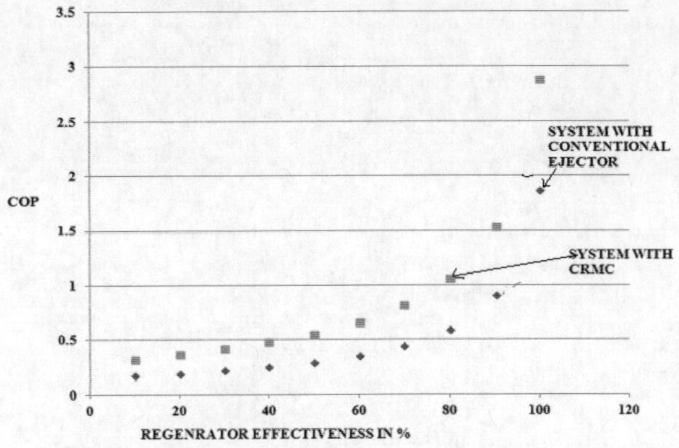

Graph 1. Variation of COP with regenerator effectiveness

8. Parametric study on different boiler temperature

8.1 Boiler temperatures effect on COP (theoretical)

The main objective of this study is to evaluate the performance of the system for various boiler temperatures. The operating conditions are condenser temperature 313K and evaporator temperature278K. The ratio of refrigeration capacity to boiler heat and pump work is used to calculate the COP. It is seen from the graph that the coefficient of performance is found to be decreasing as the heat that is supplied to the boiler increases with increase in temperature.

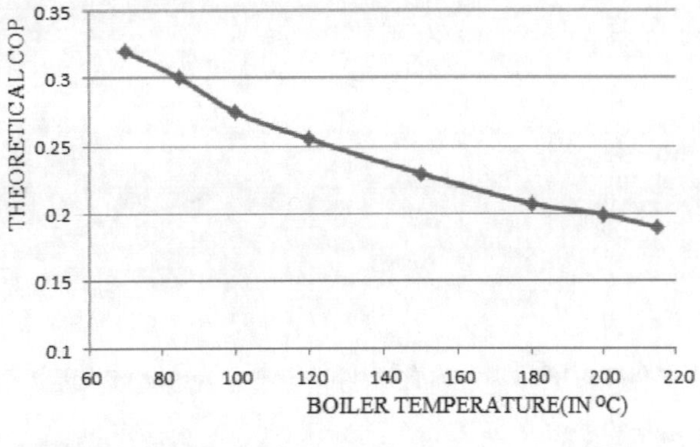

Graph 2. Effect Of boiler temperature on (COPth)

8.2 Effect of boiler temperature on (COPth)RE

It is found that when the boiler temperature exceeds an optimum value, an increase in primary mass flow rate leads to an increase in the heat supply in the boiler consequently which decreases the coefficient of performance. When the effectiveness of the regenerator varies the coefficient of performance also varies. The boiler temperature is a function of heat supplied. Hence it affects the coefficient of performance.

IOP Conf. Series: Earth and Environmental Science **312** (2019) 012010 doi:10.1088/1755-1315/312/1/012010

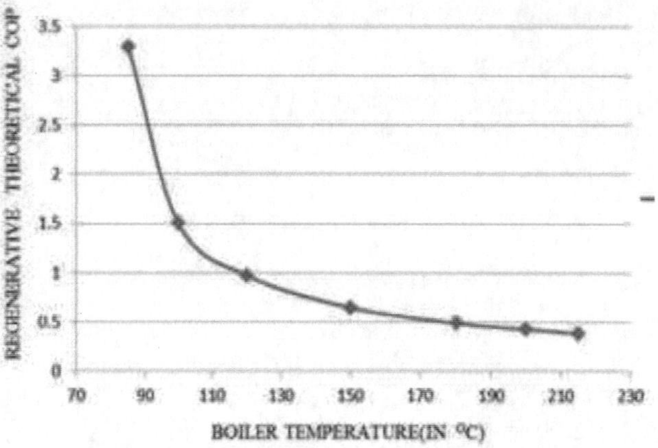

Graph 3. Effect of boiler temperature on (COPth)RE

8.3 Effect of boiler temperature on pumping power

Pumping power is plotted against different temperatures of boiler. As the temperature is increased, then there is an increase in the entrainment ratio hence the primary mass flow decreases, which in turn increase the pumping power. Pump work is negligible on comparing to heat supplied. The coefficient of performance is not much varied by the variation of the pumping power.

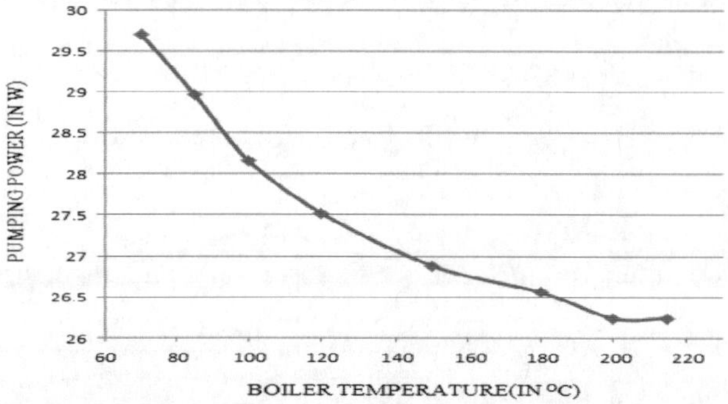

Graph 4. Effect of boiler temperature on pumping power

8.4 Effect of the temperature of boiler on ER

The entrainment ratio obtained on the various boiler temperatures is plotted. It is observed that the increase in boiler temperature leads to decrease in the flow rate through the motive nozzle. Hence power required to pump the working fluid becomes less. As boiler temperature is increasing, the heat that has to be supplied is high thereby it decreases the co-efficient of performance is decreased.

IOP Conf. Series: Earth and Environmental Science **312** (2019) 012010 doi:10.1088/1755-1315/312/1/012010

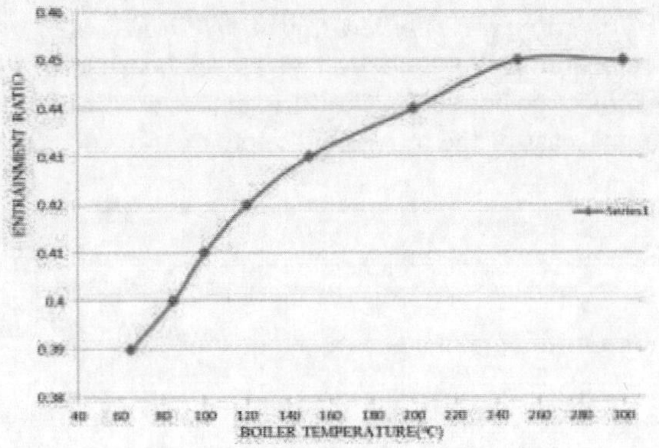

Graph 5. Effect of boiler temperature on Entrainment Ratio

8.5 Effect of solar flat plate collector temperature on Diffuser Outlet Temperature

It is observed that the increase in boiler temperature leads to decrease in amount of mass flow rate through primary nozzle. Hence the temperature of the fluid entering the ejector is found to be high which entrains the secondary flow from evaporator. The fluids mix together and diffuse in the ejector. The temperature at the outlet of diffuser is directly proportional to the boiler temperature.

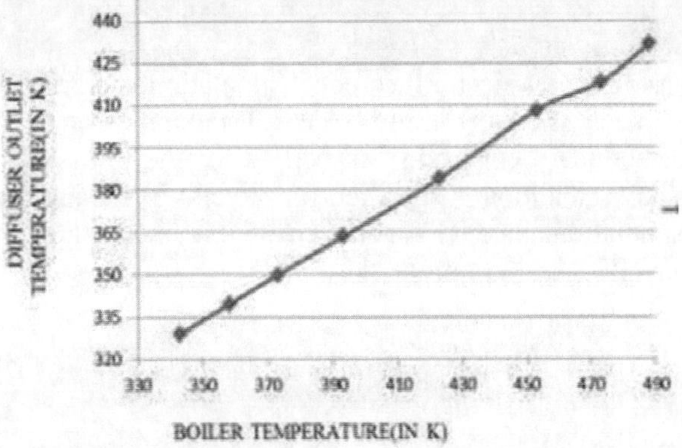

Graph 6. Effect of boiler temperature on Diffuser Outlet Temperature

8.6 Effect of boiler temperature on heat supplied at boiler

It is observed that the increase in boiler temperature leads to the heat supply at the boiler is high to attain such high temperatures. More amount of heat supply decreases the coefficient of performance.

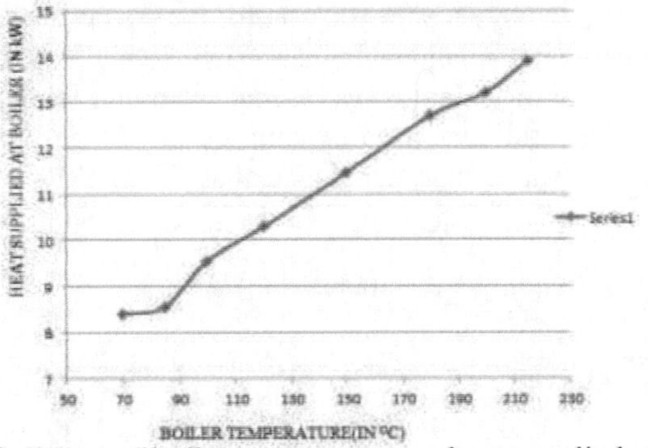

Graph 7. Effect of boiler temperature on heat supplied at boiler

8.7 Effect of solar flat plate collector Temperature on Heat Rejected at Condenser

It is seen that the heat rejection at the condenser increases because of the increase in diffuser outlet temperature, which demands large condenser. For the same size of condenser, the heat that has to be given out to liquefy the total mass of the refrigerant, increases.

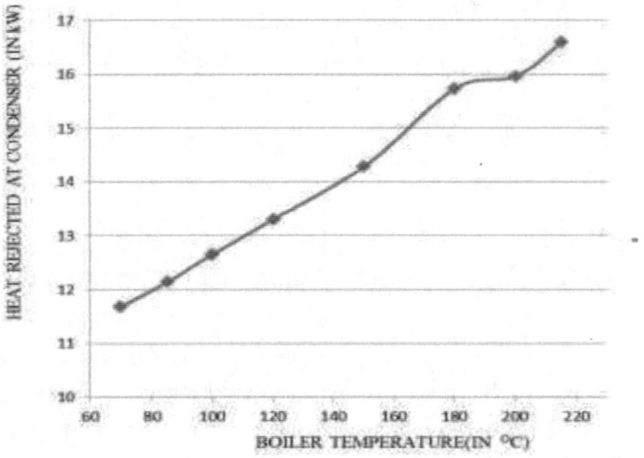

Graph 8. Effect of boiler temperature on Heat Rejected at Condenser

9. Computational fluid analysis using FLUENT

Using the design dimensions of the jet ejector theoretical simulation is carried out using FLUENT software. The jet ejector geometry is created using GAMBIT and a mesh size of 0.125 mm is used for meshing. Pressure inlet boundary condition is specified for the primary and secondary flow inlets. Pressure outlet boundary condition is specified for the exit of the jet ejector. To define the working fluid used in the simulation since R134a is not available in fluent database, the properties have been incorporated from NIST real gas properties.

NIST Real Gas:

Material name: R134a--1, 1, 1, 2-tetrafluoroethane Molecular Weight: 102.032
Critical properties:

- Temperature: 374.21(K)
- Pressure : 4.05928e+006 (Pa)
- Density : 5.01705 (mol/L) 511.9 (kg/m3)

The simulation is carried out and the results of FLUENT simulation matched well with the theoretical analysis of the jet ejector.

10. Conclusion

The vapour ejector refrigeration system for 0.75ton for the purpose of automobile air-conditioning been designed theoretically. A regenerator has been incorporated to increase the COP. A parametric study on the effect of variables by varying the boiler temperatures at constant pressure has been reported. The computational fluid dynamics analysis on the designed system had been carried out using fluent software. It is concluded that the CRMC ejector is found to be more efficient than the conventional ejector and COP of the system increased with addition of regenerator in the cycle.

11. References

[1] Zhu Y and Jiang P 2012 Hybrid vapor compression refrigeration system with an integrated ejector cooling cycle *International Journal of Refrigeration* **35** 68–78.

[2] Sun D-W 1998 Evaluation of a combined ejector-vapour-compression refrigeration system *International Journal of Energy Research* **22** 333–42.

[3] Kumar V, Singhal G and Subbarao P M V 2013 Study of supersonic flow in a constant rate of momentum change (CRMC) ejector with frictional effects *Applied Thermal Engineering* **60** 61–71.

[4] Selvaraju A and Mani A 2006 Experimental investigation on R134a vapour ejector refrigeration system *International Journal of Refrigeration* **29** 1160–6.

[5] Cizungu K, Mani A and Groll M 2001 Performance comparison of vapour jet refrigeration system with environment friendly working fluids *Applied Thermal Engineering* **21** 585–98.

[6] Kashyap S and Gupta R C 2011 Theoretical study of ejector refrigeration system with working fluid R410a *International Journal of Engineering Science and Technology* **3** 6508-6513.

[7] Kitratana B, Aphornratana S, Thongtip T 2017 comparison of traditional and CRMC ejector performance used in a steam ejector refrigeration *Energy Procedia* **138** 476-481.

Effect of RF/DC Magnetron Sputtering Process Parameters for Improving Efficiency of Solar PV Cell

Natarajan Purushothaman[1], R Elansezhian[2], A V Raviprakash[3]

[1]Research Scholar, Department of Mechanical Engineering, Pondicherry Engineering College, Puducherry, 605014, India

[2,3]Professor, Department of Mechanical Engineering, Pondicherry Engineering College, Puducherry, 605014, India

E-mail: natrajpurushothaman@gmail.com

Abstract. The biggest disadvantage of solar cells is that it has low solar power conversion efficiency typically 14%. So as to improve the power conversion performance, several attempts was made by researchers and once such technique is to reduce the reflection by providing textures on the solar cells. This texture will reduce the reflectance and thereby increase the solar power conversion efficiency. In this work, an investigational study was conducted to improve an energy efficiency of solar PV cells. Here Cu, Ag, Al was deposited for metallization purpose at specific places by using RF and DC magnetron sputtering process. The input parameters such as power, deposition time and substratum temperature were optimized during the sputtering process and the corresponding coating thickness, optical properties and surface topography were investigated. The coating thickness and surface roughness were increased as power increased from 15W to 45W. Average surface roughness (Ra) of 6nm and was obtained at the optimum power of 45W and deposition rate of 5-7 mins. The surface morphology and topography was characterized by using SEM and AFM which shows uniform coating with spherical shaped with densely packed nanostructures with increase in grain size were observed and elemental composition were confirmed by using EDAX. The absorption and reflection were investigated by using UV-VIS-NIR spectrometer indicates that the optical absorption increased from 1.2% to 1.68% at the wavelength range from 250 to 314nm. The reflection was reduced from 13.4% to 3.67% and the conversion efficiency was increased from 14.22% to 16.63%. This paper reports the complete experimental details, results and analysis.

Keywords: Solar PV cell; Magnetron sputtering; Nano coatings; Nano materials; Solar Efficiency.

1. Introduction

A solar cell is a solid state semiconductor device it can convert light energy into electrical energy based on the principle of photovoltaic effect. When compare to the other readily available energy sources conversion of electricity was more expensive. Yet, thin film sunlight based solar cell is cheap and could utilize environmentally friendly power vitality which is accessible abundant and give sun oriented power. The thin film metallic coating has been carried out with the process like electro

IOP Conf. Series: Earth and Environmental Science **312** (2019) 012011 doi:10.1088/1755-1315/312/1/012011

deposition, chemical deposition, thermal oxidation, spray pyrolysis and magnetron sputtering. Contrast with electro and chemical deposition method, magnetron sputtering gotten significant consideration since it has less ecological contamination and it is utilized for high space deposition with comparatively high deposition rate will be achieved [1]. This methodology is well appropriate for depositing oxides, nitrides, and carbides for surface protection and used as optical film [2]. Over 10.2% efficiency has been accomplished in $Cu_2ZnSn(S,Se)_4$ flimsy film which is which is manufactured by utilizing magnetron sputtering process [3]. Since copper and silver exhibit good properties like high thermal conductivity, low electrical resistivity and good antibacterial agent so the Metallization of copper and silver to the silicon wafer can improve the overall performance of the solar cell [4,5]. Aluminium films generally utilized in numerous application like optical layers, dielectric layers, electronic seals and insurance coatings in light of the fantastic warm, electrical, mechanical and concoction properties [6]. The properties of the film including its thickness is depend upon the input parameters employed in magnetron sputtering such as power, deposition time, argon/oxygen flow rate, base pressure, temperature. The vacuum degree plays the major effect on tungsten-rhenium adherence towards the alumina ceramic substrate have been demonstrated at optimized combination of argon flow rate 30sccm, power 400W and vacuum of 0.7×10^{-6} torr [7]. The impact of oxygen fractional weight and substrate temperature can improve the stoichiometry of the film [8]. The affidavit time can modify the morphology of the structure because of variety in the molecule measure appropriation. Longer statement indicates diminish the surface coefficient and shorter affidavit time gives the film unpredictable grains with high harshness [9]. The normal surface roughness and the power of the film get expanded with increment surface temperature 200 to 300^0 C and decrease of Ar has harmed the crystallinity yet produce smooth surface and thick morphology [10]. The aim of this work is to developing Nano coating on solar PV cell by using RF/DC magnetron sputtering unit and to study the impact of process parameters in material properties and enactment of solar cell.

2. Experiment

The substrate utilized in this research was multicrystalline solar Photovoltaic cell of 0.5v capacity. For trialling it was turned into size of 20 x 40mm. Surface cleaning is essential to achieve efficient coating to avoid contaminations. First the surface was washed with acetone and then rinses with distilled water and ethanol. The sputtering unit and vacuum chamber shows in figure 1. The target materials of 50mm diameter and 20mm thickness Cu, Ag, Al (99.99%) was purchased from HHV, Bangalore as cathode material was placed at a distance of 4cm from the substrate, metallization of Cu, Ag, Al were performed by varying the input parameters that are employed in the sputtering process such as power, deposition time and substrate temperature are tabulated in table 1. The cathode materials were pre - sputtered before deposition of about 5minutes to dispose of any defilement on its surface. All the coating process was performed at a constant base pressure of 3×10^{-6} mbar at argon (99.999%) rate of 10sccm in the vacuum chamber.

Table 1. Experiments by Varying Process Parameters.

Experiments	Power (W)	Deposition Time (Minutes)	Temperature (Degree Celsius)
1	15	3	100
2	15	5	200
3	15	7	300
4	30	3	200
5	30	5	300
6	30	7	100
7	45	3	300
8	45	5	100
9	45	7	200

IOP Conf. Series: Earth and Environmental Science **312** (2019) 012011 doi:10.1088/1755-1315/312/1/012011

Figure 1. Sputtering unit.

Coating properties such as thickness and the surface roughness was measured by using STYLUS PROFILOMETER. The grain size and the cross sectional view was measured by FEI Quanta FEG 200 HRSEM and elemental composition were confirmed by EDX. An optical property such as absorption and reflection was measured by UV visible spectrophotometer (Model – UV-3600 plus) with wavelength range from 200 to 2500nm. I-V characteristics were studied by using class AAA solar simulator - ORIEL SYSTEM.

3. Results and discussion

3.1. Coating properties

figure 2(a) & (b). indicate the graphical representation of coating thickness and surface roughness against process parameters involved in sputtering process and their values are tabulated in table 2. The process parameters are influencing the coating properties. When the power is increasing the coating thickness and surface roughness is also increasing. As power increases the energy available to eject atoms from target increases and hence more number of atoms will be ejected and gathered on to the substrate [11]. So the maximum surface roughness was obtained as 6nm at the power of 45W. Amount of roughness can create texture on to the surface of the substrate.

Table 2. Measured Coating Properties by Varying Process Parameters.

Experiments	Power (W)	Deposition Time (minutes)	Temperature (Degree Celsius)	Coating thickness (nm)	Surface roughness Ra (nm)
1	15	3	100	49.640	1.33
2	15	5	200	70.466	2.37
3	15	7	300	61.359	3.19
4	30	3	200	150.620	2.55
5	30	5	300	138.286	2.99
6	30	7	100	160.079	5.01
7	45	3	300	310.670	5.03
8	45	5	100	227.415	5.90
9	45	7	200	203.500	6.76

IOP Conf. Series: Earth and Environmental Science **312** (2019) 012011 doi:10.1088/1755-1315/312/1/012011

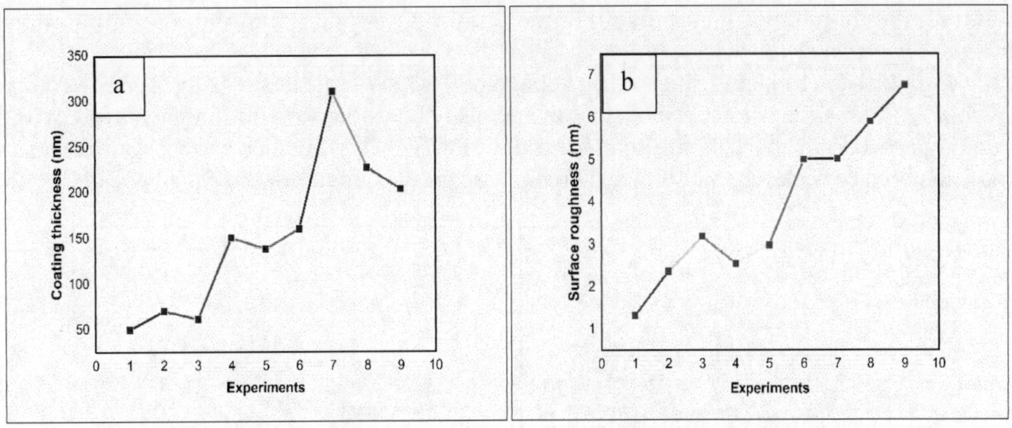

Figure 2. Shows the graphical representation of (a) coating thickness, (b) surface roughness

3.2. Surface morphology

The surface morphology of Cu, Ag, and Al coated samples with different magnifications of 1μm, 2μm, 400nm and 500nm shown in figure 3(a), (b) &(C). The image shows non-uniformly distributed surface under 15w and 30W. From Fig. 3(c) at 45W spherical shape nanostructure morphology with densely packed throughout the surface with increase in grain size from 80nm to 150nm respectively. Coalescence of atoms was observed due to increase in anode temperature [12].

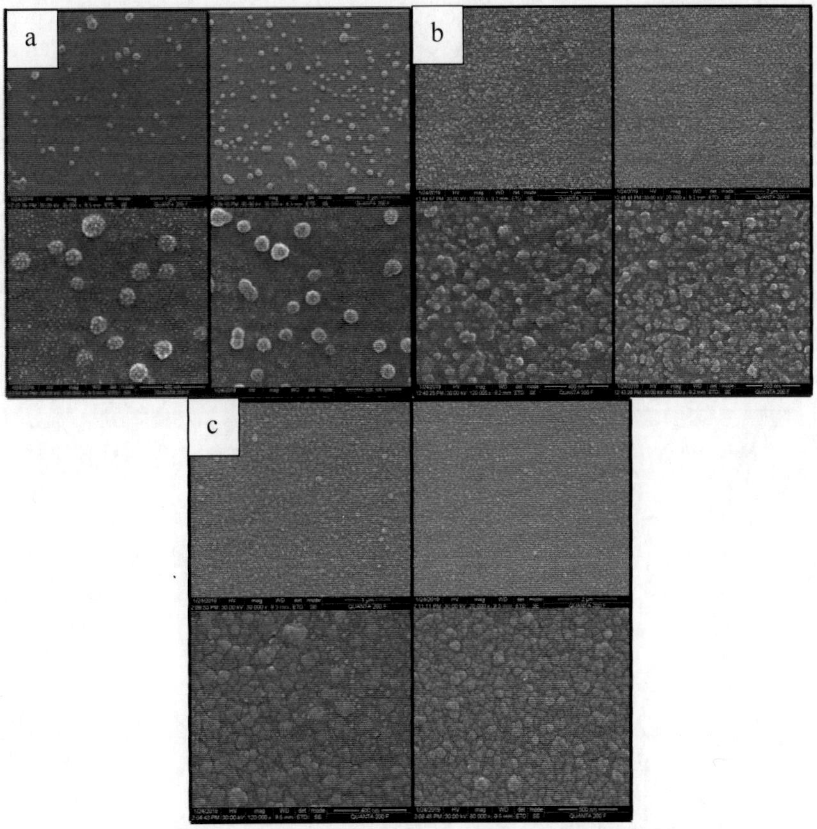

Figure 3. Shows the SEM image of the samples (a) 15W, (b) 30W & (c) 45W

3.3. Surface topography

From the Fig. 4(a) & (b) shows the surface topography which indicates that the surface has many spikes which confirms the surface is not smooth and also shows the variation in the surface with less inter grain space due to increase the sputtering power [13]. This rougher surface can create more reflection and bounce back into the surface thereby increase the output power of the PV cell.

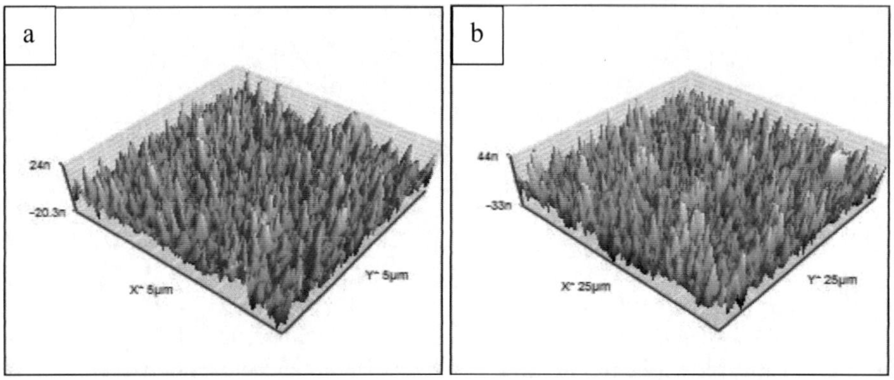

Figure 4. Shows the AFM image of the (a) sample 6 & (b) sample 8

3.4. Optical properties

The absorption spectra of the coated and non-coated samples are presented in the figure 5(a), (b), (c) & (d). The absorbance of non-coated solar PV cell was about 1.22% in the wavelength of 280nm. After metallization the absorption was increased to 1.56% for the sample 1 and 1.61% for sample 6 in the wavelength shifted to 320nm and the maximum absorption of 1.68% was observed at sample 8 in the wavelength range of 312nm so the optimum combination of process parameters maintained at 45W and 5 mins of deposition. figure (e), (f), (g) & (h). shows the reflection spectra, it shows the reflection of 23% for non-coated sample and the reflection was decreased to 13.4% in the wavelength range of 360nm for the sample 1 at 15W and 5.4% of reflection for sample 6. For sample 8 we observed that the reflection reduced to 3.67% in shifted wavelength range of 270nm at 45W. Since at 45W the amount of deposition of Nano particles was more compare to the 15W and 30W such that less amount of photons was reflected back due to multiple reflection with cu, Ag, & Al Nano particles.

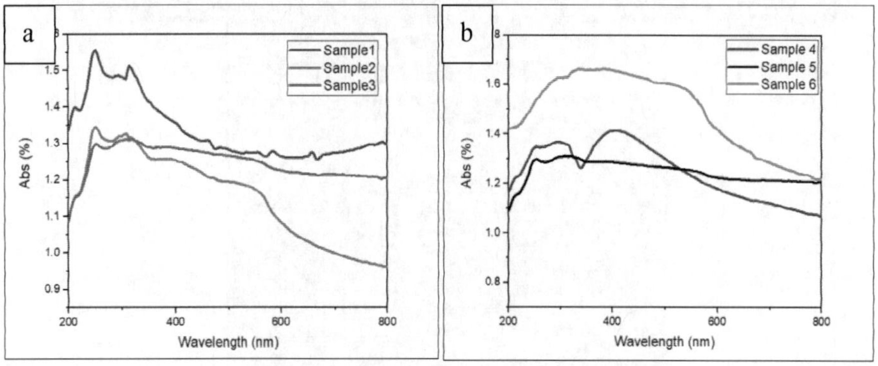

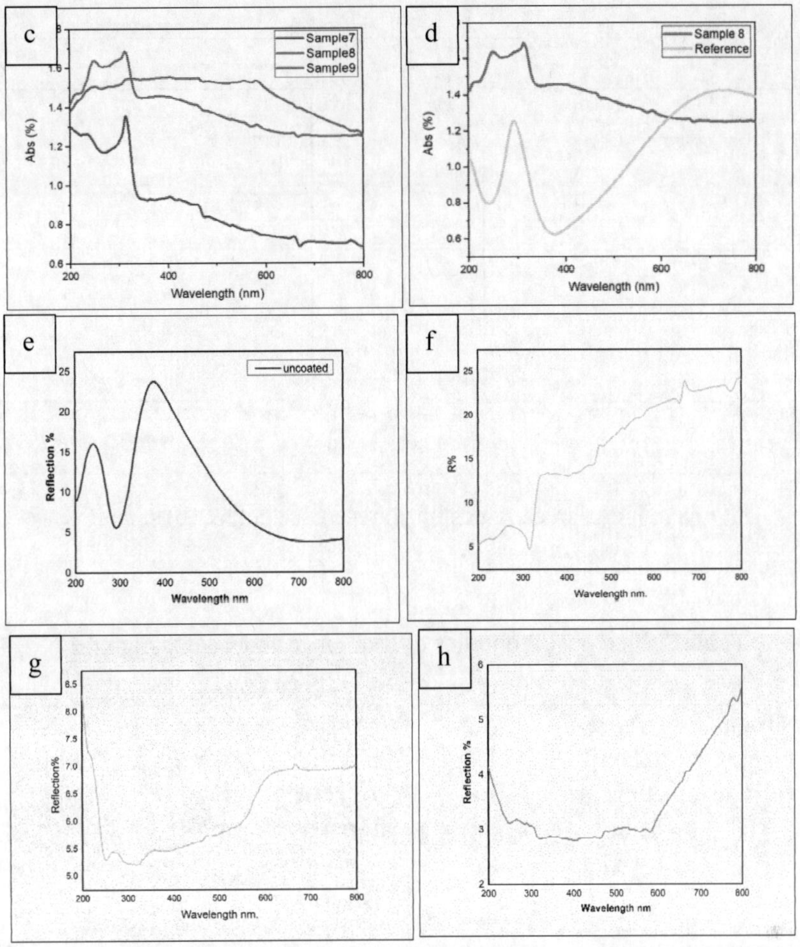

Figure 5. Shows the absorption spectra of (a) sample 1-3, (b) sample 4-6, (c) sample 7-9 and reflection spectra of (e) reference cell, (f) sample 1, (g) sample 6, (h) sample 8

3.5. I-V characteristics

The performance of multicrystalline solar cell without coating (Reference sample) and the maximum absorption sample were measured and the outcomes are potted in the table 3. And the corresponding I-V curve is presented in figure 6(a) & (b). From the table 3. the electrical properties such as V_{oc} and P_{max} are slightly increased to 0.6283 volts and 133.11mW compare to the reference cell respectively. The fill factor of the coated sample is 73.57 % compared to 67.04% of the reference cell which leads to increase of 23% efficiency. Since the sample 8 has maximum absorption of 1.68% which shows the maximum efficiency.

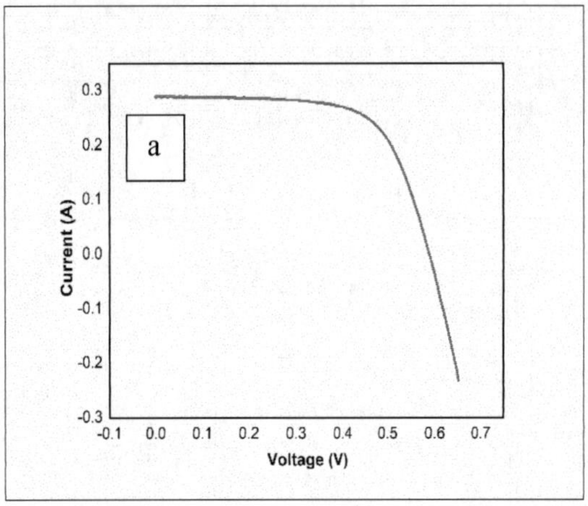

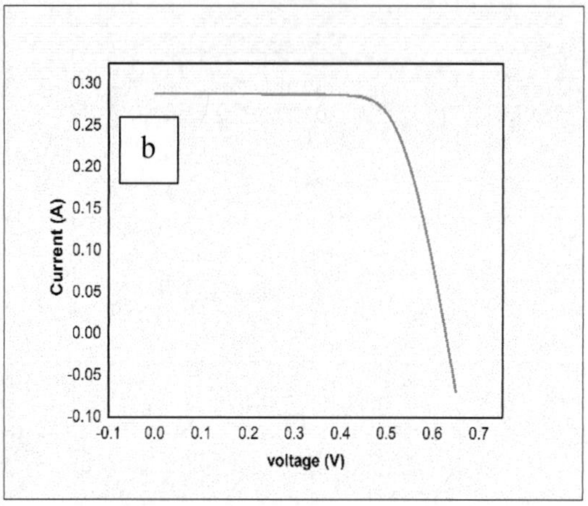

Figure 6. I-V characteristics of (a) reference cell & (b) sample 8

Table 3. Electrical Properties of Reference and Coated Sample

Sample no	**Open circuit voltage V_{oc} (v)**	**Short circuit current I_{sc} (A)**	**Max voltage Vmax (V)**	**Short circuit current density J_{sc} (A/cm^2)**	**Max Power P_{max} (mW)**	**Fill Factor (%)**	**Efficiency (%)**
Reference cell	0.5877	0.2887	0.4534	36.09	113.88	67.04	14.22
Sample 8	0.6283	0.2879	0.2712	35.99	133.11	73.57	16.63

4. Conclusion

Copper, silver and aluminium was deposited by using Magnetron sputtering process and the input parameters were varied as well as the optimum combination was found to be 45W, 5mins and 100^0C. The coating properties were analysed by using stylus Profilometer that shows the coating thickness and surface roughness was increased with increase in power and deposition time and the maximum roughness of about 6nm. SEM and AFM images show the Spherical shape with tightly packed nanostructure due to increase substrate temperature. Optical properties were characterized by using UV spectrometer which shows the maximum absorption of 1.68% and reflection reduced to 3.67% and I-V measurement shows 14% of increased efficiency.

References

[1] Donald L Smith 1995. Thin-Film Deposition Principles and Practice, , McGraw-Hill, Inc.,
[2] K. Ellmer, J. H. 2002. Copper indium disulfide solar cell absorbers prepared in a one-step. *Thin Solid Films*, **413**, 92-97.
[3] Jianjun Li, H. W. 2016. 10% Efficiency Cu2ZnSn(S,Se)4 thin film solar cells fabricated by magnetron sputtering with enlarged depletion region width. *Solar Energy Materials &*

SolarCells, **149**, 242-249.

[4] B. Giroire, M. A.-G.-Q. 2017. A comparative study of copper thin films deposited using magnetron sputtering and supercritical fluid deposition techniques. *Thin Solid Films*, **643**, 53-59.

[5] Paolo Nicolas Catalano, M. P.-I. 2016. Optically transparent silver-loaded mesoporous thin film coating with long-lasting antibacterial activity. *Microporous and Mesoporous Materials*, **236**, 158-166.

[6] Chien-Ming Chiang, L.-S. C. 2005. Microstructure and characterization of aluminum oxide thin films prepared by reactive RF magnetron sputtering on copper. *Surface & Coatings Technology*, **198**, 152-155.

[7] Bian Tian, Q. Y. 2018. Effect of magnetron sputtering parameters on adhesion properties of tungsten-rhenium thin film thermocouples. *Ceramics International*, **44**, S15-S18.

[8] T.K. Subramanyam, G. M. 2001. Process parameter dependent property studies on CdO films prepared by DC reactive magnetron sputtering. *Materials Chemistry and Physics*, **69**, 133-142.

[9] R.S. Gonçalves, P. B. 2018. The effect of thickness on optical, structural and growth mechanism of ZnO thin film prepared by magnetron sputtering. *Thin Solid Films*, **661**, 40-45.

[10] Vipin Chawla, R. J. 2008. Morphological study of magnetron sputtered Ti thin films on silicon substrate. *Materials Chemistry and Physics*, **111**, 414-418.

[11] Ellmer, K. 2000. Magnetron sputtering of transparent conductive zinc oxide: relation between the sputtering parameters and the electronic properties. *Journal of physics D: applied physics*, **33**, R17- R32.

[12] B. Subramanian, R. A. 2011. Influence of substrate temperature on the materials properties of reactive DC. *Materials Science and Engineering,* **176**, *1-7*.

[13] Shuyun, C. C. 2010. Orthogonal analysis for perovskite structure microwave dielectric ceramic thin films fabricated by the RF magnetron-sputtering method. *Mater Sci: Mater Electron*, **21**, 349–354.

Acknowledgement

The authors would like to thank Centre for Nanoscience and nanotechnology, Sathyabama institute of science and Technology for providing Magnetron sputtering Coating facility and Stylus Profilometer instrument for measuring Surface roughness and coating thickness. We extend our thanks to IIT Madras for SEM facility and Vellore Institute of technology, Vellore for AFM facility.

IOP Conf. Series: Earth and Environmental Science **312** (2019) 012012 doi:10.1088/1755-1315/312/1/012012

A succinct review on fuel cells

Subbarama Kousik Suraparaju[a], Sendhil Kumar Natarajan[a*] and P. Karthikeyan[b]

[a] Department of Mechanical Engineering,
National Institute of Technology – Puducherry, Karaikal 609609
U.T of Puducherry, INDIA.

[b] Fuel Cells and Energy Systems Laboratory,
Department of Automobile Engineering,
PSG college of Technology, Coimbatore, Tamilnadu – 641004.

Email: drsendhil1980iitmuk@googlemail.com

Abstract. The role of renewable energy is becoming significant in the current scenario due to increasing power demand, instability of fuel prices and economic-environmental glitches. Fuel cell stood as one of the emerging and leading contributors in the realm of renewable energy because of its higher efficacy, cleanliness and economic supply of demanded energy to the consumers. This article gives a brief knowledge on the different types of fuel cell technologies along with their working principles, pros, cons and application of fuel cells in various sectors. It also discusses the techno-economic issues of this current technology. Also briefed about the future recommendations to wide spread this technology for better economic and environmental solutions.

Keywords: Fuel cell, Renewable energy, Socio-economic solutions, I-V Characteristics.

1. Introduction

The consumption of energy is increasing day by day in all forms but at the same time sources of energy are depleting at the same rate. The context of renewable energy is evolved in this situation due to increased awareness in public for safeguarding environment and current status of fossil fuels. Among all the energy generating systems, the world is shifting more towards fuel cells as equal as solar energy because of fuel cells are capable of generating heat and electricity at an operation. Fuel cells directly transform chemical energy into electricity and generates water as its byproduct [1,2]. Even though, the conventional devices generate electricity from chemical energy but ended up with less conversion efficiency when compared to fuel cells. The fuel cells are literally seen as the device with collective features of heat engine and batteries; similar to an engine, fuel cells can function for longer duration as fuel is accessible without any transitional mechanical energy transformation and the features of fuel cell are almost similar to batteries [3]. It is intended that current article may be beneficial for forthcoming investigators to know the concept of fuel cell in a broader manner.

2. Fuel cell approach

Christian Friedrich Schönbein, a Swiss scientist, discovered the fundamental principle of fuel cell in 1838. First fuel cell accidentally developed by Sir William Grove by reversing the electrolysis of water [2] in 1839. A 5kW alkaline fuel cell was first illustrated by Francis Bacon in 1950 at Cambridge University. After the fruitful evolution of alkaline fuel cells, NASA required a compact system to produce electricity. In 1970s, an alkaline fuel cell of 12kW capacity was developed for space shuttle orbiter of NASA to offer consistent power without any backup usage. Later in the mid-1960s the studies on the fuel cells was more elaborately concentrated on the further advancement of different fuel cells for stationary power and transportation applications. Further the international government agencies have predominantly increased the opportunities for further research and development in fuel cells [3]. But in all the countries it took almost after 50 years for development due to its high installation cost. Later, the development of power conversion devices leads to the reduce in the decreased the installation cost.

2.1. *Principle of Fuel cells*

Fuel cells are direct energy conversion devices that transform the chemical energy into electricity and generates water and heat as its byproduct as shown in Fig.1 [4] and the fuel cell comprises of electrodes immersed into electrolyte. The oxygen from air is continuously fed to cathode and the hydrogen is continuously fed to anode of fuel cell. At anode, hydrogen split into positive ions and electrons are released. The electrolyte allows only the positively charged hydrogen ions and they flow from anode to cathode. These free electrons moved to cathode from anode through the external load connected.The oxidant reacts with cathode to form water. The chemical reactions involved are given as:

At Anode:

$$H_2 \rightarrow 2H^+ + 2e^- \quad (1)$$

At Cathode:

$$\frac{1}{2}O_2 + 2H^+ + 2e^- \rightarrow H_2O \quad (2)$$

Overall Reaction:

$$H_2 + \frac{1}{2}O_2 \rightarrow H_2O \quad (3)$$

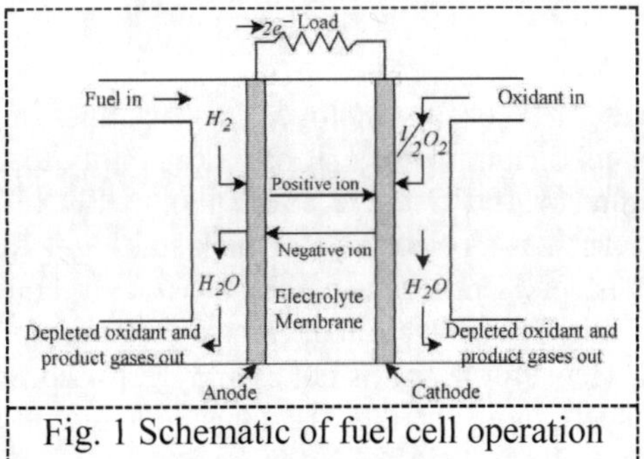

Fig. 1 Schematic of fuel cell operation

2.2. *Fuel cells categorization*

The fuel cells are categorized on the basis of electrolyte and fuel choice. Currently there are six types of fuel cells are serving in the realm of renewable energy.

i. Proton exchange membrane fuel cell (PEMFC)
ii. Alkaline fuel cell (AFC)
iii. Phosphoric acid fuel cell (PAFC)
iv. Molten carbonate fuel cell (MCFC)
v. Solid oxide fuel cell (SOFC)
vi. Direct methanol fuel cell (DMFC)

They are again categorized on the basis of operating temperature.

i. Low Temperature (50-250 °C) for PEMFC, AFC and PAFC.
ii. High Temperature (650-1000 °C) for MCFC and SOFC

2.2.1. Proton Exchange Membrane Fuel Cell (PEMFC)

The proton exchange membrane fuel cell comprises of a solid polymer electrolyte for swapping the ions between the electrodes, which is an insulator for electrons and conductor for protons. The fuel cell operates at a temperature of as low as 100 °C. The chemical reactions involved in the fuel cell are given in Eqns (1) – (3) [5]. The key benefits of PEMFCs are its quick startup power and higher densities for automobiles. Also, its low operating temperature makes it more viable in applications like transportation, mobile phones, bicycles and laptop computers etc. The main disadvantages of fuel cell are its low operating efficacy about 40%-45% and the other lagging factor is that use of platinum which is not cost effective. The PEMFC based sources of power are also being advanced for the household, hot water and building electricity applications [6]. The PEMFC is subcategorized into Direct Ethanol Fuel Cell (DEFC) and Direct formic acid fuel cell (DFAFC).

2.2.1.1. Direct Formic Acid Fuel Cell (DFAFC): The inlet fuel of DFAFC is formic acid (HCOOH) comprises of minute organic molecules fed directly to anode. It has a key benefit that the formic acid doesn't cross over the polymer which results the high concentration efficiencies and low power density compared to Methanol. This particular fuel cell at 60 °C operating temperature generates an open circuit voltage of 0.55V which is comparatively low than the theoretical value given by Gibbs free energy. So as to improve the efficiency, the Richard Masel's group at the University of Illinois utilizes catalyst as palladium [7–9].

2.2.1.2. Direct Ethanol Fuel Cell (DEFC): The main difference between the PEMFC and DEFC is input fuel. In PEMFC, Hydrogen is used as an input fuel whereas in DEFC, Ethanol is used as input fuel. In the DEFC a Nafion catalyst is used. The input fuel Ethanol is easily accessible from biomass through fermentation process. In this cell, at anode, the liquid ethanol is oxidized with the mixture of water and hydrogen ions, electrons and CO_2 are generated. The chemical reaction entangled is similar to PEMFC and the generated voltage is of 0.5V to 0.9V. The first DEFC powered vehicle is presented by University of Applied Sciences in France [10–13]

2.2.2. Alkaline Fuel Cells (AFC)

It is one of the earlier approaches of fuel cells employed in space missions of NASA. Previously, after its British inventor, it is coined as Bacon fuel cell. It has a capability to obtain the efficacy of 60% to 70% at an operating temperature of 100°C range as similar to PEMFC. Here aqueous solution of Potassium Hydroxide (KOH) is used as an electrolyte. Quick starting is the main advantage of this fuel cell but the disadvantage is that is very sensitive to carbon dioxide. It requires a distinguished system to remove the carbon dioxide from air. The utilization of electrolyte of corrosive nature is also a drawback that its life span is also shorter. Henceforth, this is not employed for commercial applications and only used in space shuttles and transportation. The AFC is subcategorized into Protonic Ceramic Fuel Cell (PCFC) and Direct Ethanol Fuel Cell (DEFC) [6,14–16].

2.2.2.1. Protonic Ceramic Fuel Cell (PCFC): It is basically a novel type of fuel cell with ceramic electrolyte material. The PCFC is capable of operating at higher temperatures of 750°C and the hydrocarbon fuel gaseous molecules supplied to anode are directly oxidized without any reformer. As the electrolyte is solid, the membrane cannot dry out like in the PEMFCs or there is no leakage of liquids as in the case of PAFCs. The open circuit voltage generated is almost equal to the given theoretical value. The only disadvantage is low current density and it can be improved by decreasing the thickness of electrolyte, optimized electrodes and improved conductivity. The studies are mainly focused to increase the efficiency to a range of 55%-65% [17–19].

2.2.2.2. Direct Borohydride Fuel Cell (DBFC): In this cell, Sodium Borohydride ($NaBH_4$) is utilized as input fuel mixed with H_2O and generates Borax ($NaBO_2$) and hydrogen ($4H_2$). This cell operates at a low temperature of 70°C. The key benefits of this particular DBFC are it is not operated with high cost platinum catalyst, high open circuit voltage and high-power density. But it has a very low efficacy of 35% and henceforth the studies are more focused on the development of efficient DBFC by utilizing the various catalysts namely Ni, Pd or Au instead of platinum to minimize the borohydride hydrolysis. This DBFC approach is still in the advancement stage and researchers are more focusing on the $NaBO_2$ recycling [20,21].

2.2.3. Phosphoric Acid Fuel cells (PAFC)
The operating temperature of this fuel cell is about 150-200°C which is almost twice of the PEMFC operating temperature. In this scenario liquid phosphoric acid is utilized as an electrolyte and it is very resistive towards the scums in hydrocarbon fuels. The chemical reactions are as same as PEMFC but only difference is input fuel [10]. The high operating temperatures lead to cogeneration of hot water and electricity. But the main disadvantage is that PAFC also uses the platinum catalyst which is not cost effective. It has been developed commercially and the 100kW, 200kW and 500kW plants are already available. A system of 1.3MW capacity is already tested in Milan and more PAFCs are installed at Japan, USA and Europe [22,23].

2.2.4. Molten Carbonate Fuel Cell (MCFC)
The operating temperature of this MCFC is about 600^0C-700°C and it comprises of two good conductive porous electrodes are in contact with molten carbonate cell. The key benefit of this fuel cell is that it has higher efficiency of 50-60% and it doesn't require any reformer and metal catalyst because of its high operating temperature [6,24]. This cell has slow startup and not resistive to Sulphur record as its disadvantages. It is mostly employed for large and medium power applications.

2.2.5. Solid Oxide Fuel Cells (SOFC)
These fuel cells are normally operated at high temperatures about 1000°C and utilize a solid ceramic material that is dense yttria stabilized zirconia as electrolyte. It has a high efficiency of 50-60% and also due to its internal reforming capability it doesn't need any separate reformer to extract hydrogen[6,25]. The high cost, slow startup and not tolerant to Sulphur content are recorded as its main drawbacks. It is mostly employed for large and medium power applications. In 1997, a Ceramic Fuel cell industry illustrated a prototype fuel cell system of 5kW capacity. Yakabe et al. [26] advanced a SOFC of 3kW and analyzed he main parameters to enhance the efficiency of SOFC in micro-grid system. Currently, the studies are focusing to develop the commercial model of 250kW capacity.

2.2.6. Direct Methanol Fuel Cell (DMFC)
This cell is totally new and innovative approach when correlated to others. Like PEMFC, it uses a polymer electrolyte but the fuel used is liquid methanol or alcohol is used in place of hydrogen. During chemical reactions, the anode attracts hydrogen by dissolving liquid methanol (CH_3OH) in water so as to eradicate the external reformer requirement. Basically, a single DFMC can supply 0.3 to 0.5V under loaded conditions and mainly employed as the substitute for batteries in cameras, notebook computers and some portable electronic devices. The key benefit of this fuel cell is that anode attracts the hydrogen from the methanol and lessens the total price because of reformer absence. Its characteristic features are identical to PEMFC, but their performance is restricted to methanol crossover to cathode from anode which drops the performance and the slow kinetics of methanol electrochemical oxidation at anode [27–29].

3. Comparison of Fuel cells
A broad and concise comparison of the above said six fuel cell approaches based on the different categories is depicted in the Table 1. It is clear that PEMFC is more appropriate for household and business applications because of its low temperature operation and quick start up but when it is for medium/large power applications, the suitable option is SOFC and MCFC. SOFCs have all the intriguing benefits but its high capital cost is its main drawback. Henceforth, the researchers are mainly

IOP Conf. Series: Earth and Environmental Science **312** (2019) 012012 doi:10.1088/1755-1315/312/1/012012

focused on decreasing the operating temperature as well as its high capital cost. These key and intriguing benefits made SOFCs as a trending approach for stationary power production from range of 2kW to some MW capacity. The AFCs are employed for few special space programs, and the PAFCs is employed for commercial applications and for transportation. [5,6,22,25–27,30–33] Currently, almost 2500 fuel cells are employed throughout the world for stationery applications and also for portable applications like mobile phones and laptops etc., [34–37]

Table 1. Comparison of Fuel cells

Parameter	Type of Fuel Cell					
	PEMFC	AFC	PAFC	MCFC	SOFC	DMFC
Electrolyte	Solid polymer membrane (Nafion)	Liquid solution of KOH	Phosphoric acid (H_3PO_4)	Lithium and potassium carbonate ($LiAlO_2$)	Stabilized solid oxide electrolyte (Y_2O_3, ZrO_2)	Solid polymer membrane
Fuel	Pure H_2	Pure H_2	Pure H_2	H_2, CO, CH_4 and other hydro carbons.	H_2, CO, CH_4 and other hydro carbons.	CH_3OH
Operating Temperature (°C)	50-100	50-200	Approx. 200	Approx. 650	800-1000	60-200
Efficiency	40-50%	Approx. 50%	40%	>50%	>50%	40%
Reformer	Yes	Yes	Yes	No	No	No
Applications	Household; UPS; transportation; emergency services such as industry; hospitals and banking; commercial	Space Shuttles; Portable Power and transportation.	Cogeneration; portable power; transportation.	Industries, Powerplants and transportation.	Household; cogeneration; utility power plants; portable power.	It is employed as replacement for batteries in laptop and mobiles.
Benefits	Quick startup, High power density and non-corrosive electrolyte.	Quick startup and high-power density.	Generate high grade waste heat.	Highly efficient and no metal catalyst required.	High grade waste heat and highly efficient.	Cost effective

Drawbacks	Sensitive to fuel impurities (CO, H_2S) and high cost catalyst	Sensitive to fuel impurities (CO, H_2S, CO_2 and H_2S) and high cost catalyst	Sensitive to fuel impurities (CO, H_2S) and electrolyte is corrosive.	Cost ineffective, slow startup, liquid electrolyte is corrosive and intolerant towards Sulphur.	Cost ineffective, slow startup, and intolerant towards Sulphur.	Low power density and low efficiency

4. I-V Characteristics

The voltage of fuel cell is very low around 1.2V. As the fuel cell having low voltage, it is required to stack as many as possible fuel cells connected in parallel and series to obtain the required power capacity. The characteristic curve of voltage and current of fuel cell is as shown in fig 2 [32]. It is depicted that a linear region exists because the rise in current density tends to drop in voltage due to its ohmic nature. This particular region is known as ohmic polarization. At low current, the ohmic loss become less significant; the enhancement in output voltage is primarily because of the chemical reactions and this region is known as active polarization. At high current, the voltage drops significantly due to the fall of efficiency of gas exchange and it is mainly because of water overflooding in catalyst. This region is known as concentration polarization. The fuel cell performance is enhanced by the electrical and thermodynamic efficiency of the system.

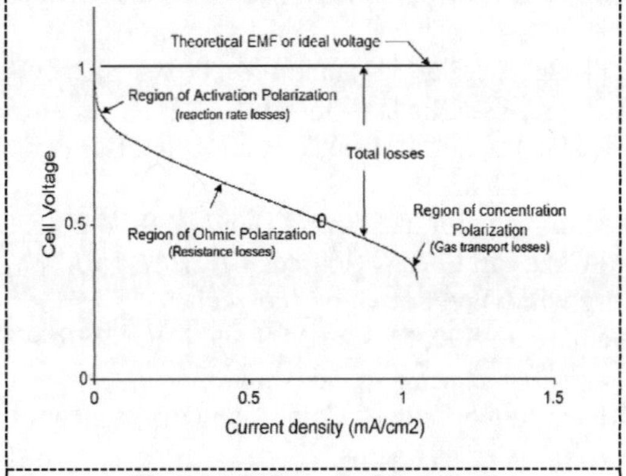

Fig 2. I-V Curve of Fuel cell

5. Recent advances in Fuel cells

The choice of optimal size of fuel cell is more significant to meet the required load demands for various applications [31]. The research is mainly concentrating on the PEM, MCFC and SOFCs to reduce the fuel cell stack cost and to enhance the life span. Currently, the development of 7kW PEMFC is being carried out for household applications by plug power and a cell with capacity of 250kW is under testing by Ballard power generation system. The extensive research on MCFCs for stationary power application is being carried out by the Department of Energy (DOE) and Fuel Cell Energy, Inc. have researched. Many MCFCs with different capacities from 250kW to 400kW are being developed for cogeneration in various parts of the world such as Holland, Italy, Germany and Spain[3–6,38] Moreover, the future is research is mainly targeting the development of PEMFC and PAFCs for cogeneration of heat and power.[39,40] The research is also concentrating on the development of 100kW to 1MW DMFC and other fuel cell types for commercial applications [7,17,20–22,26,41–43].

6. Conclusions

This review paper gives a succinct information on the principle, working, chemical reactions involved and broad categorization of different fuel cell approaches. It also concisely discussed about the type of electrolyte, type of fuel, operating temperatures, key benefits and drawbacks of particular fuel cell technologies and their appropriate applications along with I-V characterization curve. The recent advances in fuel cells have been elucidated and compared the different types of fuel cell approaches. It is depicted that from this discussion, the PEMFC is suitable for almost all applications because of its high-power density, long life span, quick start up and lower cost. But MCFCs and SOFCs are best suited for large and medium power applications due to their promising efficacy, internal reforming and cogeneration of power and heat. A more focused research on the development of long life and cost-efficient fuel cell approaches is required for the benefit of the society.

References:

[1] Wang C and Nehrir M H 2006 Distributed generation applications of fuel cells *Power Syst. Conf. 2006 Adv. Metering, Prot. Control. Commun. Distrib. Resour. PSC* 244–8
[2] Boudghene Stambouli A and Traversa E 2002 Fuel cells, an alternative to standard sources of energy *Renew. Sustain. Energy Rev.* **6** 295–304
[3] Cook B and Anonymous 2002 Introduction to fuel cells and hydrogen technology *Eng. Sci. Educ. J.* **11** 205–16
[4] Xinhong Huang 2006 Fuel Cell Technology Distributed Generation : An Overview *ISIE 2006, July 9-12, 2006, Montr. Quebec, Canada Fuel*
[5] Ellis M W, Von Spakovsky M R and Nelson D J 2001 Fuel Cell Systems: Efficient, Flexible Energy Conversion for the 21st Century *Proc. IEEE* **89** 1808–17
[6] Farooque M and Maru H C 2001 Fuel Cells - The Clean and Efficient Power Generators *Proc. IEEE* **89** 1819–29
[7] Mozsgai G, Yeom J, Flachsbart B and Shannon M 2003 A SILICON MICROFABRICATED DIRECT FORMIC ACID FUEL CELL 1738–41
[8] Rice C, Ha S, Masel R I, Waszczuk P, Wieckowski A and Barnard T 2002 Direct formic acid fuel cells *J. Power Sources* **111** 83–9
[9] Aslam N M, Masdar M S, Kamarudin S K and Daud W R W 2012 Overview on Direct Formic Acid Fuel Cells (DFAFCs) as an Energy Sources *APCBEE Procedia* **3** 33–9
[10] https://en.wikipedia.org/wiki/Direct-ethanol_fuel_cell
[11] Kamarudin M Z F, Kamarudin S K, Masdar M S and Daud W R W 2013 Review: Direct ethanol fuel cells *Int. J. Hydrogen Energy* **38** 9438–53
[12] Akhairi M A F and Kamarudin S K 2016 Catalysts in direct ethanol fuel cell (DEFC): An overview *Int. J. Hydrogen Energy* **41** 4214–28
[13] Ma K B, Kwak D H, Han S B, Park H S, Kim D H, Won J E, Kwon S H, Kim M C, Moon S H and Park K W 2018 Direct Ethanol Fuel Cells with Superior Ethanol-Tolerant Nonprecious Metal Cathode Catalysts for Oxygen Reduction Reaction *ACS Sustain. Chem. Eng.* **6** 7609–18
[14] Gülzow E 2004 Alkaline fuel cells *Fuel Cells* **4** 251–5
[15] Tomantschger K, McClusky F, Oporto L, Reid A and Kordesch K 1986 Development of low cost alkaline fuel cells *J. Power Sources* **18** 317–35
[16] Burchardt T, Gouérec P, Sanchez-Cortezon E, Karichev Z and Miners J H 2002 Alkaline fuel cells: Contemporary advancement and limitations *Fuel* **81** 2151–5
[17] Coors W G 2003 Protonic ceramic fuel cells for high-efficiency operation with methane *J. Power Sources* **118** 150–6
[18] https://en.wikipedia.org/wiki/Protonic_ceramic_fuel_cell
[19] Duan C, Tong J, Shang M, Nikodemski S, Sanders M, Ricote S, Almansoori A and O'Hayre R 2015 Readily processed protonic ceramic fuel cells with high performance at low temperatures *Science (80-.).* **349** 1321–6
[20] Jamard R, Salomon J, Martinent-Beaumont A and Coutanceau C 2009 Life time test in direct

borohydride fuel cell system *J. Power Sources* **193** 779–87
[21] Colominas S, McLafferty J and Macdonald D D 2009 Electrochemical studies of sodium borohydride in alkaline aqueous solutions using a gold electrode *Electrochim. Acta* **54** 3575–9
[22] O'Sullivan J B 1999 Fuel Cells in Distributed Generation 568–72
[23] Kanuri S V and Motupally S 2013 Phosphoric Acid Fuel Cells for Stationary Applications *Encyclopedia of Sustainability Science and Technology*
[24] Watanabe T 2012 *Molten Carbonate Fuel Cells* vol 45
[25] Swider-Lyons K E, Carlin R T, Rosenfeld R L and Nowak R J 2003 Technical issues and opportunities for fuel cell development for autonomous underwater vehicles 61–4
[26] Yakabe H, Sakurai T, Sobue T, Yamashita S and Corporation K 2006 Solid Oxide Fuel Cells as Promising Candidates for Distributed Generators H. Yakabe and T. Sakurai, **00** 369–74
[27] García B L, Sethuraman V A, Weidner J W, White R E and Dougal R 2004 Mathematical Model of a Direct Methanol Fuel Cell *J. Fuel Cell Sci. Technol.* **1** 43
[28] http://www.fuelcelltoday.com/technologies/dmfc
[29] Hogarth M P and Hards G A 1996 Direct methanol fuel cells: Technological advances and further requirements *Platin. Met. Rev.* **40** 150–9
[30] Canha L N, Popov V A and Farret F A 2005 Optimal characteristics of fuel cell generating systems for utility distribution networks *2002 37th Intersoc. Energy Convers. Eng. Conf. IECEC* 597–602
[31] Cheng K W E, Sutanto D, Ho Y L and Law K K 2001 Exploring the power conditioning system for fuel cell *PESC Rec. - IEEE Annu. Power Electron. Spec. Conf.* **4** 2197–202
[32] Soltani M and Bathaee S M T 2008 A new dynamic model considering effects of temperature, pressure and internal resistance for PEM fuel cell power modules *3rd Int. Conf. Deregul. Restruct. Power Technol. DRPT 2008* 2757–62
[33] Choi W, Howze J W and Enjeti P 2006 Development of an equivalent circuit model of a fuel cell to evaluate the effects of inverter ripple current *J. Power Sources* **158** 1324–32
[34] Pasricha S, Keppler M, Shaw S R and Nehrir M H 2007 Comparison and identification of static electrical terminal fuel cell models *IEEE Trans. Energy Convers.* **22** 746–54
[35] del Real A J, Arce A and Bordons C 2007 Development and experimental validation of a PEM fuel cell dynamic model *J. Power Sources* **173** 310–24
[36] Corrêa J M, Farret F A, Popov V A and Simões M G 2005 Sensitivity analysis of the modeling parameters used in simulation of proton exchange membrane fuel cells *IEEE Trans. Energy Convers.* **20** 211–8
[37] Jacobs T and Beukes J 2006 Suitability of fuel cell technology for electricity utility standby power applications *INTELEC, Int. Telecommun. Energy Conf.*
[38] Meng J 2004 A distributed power generation communication system 483–6
[39] Lakshminarayanan V and Karthikeyan P 2016 Optimization of flow channel design and operating parameters on proton exchange membrane fuel cell using MATLAB *Period. Polytech. Chem. Eng.* **60** 173–80
[40] Arun Saco S, Thundil Karuppa Raj R and Karthikeyan P 2016 A study on scaled up proton exchange membrane fuel cell with various flow channels for optimizing power output by effective water management using numerical technique *Energy* **113** 558–73
[41] García-Rodríguez L 2002 Seawater desalination driven by renewable energies: A review *Desalination* **143** 103–13
[42] Thiagarajan V, Karthikeyan P, Thanarajan K, Neelakrishnan S, Manoharan R, Chen R, Fly A, Anand R, Karuppa Raj T R and Sendhil Kumar N 2019 Experimental investigation on DMFCs using reduced noble metal loading with NiTiO 3 as supportive material to enhance cell performances *Int. J. Hydrogen Energy*
[43] Thiagarajan V, Manoharan R, Karthikeyan P, Nikhila E, Hernández-Ramírez A and Rodriguez-Varela F J 2017 Pt nanoparticles supported on NiTiO 3 /C as electrocatalyst towards high performance Methanol Oxidation Reaction *Int. J. Hydrogen Energy* **42** 9795–805

A short review on recent trends and applications of thermoelectric generators

Subbarama Kousik Suraparaju, Gudapati Kartheek, Gummalla Venkata Sunil Reddy and Sendhil Kumar Natarajan*

Department of Mechanical Engineering,
National Institute of Technology – Puducherry, Karaikal 609609
U.T of Puducherry, INDIA.

Email: drsendhil1980iitmuk@gmail.com

Abstract. The basic needs of the society were food, water and energy. The nexus between food, water and energy is unavoidable. Energy is the key element for the survival of all the living beings on the Earth. Energy crisis and environment deterioration are the two most faced glitches of the current scenario. Thermo electricity is one of the promising solutions for energy crisis that is experiencing in this 21st century. This particular article gives the brief description about the principle of thermoelectric devices and their necessity, classification and the applications of thermoelectric generators (TEG's). Also, this paper gives a short description on materials, design and optimization of TEG's. It concludes with the future research direction in the realm of thermo electricity generation.

Keywords: Thermo electric power generation; thermo electric generators; thermo electric materials; Applications.

1. Introduction

Extensive usage of fossil fuels for sustenance of human life has led to severe atmospheric and ecological problems. Subsequently, the terminologies like greenhouse gas, global warming, ozone depletion, climate change, acid rains, etc., were evolved and now they are the trending glitches of our world, in which the solutions for them are not up to the mark. One of the most reasonable solutions for lessening the fossil fuels usage is thermoelectricity, where it is gaining its popularity because of its impeccable advantages. Thermo electricity is an approach that converts the heat available from automobiles, industries etc., into electricity. Likewise, thermoelectric technology, also converts the electricity into required thermal energy for the demanded heating or cooling applications. The key benefits of this approach are that it converts thermal energy into electric energy directly without any intermediate conversion, it has no moving parts or no working fluids inside the equipment, the operation is noiseless and maintenance free, it is emission and chemical reaction free, it is harmless to the environment and it has long reliable operation. Regardless of these benefits, the application of this approach is limited mainly because of the low conversion efficiency and high cost which have been a hurdle to its advancement for most common applications from over decades. Accordingly, wide investigations into the TE technology and its materials have been carried out for attaining the high conversion efficiency and to expand its wings for all applications. The phenomenon of Thermoelectricity was discovered in the early 18th century, which generates the voltage between two dissimilar materials and was most

IOP Conf. Series: Earth and Environmental Science **312** (2019) 012013 doi:10.1088/1755-1315/312/1/012013

commonly utilized as thermocouples. Even though, novel materials are yet under the development, the elementary concept behind the TE approaches are all rely upon the principles of Seebeck effect and Peltier effect. Thermo Electric Generators (TEGs) generally comprises of a set of thermoelectric modules embedded between two heat exchangers. Each thermoelectric module then comprises of some hundreds of thermoelectric couples connected together thermally in parallel and electrically in series which directly converts the thermal energy that passes through it into electrical energy. This review thus gives the brief discussion on the thermoelectric principles, materials and applications. It would help the fellow researchers to demonstrate the thermoelectric approach and identification of techno-glitches in the technology along with the future directions of research.

2. Working Principle

The Seebeck effect came into existence in 1821 which revealed that two dissimilar metals have the distinct temperatures at the junctions, and the respective electromotive force (emf) and current in the junction circuit are called as thermo-emf and thermo current. Enhancing the temperature difference increases the temperature difference between two junctions. The proportional constant associated to material property is acknowledged as Seebeck coefficient. This coefficient is low for metals and high for semiconductors i.e., for metals it is approximately 0 µV/K and for semiconductors it is about ±200 µV/K

$$\alpha = \frac{\Delta V}{\Delta T} \tag{1}$$

Where
α = Seebeck Coefficient,
ΔV = voltage difference,
ΔT = temperature difference.
Peltier effect came into existence in 1834 that when there is current in the circuit, the junction rejects or absorbs the heat depends on the current direction. This is mainly due to the Fermi energy difference between two materials. The capability of heat rejection or absorption is mainly correlated to the temperature of junction and property of different conductors. For defining the absorbed heat per area of junction per second, ZT, a dimensionless parameter is most commonly used to evaluate the Peltier performance of a thermoelectric material.

$$ZT = \frac{\alpha^2}{k}\sigma T \tag{2}$$

Where
α = Seebeck coefficient,
σ = electric conductivity,
k = thermal conductivity. Accordingly, the thermoelectricity approach is divided into two categories according to Peltier and Seebeck,

1. Thermo electric generators (TEGs) for generating electricity when two materials are disclosed in two distinct temperatures,
2. Thermo electric coolers (TECs) for cooling applications when voltage is added onto the materials used. The TEG efficacy can be evaluated by:

$$\eta_{max} = \frac{T_H - T_C}{T_H} \times \frac{\sqrt{1+ZT}-1}{\sqrt{1+ZT}+\frac{T_C}{T_H}} \tag{3}$$

Where T_H is temperature on the hot side and T_C is temperature on the cold side and the maximum coefficient of performance of the TEC is evaluated by:

$$COP = \frac{T_C}{T_H - T_C} \times \frac{\sqrt{1+ZT}-\frac{T_H}{T_C}}{\sqrt{1+ZT}+1} \tag{4}$$

Corresponding to equations (3) and (4) it is seen that when the value of ZT is infinite, the efficacy or COP is Carnot. Thus, it was said that, TEGs are Carnot hot engines, where electrons can work as working medium. On contriving Thermoelectric modules, both p- and n-type materials are essential [1,2].

3. Materials

Meanwhile the later 1950s, the semiconductor TE appliances have been employed for cooling and power generation and subsequently for space power production because of their impeccable energy conversion and some interesting capabilities compared to the conventional power generators. During early 90's, there was an enthusiastic curiosity in the stream of thermoelectricity which is greatly powered by the necessity for more distinct materials for electricity generation and electronic refrigeration. In this regard, thermoelectric materials are broadly classified into three types, namely:

1. Semiconductors,
2. Polymers and
3. Ceramics [3].

3.1. Semiconductors.

The semiconductor materials are best suitable and capable for developing the thermocouples because they have higher Seebeck coefficients and also the way to reduce k without influencing α and s in bulk materials, thus increasing the value of ZT, is to utilize the semiconductors for its greater atomic weight like Bi_2Te_3 and the alloys with Sn, Sn and Pb. A semiconductor or solid-state electronics component can perform good and with greater reliability for many years when it is operating at near ambient temperatures. The materials with best ZT are found as heavily doped and small bandgap semiconductors [4]. The highest value of ZT for Bi_2Te_3 has been noticed as 2.4 in p-type Bi_2Te_3/Sb_2Te_3 at 300K [5]. Pei et al. [6] found the transport facilities of PbTe alloyed with MnTe resulted in a high ZT value of 1.6 at 700K. As said earlier, decreasing the thermal conductivity can enhance the value of ZT, Pei et al. [7] found that Ca – doped BiCuSeO can lower the thermal conductivity and can rise the ZT to 0.9 at 923K for $Bi_{0.925}Ca_{0.075}CuSeO$. Rhyee et al [8] discovered the ZT value of 1.48 at 705K for binary crystalline n-type material $In_4Se_{3-\sigma}$ and Hsu et al. [9] noticed the ZT value of 2.2 at 800K for $AgPb_mSbTe_{2-m}$.

3.2. Ceramics

Thermoelectric materials in real-time operations are invariably based on the alloy materials like Bi_2Te_3 and SiGe. In comparison with other thermoelectric alloys, metal oxides have some benefits such as chemical stability, less toxic, low cost and oxidation resistance so their use enables the manufacturing of more reliable and durable equipment [10]. Ceramic is a chief thermoelectric material for thermoelectric energy transformation to recover waste heat of high temperature from engines and incinerators [11]. However, these oxides are not considered as thermoelectric materials due to their low carrier mobility till the high operative thermo electric oxide of $Na_xCo_2O_4$ came into view. Presently, cobalt based oxides like $Ca_3Co_4O_9$, $NaCo_2O_4$, have been manufactured as p-type legs in thermoelectric modules. Parallelly, n-type ZnO, $SrTiO_3$ and $CaMnO_3$ ceramics also been considered. Among them, CaMnO3 can be synthesised at ambient conditions showed the extraordinary thermoelectric properties which in turn made it as a potential n-type oxide material [12]. Tsubota et al. [13] examined the TE properties of $LaCoCO_3$ with addition of B_2O_3-CuO and resulted that the material ha value of ZT = 0.073 at 373K, which was almost 1.5times greater than that of pure $LaCoCO_3$ sintered at 1473K. Butt et al. [14] evaluated the thermoelectric properties of (La, Fe) co-doping of $Ca_3Co_4O_9$ and results revealed that the value of ZT= 0.32 at 1000K which is 70% greater than that of pure $Ca_3Co_4O_9$.

3.3. Polymers

The broadly examined, developed and utilized inorganic materials associated with complicated issues such as dearth of natural resources, toxicity and manufacturing methods involved with high economy. Therefore, there exists a great need of developing a new type of material for further enhancement of properties[15]. The group of conductive polymer composites comprising conducting fillers and

IOP Conf. Series: Earth and Environmental Science **312** (2019) 012013 doi:10.1088/1755-1315/312/1/012013

insulating polymer matrices have been examined for the benefits of inexpensive, solution processability, flexibility, lightweight, low cost synthesis and more eco-friendly substitutes to conventional thermoelectric modules. Further investigations on the behaviour behavior of segregated conductive polymer composites with hybrid fillers of carbon nanotube and bismuth telluride resulted that segregated composite containing 2.6 vol% CNTs and 5.1 vol% Bi_2Te_3 exhibit the value of $ZT = 3 \times 10^{-5}$ at room temperature [16]. However, currently there is advancement in the polymers ZT, Kim et al. [17] stated that the value of ZT of dopant poly (styrenesulphonate) in poly (3,4-ethylenedioxythiophene) as 0.42 at room temperature. For polymers, ceramics and some semiconductors, though the value of ZT is low, other influencing properties like temperature resistivity and flexibility made them as intriguing thermoelectric materials.

Among all, semiconductor is widely used but still the value of ZT is not so as high enough to make them as competitive among conventional generator and cooler materials, Other major disadvantage of semiconductors is that they are resistive of high temperatures and they do lack of flexibility. Hence, ceramics and polymers are alternatives for them.

4. Applications of Thermo Electric generators (TEGs)

Thermoelectric generators are basically categorized according to the heat source type and applications where it is utilized. Some of them are as follows:

1. Waste Heat Recovery
2. Space applications
3. Solar Thermo Electric Generators (STEGs)

4.1. Waste Heat Recovery:

4.1.1. Automobiles: One of the major challenges experiencing currently is to reduce the energy that is being wasted in the form of heat energy. Automobile industry is one among them for the utilization of TEGs. Almost 65 percent of the heat generated in the internal combustion engines is being wasted as of the survey noted [18]. However, for the installation of the TEGs in automobiles, it has to meet some conditions. They are: it should not affect the operation of the vehicle. The materials should be recyclable and eco-friendly. It should be economically viable [19]**.** the ford group in association with the department of energy (DOE) has conducted a research on the v 6 engine equipped with TEGs with an objective to produce 500 W but the results published by Maranville in 2012 showed that the generated output was 250 W which is 50 percent of the target [20,21]. Generator is protected by installing a bypass in the exhaust system. Also, Volvo, Renault trucks worked on the RENOTER (recovery of energy from exhaust of an engine from 2008 to 2011). This project inferred that TEG can generate up to 130 W from a passenger car exhaust. Renoter 2 project started in year 2013 targets the usage of TEGs in industrial gas exhaust recirculation and on the hybrid gasoline engines [22]. Similarly, for aircrafts, a study conducted by boeing research and technology showed that by using TEGs, fuel reductions of 0.5% can be achievable which will reduce the operating costs of the flights significantly[19]. Subsequently, shipping is one of the largest sources of greenhouse gas emissions. Kristiansen et al. [23] studied the possibilities of using TEGs for large ships. The heat that is released by the engine is used to purify the water and to heat the accommodation areas.so the temperatures of the waste heat will get significantly reduced. Also, the weight is one of the main concerns in the heavy ships. TEGs are also an excellent opportunity in the smaller ships since the weight is not a problem, cold source (water) is free [24].

4.1.2. Industries: Heat is a by-product in almost all the industries.in some of the industries, this heat is used for some heating networks and in some industries, it is converted into electricity by using steam turbines. KELK LTD. did the experiment on the usage of the thermo electric generation at komatsu ltd., Awazu plant. The results shown that the efficiency is 5% [25]. Aranguren et al. [26] had conducted experiments on a TEGs dedicated to convert the waste heat from a combustion chamber. TEGs are being used in some of the steel industries to convert the radiant heat from molten metal into electricity.

IOP Conf. Series: Earth and Environmental Science **312** (2019) 012013 doi:10.1088/1755-1315/312/1/012013

In japan, JFE steel corporation did the studies on TEGs using radiant heat from slabs. TEG is connected to the grid [27]. Luo et al. [28] studied the usage of TEGs in Portland cement manufacturing and reported that around 10-15% of the energy is dissipating into atmosphere. This kind of wasted heat is difficult to recover as the kiln rotates permanently. The heat loss from a kiln of length 72 m and diameter 4.80 m is approximately 10000 KW. They have taken a kiln which is surrounded by a coaxial shell having thermo electric modules and performed the experiments. The results showed that TE system can produce 210 kw and it also saves approximately 3280 KW due to insulation provided by the shell around the kiln. Hesham Khalil et al. [29] investigated about the influence of the tilt angle on waste heat recovery system from a chimney placed vertically by employing TEGs cooled at their cold sides through natural convection and also the aspect ratio performance of Thermo Electric Generator has been studied. Later, the whole scenario was studied numerically in ANSYS-Fluent software and results were best fitted with the experimental outcomes. The tilting of chimney wall towards the flue gases had a positive influence on the output power of TEGs and an angle of 15° degree augments the power of upper TEGs by 5% with a nominal influence on the power of TEGs equipped at lower side. Souvik Biswas et al. [30] numerically investigated on the design and development of waste heat recovery utilization system through a heat exchanger made of Bi_2Te_3 material and got the best fit of results for industrial use.

4.2. Space Applications

Electricity production in some of the drastic environments should meet strict specifications. Weather conditions can be extreme, either very hot or cold etc. also the maintenance should be minimum as it is difficult in such type of environments. the space industries are using the radio isotope thermo electric generators (RTGs). Heat is generated by the natural radioactive decay plutonium-238 [31]. US NAVY'S transit navigation satellite first used this technology back in 1961. They equipped the satellite with the SNAP-3 nuclear auxiliary generator which generated the power of about 2.7 W only but worked for about 15 years [32]. RTGs are used in space applications because of their low weights, good reliability and also, they can work even for several years. The voyager I and II are launched in 1997 also equipped with these RTGs because of their extreme reliability to power the instruments and systems in the craft. The capacity of generation decreases gradually by 7W a year due to decay of the plutonium. NASA launched the Cassini-Huygens orbital probe in association with European space agency in 1997 to explore about Saturn and its satellites. This is also powered by RTGs [33,34].

4.3. Solar Thermo Electric Generators (STEGs)

In these generators, the heat from the sun is used as heat source. the maximum efficiency which can be obtained by using TEGs are around 5-10%. But the solar cell will have an efficiency more than that of TEGs by 3-5 times. The efficiency of the solar TEGs can be increased by increasing the temperature difference across the thermo elements. This can be achieved if a concentrating lens is used to concentrate the solar beams. [3]. The device consists of thermal concentration absorber and optical concentration system. To reduce the losses, they created a vacuum by enclosing the STEG with glass.[35,36]. Kraemer et al. [37] did the research to integrate STEGs in solar hot water vacuum tube system for co-generation application to create a device that has the potential to provide cost effective hot water and electricity. Baranowski et al. [38] did the research and proved that by using the available thermo electric materials an efficiency of 14.1% can be achieved with a temperature difference of 1000[degree centigrade] and concentration of solar intensity by 100 .Wei He et al. [39] did the research on coupling of the solar water heating with a TEG. A heat pipe is used to transfer the heat and the TEG is used to exchange the heat from heat pipe to water. Electric efficiency is around 1-2%, the hot water production efficiency is around 55%. Khanmohammadi [40] investigated the novel model of solar-based integrated energy systems (IES) for hydrogen yield with and without thermoelectric generators (TEGs) waste heat recovery system (WHRS). The proposed approach has an increase in out power by 17.6%, efficiency by 1.2% and hydrogen production is increased by 16% when compared to conventional system.

5. Future Scope

So as to be compete with conventional devices, the analysis revealed that the value of ZT should be greater than 3 [41]. Though the highest ZT ever found at the normal temperature is about 2.4. Therefore, it is even crucial to invent the highly efficient thermoelectric materials among several materials like ceramics, semiconductors and polymers. Also, it is also very essential to augment the thermoelectric devices by enhancing and optimizing the structure of heat exchangers, the operating parameters and the geometry design, because these aspects obviously influence the whole system efficiency. The field of applications can be broadening such as heat load reduction in automobiles, energy storage, thermal energy management and recovery from the manufacturing. Lastly, the long-term operation and stability assessment can be more investigated for better potentiality.

6. Conclusions

In the current discussion, a brief elucidation on the principle, working and applications of thermoelectric materials and thermoelectric generators has been presented. Also, novel materials with improved ZT, high temperature resistive materials with greater reliability also been discussed. Later the application in sector wise and the developments of TEGs was reviewed which allows the researcher to get a concise knowledge on the TEGs. At the end, the future direction in this allows the reader to enhance the researching path in the particular gaps discussed.

References.

[1] Rowe E D M, Ph D, Sc D and Group F 2006 *Thermoelectrics Handbook*

[2] Wang S, Lü Q, Li L, Fu G, Liu F, Dai S, Yu W and Wang J 2013 High-temperature thermoelectric properties of Cd 1-x Pr x O ceramics *Scr. Mater.* **69** 533–6

[3] Hamid Elsheikh M, Shnawah D A, Sabri M F M, Said S B M, Haji Hassan M, Ali Bashir M B and Mohamad M 2014 A review on thermoelectric renewable energy: Principle parameters that affect their performance *Renew. Sustain. Energy Rev.* **30** 337–55

[4] Tani J I and Kido H 2007 Thermoelectric properties of P-doped Mg2Si semiconductors *Japanese J. Appl. Physics, Part 1 Regul. Pap. Short Notes Rev. Pap.* **46** 3309–14

[5] Venkatasubramanian R, Siivola E, Colpitts T, The K, Kanatzidis G, Zt T and Vining B 2001 Thin-®lm thermoelectric devices with high room-temperature ®gures of merit *October* 597–602

[6] Pei Y, Wang H, Gibbs Z M, LaLonde A D and Snyder J G 2012 Thermopower enhancement in Pb1-xMnxTe alloys and its effect on thermoelectric efficiency *NPG Asia Mater.* **4** 1–6

[7] Pei Y L, He J, Li J F, Fuli, Liu Q, Pan W, Barreteau C, Berardan D, Dragoe N and Zhao L D 2013 High thermoelectric performance of oxyselenides: Intrinsically low thermal conductivity of Ca-doped BiCuSeO *NPG Asia Mater.* **5**

[8] Rhyee J S, Lee K H, Lee S M, Cho E, Kim S Il, Lee E, Kwon Y S, Shim J H and Kotliar G 2009 Peierls distortion as a route to high thermoelectric performance in In 4Se3-) crystals *Nature* **459** 965–8

[9] Hsu K F, Loo S, Guo F, Chen W, Dyck J S, Uher C, Hogan T, Polychroniadis E K and Kanatzidis M G 2004 Cubic AgPbmSbTe2+m: Bulk Thermoelectric Materials with High Figure of Merit *Science (80-.).* **303** 818–21

[10] Zhu Y, Wang C, Wang H, Su W, Liu J and Li J 2014 Influence of Dy/Bi dual doping on thermoelectric performance of CaMnO 3 ceramics *Mater. Chem. Phys.* **144** 385–9

[11] Yasukawa M, Kono T, Ueda K, Yanagi H and Hosono H 2010 High-temperature thermoelectric properties of La-doped BaSnO3 ceramics *Mater. Sci. Eng. B Solid-State Mater. Adv. Technol.* **173** 29–32

[12] Wang H and Wang C 2013 Thermoelectric properties of Yb-doped La0.1Sr 0.9TiO3 ceramics at high temperature *Ceram. Int.* **39** 941–6

[13] Song Y, Sun Q, Lu Y, Liu X and Wang F 2012 Low-temperature sintering and enhanced thermoelectric properties of LaCoO 3 ceramics with B 2O 3-CuO addition *J. Alloys Compd.*

536 150–4
[14] Butt S, Liu Y C, Lan J Le, Shehzad K, Zhan B, Lin Y and Nan C W 2014 High-temperature thermoelectric properties of la and Fe co-doped Ca-Co-O misfit-layered cobaltites consolidated by spark plasma sintering *J. Alloys Compd.* **588** 277–83
[15] Wang L, Wang D, Zhu G, Li J and Pan F 2011 Thermoelectric properties of conducting polyaniline/graphite composites *Mater. Lett.* **65** 1086–8
[16] Pang H, Piao Y Y, Tan Y Q, Jiang G Y, Wang J H and Li Z M 2013 Thermoelectric behaviour of segregated conductive polymer composites with hybrid fillers of carbon nanotube and bismuth telluride *Mater. Lett.* **107** 150–3
[17] Kim G H, Shao L, Zhang K and Pipe K P 2013 Engineered doping of organic semiconductors for enhanced thermoelectric efficiency *Nat. Mater.* **12** 719–23
[18] Debbie Sniderman ASME.org Using Waste Engine Heat in Automobile Engines
[19] Champier D 2017 Thermoelectric generators: A review of applications *Energy Convers. Manag.* **140** 167–81
[20] Maranville 2011 Overview of ford-DOE thermoelectric programs: waste heat recovery and climate control *2nd Thermoelectrics Applications Workshop 2011* pp 1–18
[21] Maranville C 2012 Thermoelectric Opportunities for Light-Duty Vehicles *Ford Mot. Co.*
[22] Cea L A 2012 Presentation of " RENOTER " Project on thermoelectric *DoE Workshop 2012 (Baltimore)* vol 2012
[23] Kristiansen N R, Snyder G J, Nielsen H K and Rosendahl L 2012 Waste heat recovery from a marine waste incinerator using a thermoelectric generator *J. Electron. Mater.* **41** 1024–9
[24] Wallace T 2011 Development of Marine Thermoelectric Heat Recovery Systems *US Dep. Energy, 2nd Thermoelectr. Appl. Work. 2011*
[25] Kaibe H, Makino K, Kajihara T, Fujimoto S and Hachiuma H 2012 Thermoelectric generating system attached to a carburizing furnace at Komatsu Ltd., Awazu Plant *AIP Conf. Proc.* **1449** 524–7
[26] Aranguren P, Astrain D, Rodríguez A and Martínez A 2015 Experimental investigation of the applicability of a thermoelectric generator to recover waste heat from a combustion chamber *Appl. Energy* **152** 121–30
[27] Kuroki T, Kabeya K, Makino K, Kajihara T, Kaibe H, Hachiuma H, Matsuno H and Fujibayashi A 2014 Thermoelectric generation using waste heat in steel works *J. Electron. Mater.* **43** 2405–10
[28] Luo Q, Li P, Cai L, Zhou P, Tang D, Zhai P and Zhang Q 2015 A Thermoelectric Waste-Heat-Recovery System for Portland Cement Rotary Kilns *J. Electron. Mater.* **44** 1750–62
[29] Khalil H and Hassan H 2019 3D study of the impact of aspect ratio and tilt angle on the thermoelectric generator power for waste heat recovery from a chimney *J. Power Sources* **418** 98–111
[30] Biswas S, Roynaskar A, Hirwani C K and Panda S K 2018 Design and fabrication of thermoelectric waste heat reutilization system—possible industrial application *Int. J. Energy Res.* **42** 3977–86
[31] L. R and L. G 2012 U.S. Space Radioisotope Power Systems and Applications: Past, Present and Future *Radioisot. - Appl. Phys. Sci.*
[32] Lloyd I. Shure and Harvey J. Schwartz 1965 SURVEY OF ELECTRIC POWER PLANTS FOR SPACE APPLICATIONS *Tech. Pap. Present. Fifty-Eight Natl. Meet. Am. Inst. Chem. Eng. Philadelphia, Pennsylvania*
[33] Voyager https://voyager.jpl.nasa.gov/mission/spacecraft/
[34] Nasa 1999 Spacecraft Power for Cassini
[35] De Leon M T, Chong H and Kraft M 2012 Design and modelling of SOI-based solar thermoelectric generators *Procedia Eng.* **47** 76–9
[36] Kraemer D, Poudel B, Feng H P, Caylor J C, Yu B, Yan X, Ma Y, Wang X, Wang D, Muto A, McEnaney K, Chiesa M, Ren Z and Chen G 2011 High-performance flat-panel solar

thermoelectric generators with high thermal concentration *Nat. Mater.* **10** 532–8

[37] Kraemer D, McEnaney K, Chiesa M and Chen G 2012 Modeling and optimization of solar thermoelectric generators for terrestrial applications *Sol. Energy* **86** 1338–50

[38] Baranowski L L, Snyder G J and Toberer E S 2012 Concentrated solar thermoelectric generators *Energy Environ. Sci.* **5** 9055–67

[39] He W, Su Y, Riffat S B, Hou J X and Ji J 2011 Parametrical analysis of the design and performance of a solar heat pipe thermoelectric generator unit *Appl. Energy* **88** 5083–9

[40] khanmohammadi S and Saadat-Targhi M 2019 Performance enhancement of an integrated system with solar flat plate collector for hydrogen production using waste heat recovery *Energy* 1066–76

[41] Majumdar A 2004 MATERIALS SCIENCE: Enhanced: Thermoelectricity in Semiconductor Nanostructures *Science (80-.).* **303** 777–8

IOP Conf. Series: Earth and Environmental Science **312** (2019) 012014 doi:10.1088/1755-1315/312/1/012014

Optimal Energy Scheduling of Renewable Energy Sources in Smart Grid using Cuckoo Optimization Algorithm with Enhanced Local Search

S Antonyraj[1], G Giftson Samuel[2]

[1]Anjalai Ammal Mahalingam Engineering College, Kovilvenni, Tamilnadu, India.
[2]Sir Issac Newton College of Engineering & Technology, Nagapattinam, Taminadu, India

antonyraj2524@gmail.com

Abstract. The emerging trend of micro grid enabled technologies with renewable energy sources are used to satisfy the high demand of energy requirement and are preferred over conventional energy sources. However, it has disadvantages like uncertainty in energy on hand with renewable energy sources and unpredictable demand. Optimal scheduling of power generation among the available renewable energy sources is necessary to achieve minimum cost of energy generation with consideration of power loss. In this work, we have used Cuckoo search optimization algorithm with enhanced local search using Tabu search for optimal energy scheduling. This approach is compared with other evolutionary algorithm and existing approaches and results show that our proposed approach performs well.

1. Introduction

Renewable energy resources are now a days more important in order to meet the energy requirement. To deal with the growing demand for electricity, various researches have been reported. Most of the approaches pay attention to improve the efficiency of conventional power generation and distributions. But this is restricted due to the existing infrastructure of power station. Hence this work is focused on micro grid optimization which also efficiently satisfies the growing demand of electricity. Power producer offered electrical energy at different rates and optimal power scheduling is necessary to meet the demand with minimum generation cost and could handle the complexity of power system [1].

In this work, Energy scheduling management is proposed which optimizes the generated power scheduling to balance between the generated and received energies. The numbers of renewable energy resources (RER) is selected to satisfy the requirement. RER sometimes may fail to generate the power to meet the load due to the stochastic nature of RER. In this case, RER uses the power from the storage systems and virtual power plants. Energy scheduling management should also handle this issue. But in our work we considered that the energy generated by RER is higher than the requested power and also considers that the power loss may occur in RER. In previous literature optimization of power scheduling is rarely investigated. Hence, cuckoo based optimization procedure is proposed for optimal sharing of power generation of RER to the requested power.

Evolutionary algorithms are effective in energy scheduling problem as the nature of energy scheduling is nonlinear and complex. Cuckoo search is used for engineering optimization problems since it can handle continuous problems and NP-hard problems well [2]. Cuckoo search often falls in local minima if the diversity of the solution is not met over the generations. So in this work, the local search is further enhanced in cuckoo optimization with the Tabu search. Smart grid with three wind plants (WP), two photo voltaic plants (PV), and one CHP plant is simulated to calculate the effectiveness of the proposed approach and energy scheduling optimization based on cuckoo with enhanced local search is applied and evaluated.

The remaining paper is organized as follows. Section 2 summarizes the related work. Section 3 explains the proposed work on energy scheduling algorithm. Section 4 discuss about the experimental results and section 5 concludes the paper.

IOP Conf. Series: Earth and Environmental Science **312** (2019) 012014 doi:10.1088/1755-1315/312/1/012014

2. Related works

The author in [3] used Memory based genetic algorithm (MGA) for optimal energy scheduling among RER in the micro grid. MGA was used to increase the local exploration ability of Genetic Algorithm (GA) by strengthening the local and global searches. Particle swarm optimization algorithm (PSO), Genetic algorithm and some variations of PSO are used for comparison. But this work does not consider the load uncertainties and power loss of renewable energy resources. Fuel cost is minimized in the micro grid [4] by optimal scheduling of the communication to aid distributed generations.

Hybrid evolutionary algorithms were also employed in optimization of maintenance scheduling in terms of both economic and reliability [5]. Authors in [6] used additive increase and multiplicative decrease algorithm for optimal RER management. Economic dispatch problem was handled in the micro grid with multi objectives in [7] by pareto concavity removal in both the grid associated and isolate modes and the results were compared with non-dominated sorting GA and PSO.

Authors in [8], considered micro grid with RER and storage devices for power scheduling. Aggregator was used to allocate the energy among the charging station and drop based controller was used in energy management of batteries. Charging stations were responsible for allocating energy among the individual batteries using droop participation factor. Multi objective with environmental, reliable and minimum cost were considered in [9], and PSO was used for optimal scheduling of power among the renewable energy resources.

3. Proposed Method

3.1. Energy Scheduling Management

Micro grid contains loads and generation units of combined heat and power, photovoltaic, and wind plants etc. Loads are supplied using generation units in microgrid and are changeable. Power availability of generation units also variable at time varies. There are so many approaches to share the produced power among renewable energy resources to meet up the requested power. Cost minimization is a powerful objective function for energy scheduling. The cost minimization objective function in microgrid is formulated [9], [10] as follows:

$$cost\ minimization = \sum_{i=1}^{n_r} c_i p_i = \sum_{i=1}^{n_r} [\alpha_i \times p_i + \beta_i \times p_i + \gamma_i]$$

Subject to

$$\sum_{i=1}^{n_r} p_i = p_l \tag{1}$$

where i denotes each available RER. c_i is the cost in hourly basis, p_i is the generated energy of ith unit. α_i, β_i, and γ_i are the function co efficient fixed numbers and assigned based on the nature of renewable energy such as fuel cost, efficiency, etc. p_l is the requested load. n_r is the number of renewable energy resources. The load should be less than or equal to the generated power and the cost of generation must be equal to the sum of cost of each RER power generation. Energy scheduling management should share the requested power optimally with the renewable energy resources available as well as minimize the cost at each hour.

In this paper, micro grid is constructed with three wind plants, two photo voltaic plants, one CHP renewable energy system and the problem is formulated as follows:

$$\sum_{i=1}^{n_r} c_i p_i = \left[\sum_{i=1}^{n_r}[\alpha_i \times p_i + \beta_i \times p_i + \gamma_i]\right] + p_f \times |p_{w1} + p_{w2} + p_{w3} + p_{pv1} + p_{pv2} + p_{CHB} - p_l| \tag{2}$$

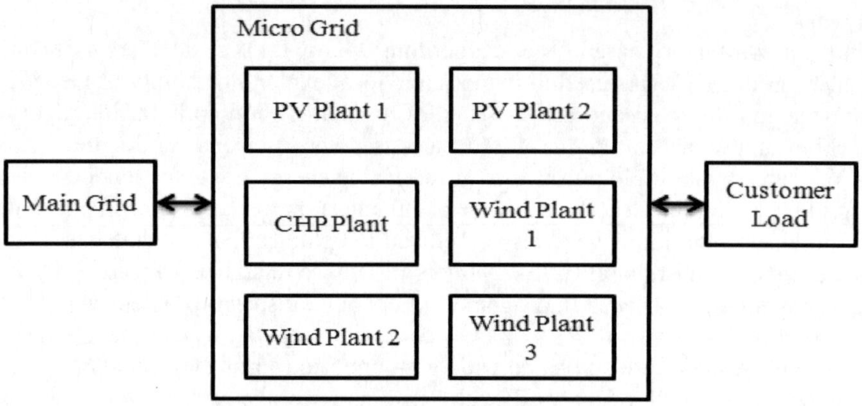

Figure 1 Micro grid Setup

In this work power loss is also included in the optimization problem due to the spasmodic nature of RERs. So the equation (3) is modified as in (4),

$$\sum_{i=1}^{n_r} c_i p_i = \left[\sum_{i=1}^{n_r}[\alpha_i \times p_i + \beta_i \times p_i + \gamma_i]\right] + p_f \times |p_{w1} + p_{w2} + p_{w3} + p_{pv1} + p_{pv2} + p_{CHB} - p_l - p_{loss}| \quad (3)$$

Power loss is calculated by kron's formula as follows:

$$p_{loss} = \sum_{i}^{n_r} \sum_{j}^{n_r} p_i a_{ij} p_j$$

where a_{ij} is the power loss co efficient.

3.2. Energy Scheduling using Cuckoo Search with Tabu based Local Search

Optimal scheduling of renewable energy resources to successfully meet the energy demand is calculated using proposed cuckoo search with enhanced local search algorithm.

3.2.1. Cuckoo Search. It is depends on characteristics of cuckoo birds laying its own eggs in other birds nest. Eggs in a nest represent solutions and cuckoo egg is the new solution. The goal of the algorithm is to work the fresh and possibly enhanced solutions (cuckoos) to interchange worst solutions in the nests with the following constraints:

1. Randomly each cuckoo lays one egg in a selected host bird during initialization
2. Best solutions only are agreed over to the following generation.
3. Host birds can remove the foreign egg or abandon their own nest with probability p_a [0, 1], then locate a new nest.

3.2.2. Cuckoo Search with Enhanced Local Search using Tabu. Cuckoo search can converge early to sub optimal solutions if the generated solutions are not diverse. This local optima problem can be solved if it is combined with some good local search algorithm. Hence Tabu search is used for local search in cuckoo search algorithm. First, the population should be initialized. The solution (individual cuckoo) is represented as follows:

Sol = [Powergeneration RER1, Powergeneration RER2, …,PowergenerationRERnr]

where nr is the number of RERs in smart gird. The solution contains the optimal generation of RERs in a smart grid with minimum production cost. The population is initialized with randomly generated solutions. Each RER is assigned real numbers in the range between minimum and maximum power that can be produced by that particular RER. Each solution in the population should be validated based on the objective function in (4). The user constrains are checked before starting of exploration. The overall procedure for validating the population and returning best solution based on cuckoo with enhanced local search is depicted in Algorithm CELTES.

Algorithm of CELTES

Input: Maximum Iteration Max_{it}, Minimum power limit of particular $RER_i min_p(i)$, Maximum power limit of particular $RER_i max_p(i)$.
Output: Optimized solution
Procedure:

1. Generate initial population $X = \{x_1, x_2, \ldots, x_{npop}\}$.
   Each RER is initialized in the maximum and minimum power range in solution space.
2. While (it <Max_{it})
3. Select randomly a solution x_i from the current population and replace it with Levy flights
4. Calculate the fitness of x_i : $F(x_i)$ as in (4)
5. Select randomly a nest j among n and calculate fitness function of $F(x_j)$
6. $q = F(x_j) - F(x_i)$;
7. T = schedule [itr];
8. if ($F(x_i)$ less than or equal to $F(x_j)$)
9. Replace x_j with x_i
10. else if ($e^{\frac{q}{T}}$ greater than randomly generated value in between [0,1])
11. $x_{j,new}$= Tabu search ($x_{j,old}$)
12. end if
13. A portion (p_a) of poorest solutions are replaced with new ones using levy flights
14. Retain the top solution
15. Rank the solutions and calculate the best
16. Increment the iteration

End while

3.2.3. Tabu Search. Tabu search uses memory to prevent the searching from revisiting the past examined solutions and exploring the new solutions that are not examined before [11]. In this work Tabu search is invoked with the power production of renewable energy resources (x_1, x_2, …,x_n) as starting point. Local optima is avoided by preventing revisit of already examined solution. Tabu search is adopted for our problem in the following way: Tabu search is started with a renewable energy configuration (Powergeneration RER1, Powergeneration RER2 …, PowergenerationRERnr). Tabu list will be maintained for storing the previsited solutions using short term memory. One RER is randomly selected using Random (1, $max_{nr}(i)$) in each iteration. The solution is updated by increasing or decreasing the value of particular RER selected early. For example, if the solution is (80, 40, 20, 70) and the random value d is selected in the minimum and maximum power range of the randomly selected index i (assume i=3, d=30), then for even number iteration d value is added and new solution is created as (80, 40, 50, 70) and for odd numbered iteration the d value decremented. The selected i value is included in the Tabu list. Whenever the RER type is modified the counter value incremented. If the counter value exceeds the tabu limit,' then the particular RER is forbidden for further modification. The above procedure is continued for maximum number of iteration in Tabu search and returns the best solution and this will be used as a new solution in cuckoo search. The overall procedure is presented in Algorithm TabuLocalSearch.

4. Experimental Results and Analysis

MATLAB R2018a is used to implement the proposed approach with the micro grid containing five generation units viz., the three wind plants, two photovoltaic plants, and one CHP plant using IEEE-37-node feeder as in [3]. The maximum population is 100 and the maximum amount of iterations is 1000. The cost co efficient values such as α_i, β_i, and γ_i are calculated as in [3].

The performance is compared with the GA, MGA, PSO, and PSO with inertia weight as discussed

IOP Conf. Series: Earth and Environmental Science **312** (2019) 012014 doi:10.1088/1755-1315/312/1/012014

Algorithm Tabu Search

Input: Maximum Iteration $TMax_{it}$, Maximum number of renewable energy resources $max_{nr}[1, 2, \ldots, nr]$, Minimum power limit of particular $RER_i min_p(i)$, Maximum power limit of particular $RER_i max_p(i)$.

Output: New solution $x_{j,new}$

Procedure:

```
While ( it <TMax_it)
   Randomly select i from the number of renewable energy resources required tuning.
   if (i is not in Tabu list)
     d = Random(min_p(i), max_p(i))
   end if
   if (iteration % 2:=0)
     x_i = x_i + d
   else
     x_i = x_i - d
   end if
   Increment count[i] // count[i] is the counter for index I;
   if(count[i] >Tabu limit )
     Add i to Tabu list
   end if
  if (f(x_i ) < f(x_j,old)
x_j,new : = x_i
  end if
end while
return x_j,new
```

in [3]. The power capacities of Each RER are represented in Table 1 and the hourly requested power is depicted in Figure 2. It can be seen from the Table 2, the proposed method produced lowest cost during a day compare to other existing approaches as the proposed method considers the power loss of renewable energy resources and avoiding local optima using Tabu search.

Moreover, the optimal power generated at each hour by RERs in microgrid during a day is presented in Figure 3. It can be seen from the Figure 3, that the proposed approach effectively meets the requested demand with the RERs generated power with the minimum cost.

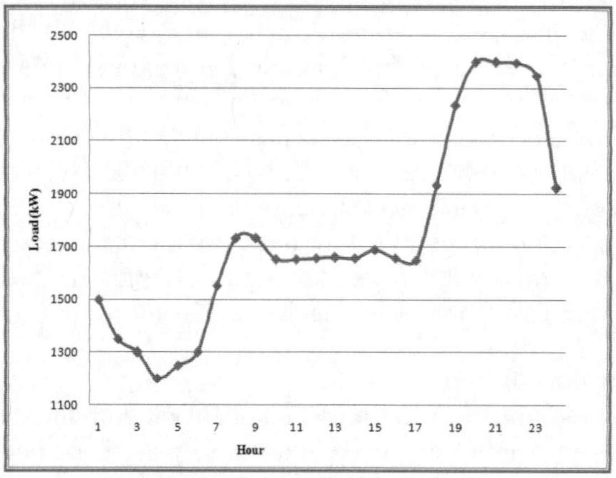

Figure 2 The Hourly Load

Table 1 Available Power of RERs

RER	Maximum Power/hour
WP	0.75 MW
PV	0.2 MW
CHP	1 MW

Table 2 Cost comparison of the proposed approach with existing approaches

Optimization Method	Cost per day
GA	1868.55
PSO	1273.87
PSO with inertia weight	1259.69
MGA	1203.88
Proposed method	1104.98

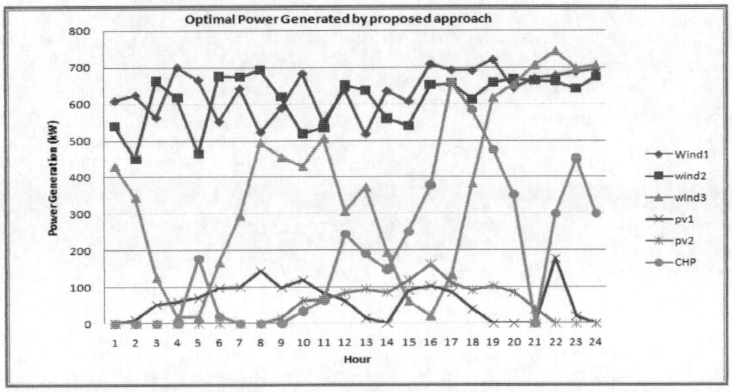

Figure 3 Optimal Power generation of proposed method

5. Conclusion

Optimal and robust energy scheduling among the renewable energy resources is required to achieve minimum generation cost. In this work cuckoo search with enhanced local search using Tabu is used to optimize the energy among the RERs. Moreover this work considers the intermittent nature of renewable energy resources. Simulation was conducted with three wind plants, two photovoltaic plants, and one CHP plant and the results proved that the proposed approach performs well compared to other evolutionary algorithm based existing power scheduling approaches in terms of cost.

References

[1] Morais H, Kadar P, Faria P, Vale Z A and Khodr H M 2010. Optimal scheduling of a renewable micro-grid in an isolated load area using mixed-integer linear programming. *Renewable Energy*, **35** pp 151-6.

[2] Yang X S, and Deb S 2014. Cuckoo search: recent advances and applications. *Neural Computing and Applications*, ***24*** pp 169-74

[3] Askarzadeh A 2018. A memory-based genetic algorithm for optimization of power generation in a microgrid. *IEEE transactions on sustainable energy*, ***9*** (3), pp 1081-89

[4] Maulik A and Das D 2017. Optimal operation of microgrid using four different optimization techniques. *Sustainable Energy Technologies and Assessments*, **21** pp 100-20

[5] Samuel G G and Rajan C C A 2015. Hybrid: particle swarm optimization–genetic algorithm and particle swarm optimization–shuffled frog leaping algorithm for long-term generator maintenance scheduling. *International Journal of Electrical Power & Energy Systems*, **65** pp 432-42

[6] Crisostomi E, Liu M, Raugi M, and Shorten R 2014. Plug-and-play distributed algorithms for

IOP Conf. Series: Earth and Environmental Science **312** (2019) 012014 doi:10.1088/1755-1315/312/1/012014

optimized power generation in a microgrid. *IEEE transactions on Smart grid*, **5** pp 2145-54
[7] Sarfi V and Livani H 2018. An Economic-Reliability Security-Constrained Optimal Dispatch for Microgrids. *IEEE Transactions on Power Systems*, **33** pp 6777-86
[8] Singh S, Singh M and Kaushik S C 2016. Optimal power scheduling of renewable energy systems in microgrids using distributed energy storage system. *IET Renewable Power Generation*, **10** pp 1328-39
[9] Azaza M and Wallin F 2017. Multi objective particle swarm optimization of hybrid micro-grid system: A case study in Sweden. *Energy*, **123** pp 108-18
[10] Caldon R, Patria A R, and Turri R 2004. Optimal control of a distribution system with a virtual power plant. *Bulk power system dynamics and control, Cortina. d'Ampezzo, Italy*, p 18
[11] Glover F 1986. Future paths for integer programming and links to artificial intelligence. *Computers & operations research*, **13** pp 533-49

IOP Conf. Series: Earth and Environmental Science **312** (2019) 012015 doi:10.1088/1755-1315/312/1/012015

Aspen HYSYS simulation of biomass pyrolysis for the production of methanol

Arul M, Dinesh Kumar M, Anand Ramanathan*

Department of Mechanical Engineering

National Institute of Technology Tiruchirappalli 620015, India

*Corresponding Author. Tel: 91-431-2503423 Fax: 91-431-2500133, email id: anandachu@nitt.edu

Abstract. In terms of supplying energy carriers for the transport sector and storing electricity outflows from intermittent sources, the importance of renewable fuel production has become significant. In this work, the production of methanol fuel from biomass is simulated. Biomass is an excellent renewable resource for the production of methanol. It is of utmost importance to make effective use of biomass resources. There are different methods available for the production of methanol from biomass. One of the best methods is pyrolysis to convert biomass into methanol. This is due to the fact that pyrolysis is an efficient conversion method compared to other thermochemical conversion practices. Pyrolysis is the process of decomposing biomass in an inert atmosphere to convert it into worthwhile products. The production of methanol from sugarcane bagasse via pyrolysis was simulated using Aspen HYSYS because of its ability to solve chemical as well as energy problems. To simulate the microwave assisted pyrolysis reactor, an Aspen HYSYS model was developed. The model is based on Gibbs free energy and it has been calibrated using the restricted equilibrium method. The model was validated and foresees the percentage of methanol yield, the predicted values very well agreed with the available data. Important parameters of the pyrolysis process such as pyrolysis temperature, sweeping nitrogen gas flow rate, heating rate, biomass moisture content were varied. It was found that pyrolysis temperature, nitrogen flow rate, heating rate have a very profound influence on the pyrolysis process and methanol yield, while the moisture content of biomass had a lesser impact.

Key Word: Aspen HYSYS; Biomass; sugarcane bagasse; methanol; pyrolysis; simulation.

1. Introduction

Methanol is a promising renewable fuel that has very less carbon emissions contrasted with traditional fossil fuels [1]. Agricultural, municipal, and biowaste produced throughout the country could be used to produce methanol and the energy produced from this methanol could supply 9-23% of the electrical energy used up in four states of the country mentioned above. The utilization of methanol could decrease carbon outflows by engine vehicles by up to 79% just as up to 31% of carbon emanations, if methanol-enhanced fuel cells were utilized to produce power [2].

There are different methods available for the production of methanol from biomass. One of the best methods is pyrolysis to convert biomass into methanol [3]. This is due to the fact that pyrolysis is an efficient conversion method compared to other thermochemical conversion practices. Pyrolysis is

IOP Conf. Series: Earth and Environmental Science **312** (2019) 012015 doi:10.1088/1755-1315/312/1/012015

the process of decomposing biomass in an inert atmosphere to convert it into worthwhile products. The creators showed that methanol was more profitable part in pyrolysis fluid and might be a conceivable substitution for traditional gas or diesel fuel. Methanol obtained as a result of pyrolysis, officially used as a fuel or part of fuel. [4]. The objective of this research was to simulate the production of methanol from sugarcane bagasse by microwave assisted pyrolysis process. This study also analyzed the effect of varying the parameters of the pyrolysis process such as pyrolysis temperature, sweeping nitrogen gas flow rate, heating rate, biomass moisture content on the yield of methanol.

Several researchers used Aspen HYSYS process simulator to simulate the pyrolysis reactor and gasifier [5,6]; Even so, the simulation made on the production of methanol from biomass pyrolysis is limited. An Aspen HYSYS model was used by Mansaray et al. [6] to predict the pyrolysis of biomass based on chemical equilibrium, energy equilibrium and equilibrium of components. Due to the presence of highly volatile material in biomass and the intricacy of kinetics in the biomass reaction rate in fluidized beds, they disregarded the formation of char and analyzed the process on the assumption of biomass pyrolysis tends to follow Gibbs equilibrium.

The ultimate purpose behind this research is to incorporate simulation to accurately predict the steady-state performance of a microwave assisted pyrolysis process by considering the reaction rate kinetics. Gibbs equilibrium defines the products of homogeneous reactions and kinetics are used to define exactly the pyrolysis products. This article illustrates the specifics of modelling strategies taken to acquire process computation for biomass pyrolysis in a microwave assisted pyrolysis reactor.

2. Modelling and simulation

The Aspen HYSYS flowsheet of the pyrolysis process for methanol production is portrayed in Fig. 1. The model is constructed based on the following key premises: Stable operation; zero-dimensional; ambient pressure; ideal gases; negligible drop of pressure; char is purely of carbon (C); pyrolysis process is prompt; the formation of tar is not taken into account; Pyrolysis reactor heat loss is neglected.

The state equation of Peng - Robinson was chosen as the property algorithm for the framework. Referring to Fig.1. non-traditional stream of sugarcane bagasse stated as 'BIOMASS' various properties of sugarcane bagasse were entered. Lower heating value of the chosen biomass was also quantified. Subsequently, the thermodynamic state of the stream (1 atm and 28 ° C) and the mass flow rate were inputted. The pressure was set to 1 atm for all the feed streams and unit operating blocks (i.e. There is no pressure drop). In this simulation R-Gibbs minimum energy reactor was used and named as 'PYRO-RE'. The different yields of the pyrolysis reactor obtained from sugarcane bagasse were determined using a calculator block.

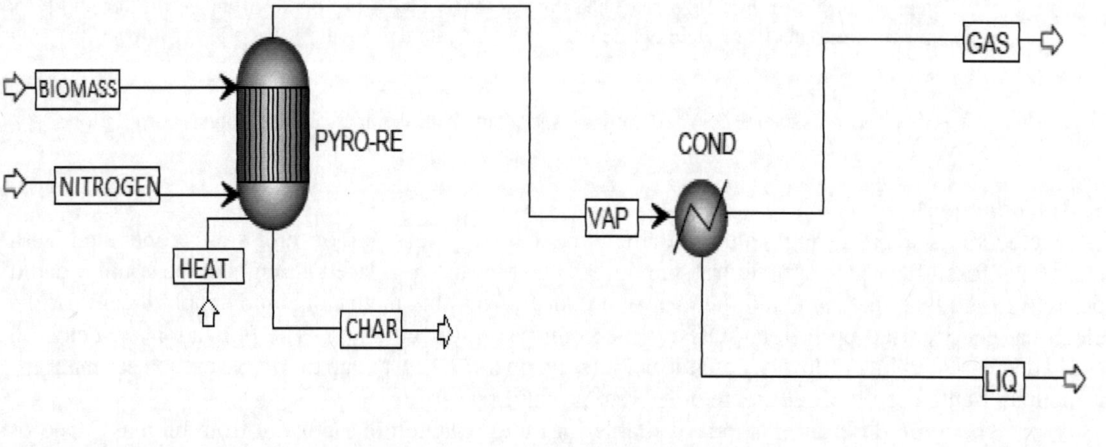

Fig .1. Pyrolysis process Aspen HYSYS flowsheet.

IOP Conf. Series: Earth and Environmental Science **312** (2019) 012015 doi:10.1088/1755-1315/312/1/012015

The outlet stream "VAP" from the pyrolysis reactor "PYRO-RE" is fed into a condenser "COND" whose purpose to condense the pyrolysis vapour into fuel oil and some non-condensable gas will also come out.

3. Results and Discussion

To analyze the susceptibility of the primary operating parameters with regard to pyrolysis reactor performance the evaluated model was used. Important parameters including pyrolysis temperature, nitrogen flow rate, heating rate, biomass moisture content. Sensitivity analysis was done by keeping input data same as used for the model validation and at any given time a single parameter was varied.

3.1. Sensitivity analysis; Pyrolysis temperature

Figure 2 shows the influence of pyrolysis temperature on the percentage of methanol yield. Pyrolysis temperature is varied from 300-590 °C and it shows that pyrolysis temperature has a very strong influence on pyrolysis yield. Fixed bed sugarcane bagasse pyrolysis reactor should be operated above 300 °C to ensure the yield formation. There is no yield formation below 300°C as it can be seen from the simulation result and maximum yield is obtained at around 450°C even if pyrolysis temperature is increased beyond 450°C, there will be no change in yield percentage so we can conclude that optimum temperature for the methanol production from biomass is around 450°C.

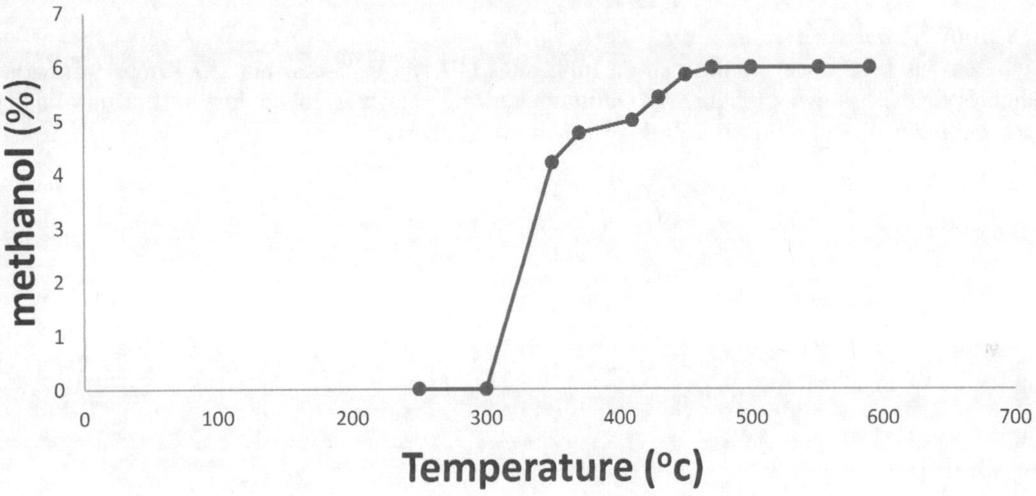

Fig.2. Variation of yield percentage with pyrolysis temperature (°C).

3.2. Sensitivity analysis; Nitrogen flow rate

After pyrolysis temperature, flow rate of nitrogen supplied for purging the reactor vessel to create inert atmosphere and to drive out the pyrolysis vapour has a great influence on pyrolysis yield. When the nitrogen flow rate is low residence time of pyrolysis vapour inside the reactor vessel increases so the methanol oil yield will be less, on the other hand if the flow rate increased above optimum range then also oil yield will decrease because of less time available for the condensation. Maximum methanol yield was around 6.42 % at the nitrogen flow rate of 200 cc/min.

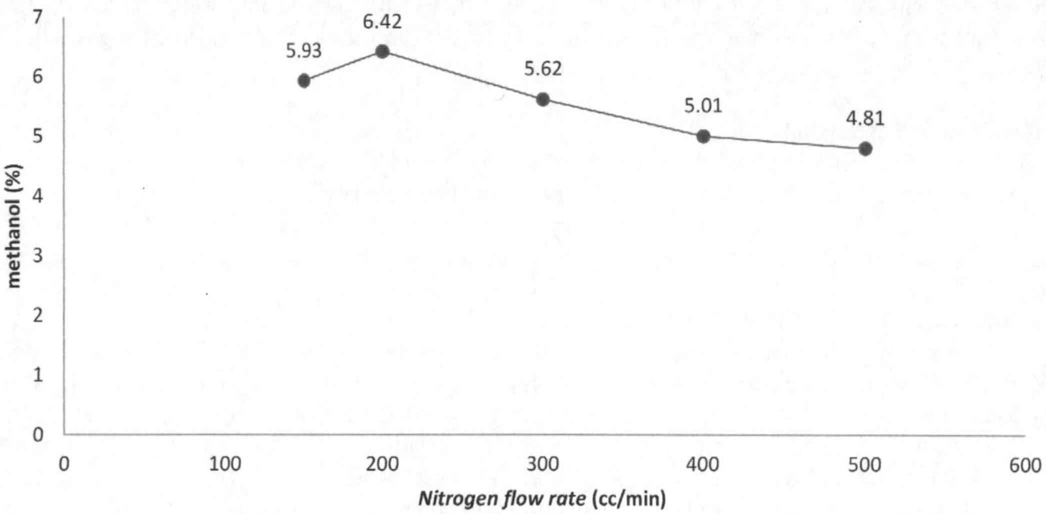

Fig .3. Methanol yield for different Nitrogen flow rates.

3.3 *Sensitivity analysis; Heating rate*
Simulation has been done for different heating rates (10 K/min, 20 K/min, 30 K/min) variation of methanol yield is shown in figure 4. Optimum heating rate was found to be 10 K/min and the corresponding methanol production was 6.36 % on weight basis.

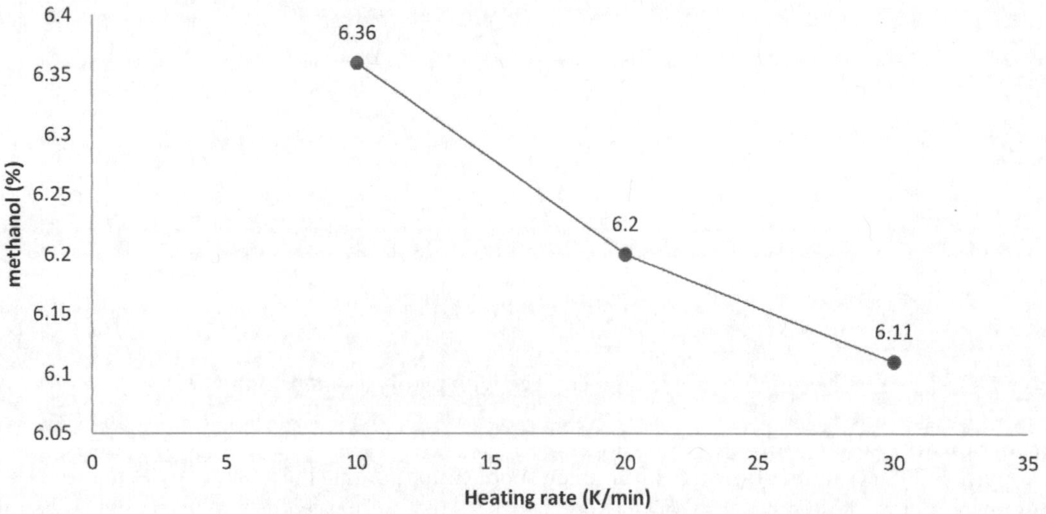

Fig .4. Methanol yield for different heating rates.

3.4 *Sensitivity analysis; moisture content of biomass*
Moisture content of the biomass has no significant effect on methanol yield as it can be seen in the figure 5 given below.

IOP Conf. Series: Earth and Environmental Science **312** (2019) 012015 doi:10.1088/1755-1315/312/1/012015

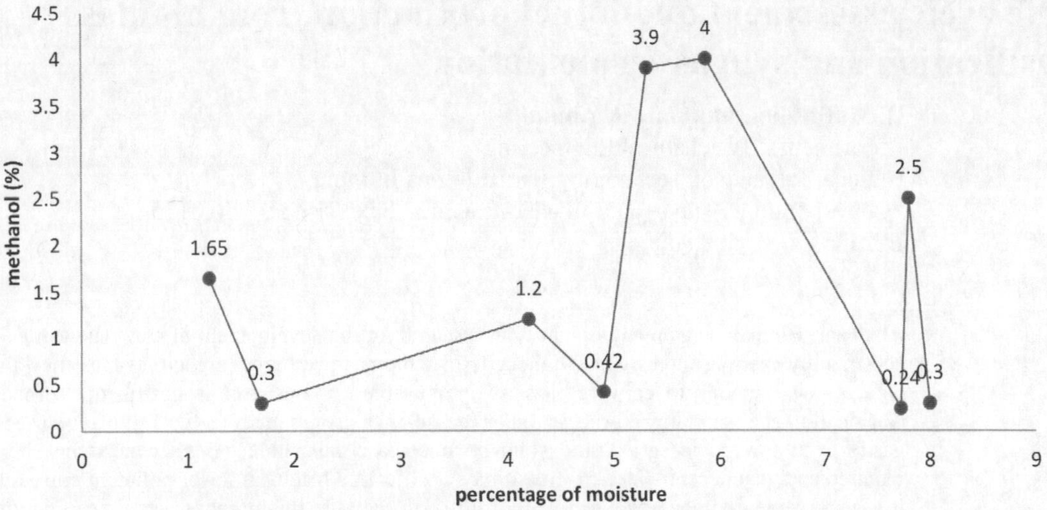

Fig .5. Methanol yield for different percentage of moistures.

4. Conclusion

The objective of the research work was to develop a pyrolysis reactor model and to use the same for investigating the impact of main operating parameters on reactor performance. A pyrolysis reactor computer simulation model was developed using Aspen HYSYS and results were obtained varying operating parameters. The findings revealed the following: pyrolysis temperature, nitrogen flow rate, heating rate are the most important methanol yield variables; the pyrolysis process must be done in the temperature range of 400 – 500°C, nitrogen flow rate of 200 cc/min, and the heating rate of 10 K/min. The simulation showed that the maximum percentage of methanol yield from sugarcane bagasse is around 6.42 % on the weight basis.

References

[1] Suntana A S, Vogt K A, Turnblom E C and Upadhye R 2009. Methanol potential in Indonesia forest biomass as a source of bio-energy that reduces carbon emissions. *Applied Energy* **86,** S215–S221.

[2] Vogt K A, Vogt D J, Weynand T P, Upadhye R., Edlund D, Edmons R L, Gordon J C, Suntana A S, Sigurdardottir R, Miller M, Roads P A and Andreu M G 2008. Methanol: how energy choices in the western United States can help mitigate global climate change. *Renewable Energy* **34,** 233–241.

[3] Gullu D and Demirbas A 2001. Biomass to methanol via pyrolysis process. *Energy Conversion and Management* **42,** 1349–1356.

[4] Hasegawa F, Yokoyama S and Imou K 2010. Methanol or ethanol produced from woody biomass: which is more advantageous? *Bioresource Technology* **101,** S109–S111.

[5] Kundsen RA, Bailey T and Fabiano LA. Experience with ASPEN while simulating a new methanol plant. *AICHE Symposium Series* **1982,**178-214.

[6] Mansaray KG, Al-Taweel, Ghaly, Hamdullahpur F and Ugursal VI. Mathematical modeling of a fluidized bed rice husk gasifier. *Energy Sources* **2000,**83–98.

[7] Mehrdokht B and Nader Mahinpey 2008. Simulation of biomass gasification in fluidized bed reactor using ASPEN PLUS. *Biomass and Bioenergy* **32,**1245–1254.

Life cycle assessment on biofuel production from biomass gasification and syngas fermentation

R. Muthudineshkumar, R. Anand*
Department of Mechanical Engineering
National Institute of Technology, Tiruchirappalli, India.
*Corresponding Author: Tel: 91-431- 2503423 Fax: 91-431- 250 0133,
E-mail: anandachu@nitt.edu

Abstract. Biomass is gaining popularity in research for the development of new renewable and eco-friendly energy sources. Since lignocellulosic biomass, resources are ample and renewable. One of the methods to convert biomass into usable gaseous fuel is gasification process. Gasification is a very simple process to convert solid fuels into gaseous fuels. It can be used to gasify even low-grade agricultural solid residues to combustible gases to meet the energy requirements in a decentralized manner. Since gasification results in fewer pollutant emissions, it is also an environmentally clean way of utilizing fuels. In this proposed study, comparative life cycle Assessment will be performed for power generation from biomass downdraft gasification and syngas fermentation for biofuel synthesis in two stages. The present investigation is to analyse the Prosopis Juliflora biomass waste intended for ultimate, calorific value and proximate analysis and to evaluate their characterization as feedstock for utilization in gasification process. In the first stage, the feedstock is first fed into downdraft gasifier from top to bottom, followed by a sequence of operation namely drying, pyrolysis, oxidation, and reduction for syngas generation. The synthesis gas derived from biomass downdraft gasification often contains additional components such as tar, acetylene, ethane and ethylene. The impurities available in the synthesis gas causes a potential scaling in the ways and by inhibiting microbial catalyst, which involves a cellular low growth and poor yield in product. Some recent studies reported that syngas has been converted into liquid fuel by using Fisher trophic synthesis, but it requires a catalyst. So, in the second stage the syngas fermenter with the capacity of 7-L would be designed and fabricated for the continuous production of 1 g/L biobutanol from purified syngas using the mixed culture mainly consists of C1 fixing homoacetogenic and C4-producing butyrate microorganisms or butyrogens. The produced syngas can also be used directly as a fuel for Solid Oxide fuel cell applications.

Keywords: Prosopis Juliflora; Gasification; Syngas; Life cycle Assessment; Fermentation

1. Introduction

In recent years the global is facing major energy crisis due to diminution of fossil fuels and mounting ecological anxieties. To overcome increasing energy demand research have been focused on the utilization of renewable energy resources [1]. Among several renewable sources, Biomass is, abundant, cheap biocompatible carbon neutral and has been used to produce chemicals, energy, and fuels. In almost 50 years worldwide potential of biomass energy will be approximately 35 to 1136 EJ/yr [2].This energy content will be equal to 200 billion of barrels of oil which will be sufficient enough to compensate global energy demand up to 2040[3].Conversion technologies of biomass can be classified into four categories: combustion, biochemical, agrochemical and thermochemical processes depending on biomass characteristics and the requirement of the end product and its applications [4]. Meanwhile, the thermochemical process further classified into gasification, pyrolysis and combustion. Combustion technique is used to provide heat and power commercially practiced all over the world. But it is not popular due to low efficiency and causes environmental pollution. Pyrolysis can be defined as thermal decomposition of any biomass feedstock without the supply of oxygen [5].Biomass gasification is the thermochemical conversion of solid biomass into combustible and non-combustible gases under partial air supply for the stoichiometric condition. The gaseous products mainly consist of hydrogen (H_2) carbon monoxide (CO) and a lesser volume of carbon dioxide (CO_2), methane (CH_4) and hydrocarbons (C2+) [6]. The quality of producer gas depends on several factors such as reactor design, operating condition, and purification techniques. According to the literature [7], the gasification process is performed when the moisture content is below 35%. When the biomass feedstock possesses the moisture content in the range of 25 to 60 % and if it is fed into the gasifier will result in energy losses in the overall system. Hence it is recommended for pre-treatment or dried before sending to the gasifier and moisture content should be in the range of 10 to 20% [8]. Tar is the major concern in the gasifier system. it may present as vapor phase in the gasifier and condense in liquid tar while cooling the gas [9]. In downdraft gasifier tar formation is less due to combustion zone follows pyrolysis zone in the reactor system [10]. The residence time is also shorter due to biomass feedstock moves faster due to gravity and drag force being aligned in the same direction. The present work focuses on using biomass species which is abundantly available in India, United States, south Arabia and African countries that is Prosopis Juliflora (PJ) [11].Pj belongs to the family Leguminosae, genus Prosopis it can capture CO2 from the atmosphere and can grow up to 10 m [12]. The zero emission and wider availability usually makes it as biomass fuel at international standard [13].LCA is described in the ISO standards 14040. It is one of the several methods for calculation of environmental impacts caused by the product or any process and service [14]. A mixture carbon, hydrogen and carbon dioxide is used as carbon and energy source and then converted into fuels and chemicals by microorganism [15].

2. Material and Methods

2.1. Selection of feedstock for investigation
The Prosopis Juliflora plant consist of a perverted trunk as well as springy branches having long and strong spikes. This plant can grow in different types of soils for example sandy, rocky, saline and alkaline soils . The roots of this plant breaches to a great depth in the soil. The stems of approximately 25 mm in diameter were collected in and around college premises of NIT Tiruchirappalli, Tamilnadu, south India.

2.2. Ultimate analysis
The elementary analyser various ELIII functions according to the principle of the combustion of catalytic tubes in oxygenated CO_2 atmosphere and high temperatures. The combustion gases are released from foreign gases (for example halogenous volatile). The preferred measuring components are separated from each other using specific and successively given columns of adsorption with thermal detector of conductivity (TCD). Helium (He) is used of rinsing and carrier gas. Win var software were used with accurate measurement of 0.05 % with analysis time of 10 to 14 min.

2.3. Proximate Analysis
The proximate analysis was performed in the tubular furnace. initially 1 mg sample was weighed and placed in the furnace for 1050 C for half an hour and the it was weighed for estimation of moisture content. Same sample was then placed in the furnace for 9250 C for 7 min at the nitrogen atmosphere. Sample was weighed to estimate the volatile matter. Finally, ash content was estimated by keeping the sample in the furnace for 5750 C for 5 hrs. Fixed carbon was calculated by the difference (100-M%+V.M%+Ash %).

2.4. Calorific value Analysis
To determine the calorific value of sample, initially the dried biomass sample is weighed for 5 mg and it is kept inside a bomb which is a heavy-duty stainless steel. The sample is exploded electrically inside bomb which is wrapped with oxygen. The temperature difference is resulted in the water bath adjoining the bomb due to complete oxidation of the compound. The rise in temperature is measured using digital sensor and constant volume heat of combustion is measured. By pelletizing sample and igniting inside

IOP Conf. Series: Earth and Environmental Science **312** (2019) 012016 doi:10.1088/1755-1315/312/1/012016

oxygen bomb calorimeter the CV of dried biomass sample (0.5 mg) was determined. Few repetitive experiments have been conducted to obtain the accuracy of ±2% maximum which is tolerable for experiments.

2.5. Syngas fermentation

An increase in energy demand, new technologies have been developed for the production of bio-butanol, which finds application as an additive in motor fuels, and several value-added products, using biological technologies. One of such technologies is the use of syngas to produce butanol using a syntrophic culture of two different anaerobic microorganisms. The syngas will be produced through down draft gasification using biomass waste. The preliminary study was conducted with 1 kW downdraft gasifier. In this, 3 kg of waste biomass was loaded into the gasifier which results in the production of 6.6 m^3/ kg of syngas having the calorific value of 16 MJ/m^3. The quality of the syngas will be analysed by the gas analyser after passed through various cleaning and filter system to remove impurities, H_2S, and other unwanted gases. The syngas fermenter with the capacity of 7-L would be designed and fabricated for the continuous production of 1 g/L biobutanol from purified syngas using the mixed culture mainly consists of C1 fixing homoacetogenic and C4-producing butyrate microorganisms or butyrogens. This methodology is preferred because of enhancement in the yield of butanol and flexibility in operation

2.6. Life cycle Assessment methodology

2.6.1. Definition and Scope of the Analysis

Description of environmental aspects of the syngas fermentation and gasification of biomass like formation of tars, emissions of heavy metals and toxic elements and emissions of polyhalogenated aromatic compounds.

2.6.2. Boundary system and Analysis of related legislation

Initially various components involved in the system must be defined before analysing the boundary of the system. Sub suits involved in the system that have productive purpose regarded as separate components. Policy and regulatory frameworks create the market push and market pull needed to develop bioenergy production currently, there is a lack of policies and incentives to support development, particularly at the smaller scale.

2.6.3. Selection of environmental performance indicators

The assessment of gasification and biofuel production process should be evaluated by using separate indicators. Indicators summarize the relevant information about the system. A precise evaluation is performed due to the collection of raw data and relation between them

2.6.4. Inventory Analysis

The relevant input and output data from gasification and biofuel production process are collected and quantified for LCA analysis. Various Inputs and outputs parameters and products data from biofuel production process and gasification of biomass. Structure the system flow of our functional units namely gasification and pyrolysis in the field of materials, energy, product output.

2.6.5. Environmental Impact Assessment

The environmental impacts and functional unit associated with the system are easily identified and evaluated by the LCA impact assessment. There are various tools and separate software are used for performing the life cycle impact assessment.

IOP Conf. Series: Earth and Environmental Science **312** (2019) 012016 doi:10.1088/1755-1315/312/1/012016

Life Cycle Assessment

Significant findings can be obtained from comparing the results with mathematical modelling and experimental results. Sensitivity analysis will be performed to identify the parameters which has major impact on the results obtained from the functional unit.

3. Results and Discussion

3.1. Ultimate analysis

The ultimate analysis of Prosopis Juliflora is given in the table. Carbon content of biomass sample is 47.96 %, an increase in the heating value may results. Some of the carbon constitutes volatile matter and char. During pyrolysis hydrogen will be as volatile matter of 6.614%. Nitrogen content is 0.293% significant source of NOx pollutants. Sulphur is also negligible with 0.539% source for SOx pollutants. The difference is the oxygen content which reduces heating value.

Table 1. Ultimate analysis of Prosopis Juliflora on dry basis

Carbon	Sulphur	Hydrogen	Nitrogen
47.96%	0.539%	6.614%	0.293%

3.2. Proximate analysis

Proximate analysis of a biomass provides the percentage of biomass that burns in a gaseous state (volatile matter) is 72.65%, ash content is 7.24%, moisture content of 3.24% and fixed carbon of 16.87%. if biomass with greater volatile content results in increased gas production.

Table 2. Proximate Analysis of Prosopis Juliflora on dry basis

Moisture	Volatile Matter%	Ash%	Fixed Carbon%
3.24%	72.65%	7.24 %	16.87%

3.3. Accessing calorific value by ultimate analysis

The heating value using ultimate analysis is determined by using dulong and Boie equation [18]. Dulong equation is[16]

$$\boldsymbol{HHV}\left(\frac{kJ}{kg}\right) = \mathbf{33823} \times \boldsymbol{C} + \mathbf{144250} \times \left(\boldsymbol{H} - \frac{O}{8}\right) + \mathbf{9419S} \quad (1)$$

The equation is not preferable since Prosopis Juliflora has less than 10% oxygen content. But in case of Prosopis Juliflora oxygen content is greater than 10%. Hence Boie equation is suitable for finding heating values of fuel and it expressed as

$$\boldsymbol{HHV}\left(\frac{kJ}{Kg}\right) = \mathbf{35160C} + \mathbf{116225H} - \mathbf{110990} + \mathbf{6280N} + \mathbf{10465S} \quad (2)$$

Where mass fraction like C, H, N, S and O obtained by ultimate analysis of respective biomass species' oxygen has negative co-efficient since it reacts with carbon and hydrogen to for carbon dioxide(CO_2)and water (H_2O).

Table 3. Comparison of calorific value between prediction and experimental values using ultimate analysis.

IOP Conf. Series: Earth and Environmental Science **312** (2019) 012016 doi:10.1088/1755-1315/312/1/012016

Biomass	Experimental calorific value (MJ/kg)	Predicted calorific value (MJ/kg)	Mean Absolute Error %
Prosopis Juliflora	17.91	17.77	-0.78

3.4. Assessing calorific value by estimating from proximate analysis

Based on proximate analysis cordea et al [19] investigated estimation of calorific value for biomass fuels on dry basis. Correlation equation used to find heating values are based on literature [16].

$$\boldsymbol{HHV = 354.3FC + 170.8\,VM} \quad (3)$$

$$\boldsymbol{M + VM + FC + ASH = 100} \quad (4)$$

Where HHV denoted as higher heating values, VM represents volatile matter, and FC represents fixed carbon these values estimated from the proximate analysis of the biomass sample Prosopis Juliflora.

Using the equation 4 HHV has been estimated and compared with investigational results. Mean absolute error has calculated using predicted and experimental results.

$$\boldsymbol{MAE = HHV\ predicted - HHV\ experimental/HHV\ experimental} \quad (5)$$

Table 4. Comparison of calorific value between prediction and experimental values using proximate analysis.

Biomass	Experimental calorific value (MJ/kg)	Predicted calorific value (MJ/kg)	Mean Absolute Error %
Prosopis Juliflora	17.91	18.38	2.65

3.5. Downdraft gasification of Biomass

Biomass gasification used for generation of producer gas by burning biomass under restricted air supply. In gasification the feed stock is first fed into downdraft gasifier from top to bottom, followed by sequence of operation namely drying, pyrolysis, oxidation and reduction for syngas generation. Design of reactor is vertical; therefore, the four zones are arranged vertically in sequence. Drying can be done inside the gasifier itself, biomass is heated up to 100 ° C. In pyrolysis process due to the unavailability of air biomass is thermally degraded into char and tar. Heat flows from the oxidation zone to drying zone and pyrolysis zone. Air is used as gasifying medium. Carbon dioxide is formed by reacting carbon in the form of char. Oxidation of hydrogen forms water. The final zone is the reduction zone were carbon monoxide is formed by reacting with carbon dioxide. Hydrogen and carbon monoxide are formed by reacting steam with carbon. Methane (CH_4) is formed by reacting char with hydrogen. During water gas shift reaction hydrogen and carbon dioxide are formed by reacting carbon monoxide with water. Combustible gases in the producer gas are formed during reduction process. The syngas comprises of hydrogen and carbon monoxide which is produced in this process.

IOP Conf. Series: Earth and Environmental Science **312** (2019) 012016 doi:10.1088/1755-1315/312/1/012016

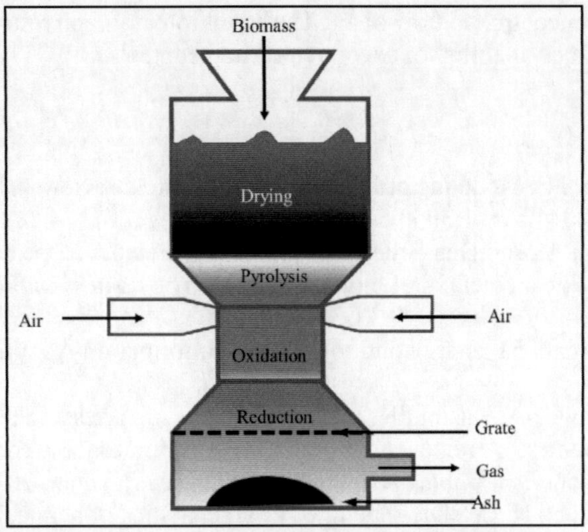

Figure 1. Schematic of Downdraft gasifier reactor

Product syngas composition

The quality of syngas produced is affected by many parameters especially feedstock and design of gasification reactor. The heating value of producer gas is in the range of 4 to 5.5 MJ/Nm^3. The small of amount of gas is collected in a tedler bag which is then analyzed in gas chromatograph using nitrogen as a carrier gas. The determined gas composition is shown in the figure 3.

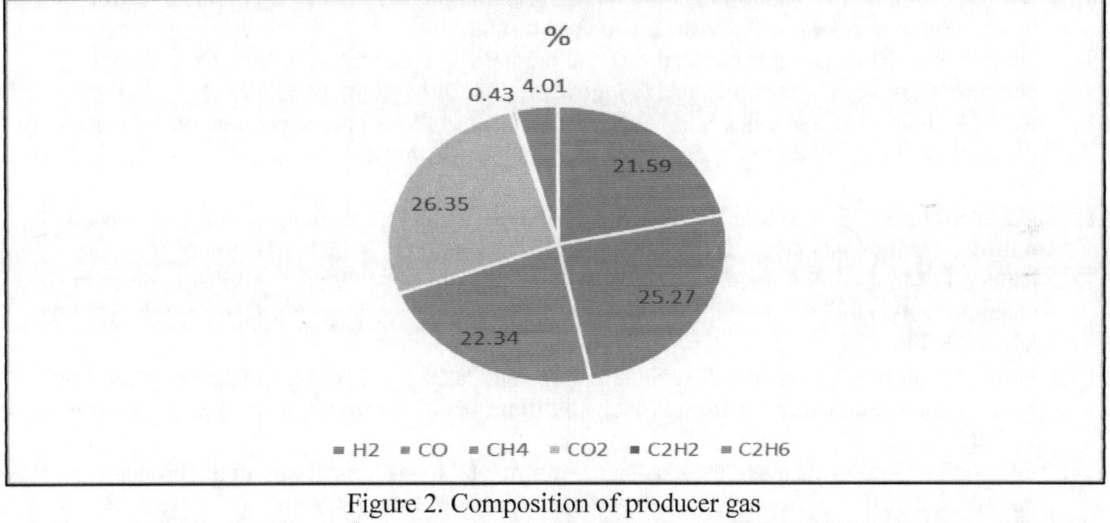

Figure 2. Composition of producer gas

The combustible gas like hydrogen gas (H_2) composition is around 21.59% and carbon monoxide (CO) is around 25.27 and methane (CH_4) is around 22.34. Due to uncontrolled supply of air carbon dioxide (CO_2) is around 26.35 %. Acetylene (C_2H_2) is around 0.43 % and ethane (C_2H_6) is around 4.01 %.

4. Conclusions

The Prosopis Juliflora having higher volatile matter content and less ash content of 72.65 and 7.25% respectively. Ultimate and proximate analysis were used to formulate the calorific values of biomass. The predicted result using Boie equation shows good agreement with experimental result. The calorific value of Prosopis Juliflora was found to be 17.91 MJ/kg which shows that it can be potential feedstock for biomass gasification. The gasification of Prosopis Juliflora in downdraft gasifier exhibits higher CO_2

of 26.35 % which is due uncontrolled flow of air. The gas chromatograph results revels that more amount of CH_4 was formed due to availability of excessive carbon content.

References

[1]. Baruah D, Baruah DC. Modeling of biomass gasification: a review. *Renewable and Sustainable Energy Reviews*. 2014 Nov 1;**39**:806-15.
[2]. Hoogwijk M, Faaij A, Van Den Broek R, Berndes G, Gielen D, Turkenburg W. Exploration of the ranges of the global potential of biomass for energy. *Biomass and bioenergy*. 2003 Aug 1;**25(2)**:119-33.
[3]. Sieminski A. International energy outlook. Energy Information Administration (EIA). 2014 Nov 18;**18**.
[4]. Ahmad AA, Zawawi NA, Kasim FH, Inayat A, Khasri A. Assessing the gasification performance of biomass: A review on biomass gasification process conditions, optimization and economic evaluation. *Renewable and Sustainable Energy Reviews*. 2016 Jan 1;**53**:1333-47.
[5]. Yang H, Yan R, Chen H, Lee DH, Zheng C. Characteristics of hemicellulose, cellulose and lignin pyrolysis. *Fuel*. 2007 Aug 1;**86(**12-13):1781-8.
[6]. Figueroa JD, Fout T, Plasynski S, McIlvried H, Srivastava RD. Advances in CO2 capture technology—the US Department of Energy's Carbon Sequestration Program. *International journal of greenhouse gas control*. 2008 Jan 1;**2**(1):9-20.
[7]. Peres AP, Lunelli BH, Fllho RM. Application of biomass to hydrogen and syngas production. *Chemical Engineering Transactions*. 2013.
[8]. Owais N. Simulation and optimization of IGCC technique for power generation and hydrogen production by using lignite Thar coal and cotton stalk.
[9]. Demirbaş A. Biomass resource facilities and biomass conversion processing for fuels and chemicals. *Energy conversion and Management*. 2001 Jul 1;**42**(11):1357-78.
[10]. Ruiz JA, Juárez MC, Morales MP, Muñoz P, Mendívil MA. Biomass gasification for electricity generation: review of current technology barriers. *Renewable and Sustainable Energy Reviews*. 2013 Feb 1;**18**:174-83.
[11]. Naseeruddin S, Desai S, Rao LV. Ethanol production from lignocellulosic substrate Prosopis juliflora. *Renewable energy*. 2017 Apr 1;**103**:701-7.
[12]. Rincón F, Muñoz J, Ramírez P, Galán H, Alfaro MC. Physicochemical and rheological characterization of Prosopis juliflora seed gum aqueous dispersions. *Food Hydrocolloids*. 2014 Mar 1;**35**:348-57.
[13]. Saravanakumar SS, Kumaravel A, Nagarajan T, Sudhakar P, Baskaran R. Characterization of a novel natural cellulosic fiber from Prosopis juliflora bark. *Carbohydrate polymers*. 2013 Feb 15;**92**(2):1928-33.
[14]. Friedrich E. Life-cycle assessment as an environmental management tool in the production of potable water. *Water Science and Technology*. 2002 Nov 1;**46**(9):29-36.
[15]. Saxena RC, Adhikari DK, Goyal HB. Biomass-based energy fuel through biochemical routes: a review. *Renewable and sustainable energy reviews*. 2009 Jan 1;**13**(1):167-78.
[16]. Basu P. Biomass gasification and pyrolysis: practical design and theory. *Academic press*;2010 Jul 19

Acknowledgement
The authors thank the Director, National Institute of Technology, Tiruchirappalli, Tamilnadu, India for extending the facilities to carry out the research work.

IOP Conf. Series: Earth and Environmental Science **312** (2019) 012017 doi:10.1088/1755-1315/312/1/012017

Energy and Life Cycle Assessment of Solar Assisted Microwave Pyrolysis of Waste Biomass

A.Santhoshkumar[1], Anand Ramanathan[1] *

[1] Department of Mechanical Engineering, National Institute of Technology, Tiruchirappalli - 620 015, TamilNadu, India

* Corresponding Author. Tel: (+91 431) 2503423, Fax: (+91 431) 2503423

E-mail: anandachu@nitt.edu

Abstract. In this proposed work focused on energy efficient solar assisted pyrolysis for biofuel synthesis from different biomass. In recent days solar-powered microwave assisted pyrolysis in increasing the research attention towards higher energy recovery from the waste biomass. Microwave-assisted pyrolysis has promoted by different researchers due to lower energy consumption, faster heating rate and uniform volumetric heating of feed. Compare to electrical and thermal pyrolysis; microwave assisted pyrolysis in increasing the process efficiency due to higher energy recovery and higher hydrogen production. Thus, Microwave-assisted pyrolysis combined with solar energy increase the economic feasibility of the pyrolysis process. To recover more energy from the waste, one must reduce the input power supply and has to improve the process efficiency. So, this proposed project concentrates on solar-powered microwave pyrolysis which will eliminate the need for input power supply. Concentrator Photo-Voltaic (CPV) systems are most likely to produce a massive amount of energy per rated power and module active area. CPV system produces around 2.5 times higher energy per module active area and around 1.3 times higher energy per rated power. Char and ash obtained from the pyrolysis process will be converted into activated carbon and further sintering at a higher temperature for converting into brick form. This reusable solid activated carbon may decrease the processing time by 50%, and raise the temperature around double the time faster compared to the conventional pyrolysis. Compared to the conventional pyrolysis, it is expected that the liquid fuel may increase by 30% due to the fast heating rate. Pyrolysis oil comprises of different chemical functional groups such as sulphur, carbonyls. The phenolic carboxyls compounds provide potentials simultaneously challenging for commercial usage. The hydrocracking and catalytic cracking are the recent technologies used in current research. The usage of hydrogen and catalyst degradation are the major challenges in current technology. To solve this problem, oxidation with ozone would be used as an efficient way for upgradation of biofuel. The life cycle assessment for the entire process is necessary to determine the carbon feet print to the environment.

Keywords: Waste to energy; Microwave pyrolysis; Solar assisted pyrolysis; Energy recovery; Life cycle assessment.

IOP Conf. Series: Earth and Environmental Science **312** (2019) 012017 doi:10.1088/1755-1315/312/1/012017

1. Introduction

The recent statistics 2016 shows, The continuous increasing fuel and oil consumption from 1999, has increased the research attention and curiosity for the energy demand as well as alternate fuel research more up-to-date than ever before. In recent days, the science community was focused on finding the new supplementary fuel from the waste to full fill society's increasing energy demand. The waste generation was increased proportionally to the population growth, with some municipal and agricultural wastes fraction was difficult to recycle; some fraction of waste may be utilized to fulfil the energy demand by direct combustion and other thermochemical conversion methods. The thermochemical conversation methods were further investigated to improve process efficiency [1].
Massive amount of energy from oil, coal, biogas, and natural gas were combusted daily to attain the ever-growing energy demand. These fossil fuels are rapidly depleting and cause environmental pollution [2]. Much waste is also generated in an ever-growing economy in various forms. Hence new sustainable technologies are needed to convert the waste generated from various sources into useful energy which is renewable. Thermo-chemical conversion is one such technology and methods like pyrolysis are providing ways to generate clean energy from waste materials, thereby avoiding pollution as well as producing alternative fuels [3].

Waste-to-energy conversion by the thermochemical process is a contemporary technology with the eco-friendly and cost-effective process. Several thermochemical conversions methods such as gasification, incineration, and pyrolysis can reduce waste volume and toxicity. However, all these conversion methods have certain limitations and drawbacks [4]. The thermochemical conversion processes like incineration, gasification, and pyrolysis were well-known methods for the management of waste. Combustion, gasification and direct incineration resulted in higher hydrocarbon emission, brominated dioxins, metal oxides and aromatic hydrocarbons. The pyrolysis is the best way of recovering both chemical and energy value of the waste without releasing any toxic oxidized species (i.e., NOx, dioxins) on the environment [5]. Direct combustion and incineration recover the only heating value of the waste materials. Different chemical values of the waste materials were not recovered in the incineration due to oxidation causes for greenhouse gas emission and toxic emissions [6].

The external heat source is necessary to convert the waste material and biomass into chemically potential biofuel. Traditional external heating source (heat delivered from the electric heaters and flue gas) has been used for the biomass pyrolysis in the last two decades [7, 8]. There are different methods that are existing for heating the waste materials in the pyrolysis. Among them, conventional pyrolysis and microwave assisted pyrolysis are currently used for waste conversion. In conventional pyrolysis process, the pyrolysis process efficiency was lower due to its slower heating mechanism based on conduction and convection heat transfer. Whereas, In modern microwave pyrolysis technique, the heat was generated equally on the pyrolysis feed due to diffused characteristics of electromagnetic field on the substance [9]. The microwave pyrolysis gives some advantage over conventional pyrolysis, which offers a faster heating rate, volumetric heating of feed, selective material heating, improve heating efficiency [10]. Paradela et al., [11] investigate the co-pyrolysis of pine waste biomass mixed with e-waste plastics to improve the maximum pyrolysis fuel yield. The prime objective behind the co-pyrolysis of biomass was to increase the pyrolysis fuel fraction and its physiochemical properties.

Traditional pyrolysis techniques are conventional and consume higher input energy for the pyrolysis. The emerging pyrolysis such as MAP has been introduced in recent days. The microwave will generate the heat depends upon the feed and susceptor used in the pyrolysis process. The activated carbon, Graphene and Silicon carbide were used as the susceptor for the different microwave pyrolysis process [12]. The microwave pyrolysis with the addition of different susceptor has decreased the energy consumption compared to the conventional pyrolysis process. Also, it decreases the production

IOP Conf. Series: Earth and Environmental Science **312** (2019) 012017 doi:10.1088/1755-1315/312/1/012017

cost while improving the quality of biofuels. The MAP technique provides rapid and selective heating of the biomass due to water vapour present in the biomass. Compared to the conventional process, the MAP was cleaner, which enhances the chemical reaction and saves energy utilization, consequently decreases the production cost [13]. So far, little research work has been carried out related to the solar thermal conversion of biomass conversion. Most of the work was carried out for the higher production of carbon monoxide and hydrogen using solar radiation. The solar pump laser and solar/ microwave hybrid system furnace has been used for this process [14-16].

In recent days hydrotreating and catalytic cracking were wildly used for fuel upgradation and enrichment of hydrogen in the fuel. Higher hydrogen pressure and high-temperature hydrocracking in the presence of metal-based catalyst will lead to the formation of useless water vapour at the end of the reaction. The demerits of catalytic cracking is that massive amount of hydrogen consumption and degradation of the catalyst after the several trails which is increasing the barrier for scale up the process. several researchers have attempted for the improving of pyrolysis fuel by different purification techniques and esterification with excess alcohol. They have identified uses of alcohol decrease the acid value and carbonyl constituents in the presence of different acid catalysts [17]. The harmful constituents of pyrolysis fuel such as aldehydes cannot be removed and it leads to affect the physicochemical characteristics of pyrolysis fuel. The aldehydes undergo oligomer reaction spontaneously to form the phenol containing molecular structure during the storage. To resolve this problem, ozone treatment can be used as an efficient way to transform the aldehydes into corresponding acids. Then, the pyrolysis fuel will be treated using alcohol in the presence of a heterogeneous acid catalyst for the conversion of aldehydes into corresponding esters at the end of the reaction [18]. The ultra-sonication treatment has used in recent days for upgradation pyrolysis fuel. The life cycle analysis is essential for the analysis of the environmental impacts of the pyrolysis process and upgradation process and its environmental impacts and carbon footprint of the pyrolysis process.

2. Experimental setup and procedure

2.1. Microwave pyrolysis process

In this experiment, the pyrolysis of waste was carried out in the microwave pyrolysis reactor using four magnetrons. Each magnetron was capable of generating the microwave frequency of 2450MHz. The pair of magnetrons was operated to heat the pyrolysis oven. The Euro-Therm microcontroller was used to control the magnetron power corresponding to the required pyrolysis temperature. Only one pair of magnetrons was powered at 3mins of operation time. Then, the power was exchanged to other pair of the magnetron to avoid magnetron failure due to overheating. The separate switch was provided to control the number of magnetrons has been used for the pyrolysis process, which can control the power rating. Each magnetron was consumed a maximum of 1.1 kW at the initial stage of the pyrolysis process. The maximum power of 2.2 kW was utilized to generate the microwave at the initial stage. The power rating was controlled by a separate switch provided for control the number of magnetrons which can be operated at the same time. The minimum power rating of 1.1 kW and the maximum power rating of 2.2 kW was used for the waste pyrolysis. The quartz vessel with 1.8 L capacity was used as a pyrolysis reactor due to excellent microwave transmittance. 1 kg of waste precisely weighted and mixed with activated carbon in the weight ratio of 40%, which facilitates the microwave to heat energy conversion. The temperature was maintained at 300 °C to 400 °C to determine the maximum yield and energy conversion of the pyrolysis process. The nitrogen was used as the carrier and inert gas and its flow rate was adjusted from 2 L/min to 5 L/min for determined the minimum required a quantity of Nitrogen flow rate. The temperature was measured using K-type thermocouple and send the signal to the Eurotherm data acquisition system corresponding to the time. A stirrer is set into the reactor at a speed of 10 rpm. The experiment is conducted at different temperatures and nitrogen flow rates. The residue (char) was quantified after the end of the experiment. The non-condensable gases (NCG) were evicted into the water to control the release of soot and particulate matters.

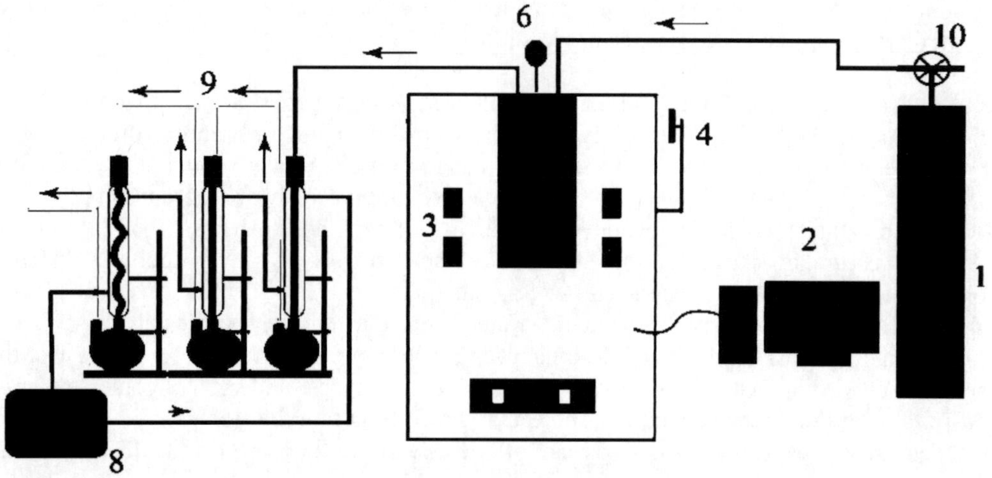

1. Nitrogen cylinder
2. Data acquisition system
3. Magnetron
4. Blower
5. Quartz vessel
6. Stirrer motor
7. Control panel
8. Water tank with pump
9. Condenser
10. Nitrogen flow regulator

Figure 1. Schematic layout of microwave pyrolysis setup.

2.2. Activated carbon preparation from char

This proposed work involves preparation of Activated carbon from char from the pyrolysis process. The microwave susceptor composites will be prepared in a brick form with the addition of different ceramics using high-pressure sintering. Thermal stability and dielectric properties of susceptor will be analysed to predict the microwave absorbing capability. Activated carbon (AC) derived from the agro waste can be used as a cleaning agent to remove tar and other impurities present in the producer gas. This proposed project overcomes the challenges associated with susceptors reusability and purification of producer gas. Activated carbon (AC) derived from the waste can be used as a cleaning agent to remove tar and other impurities present in the producer gas and pyrolysis fuel. Char, ash, and slag collected from the pyrolysis and gasification process will be dried in a furnace at 150 ˚C for 24 hrs. The dry powder will be activated with 66% H_3PO_4 for 2 hrs at 80 ˚C and dried at 110 ˚C for 24 hrs after that calcinated at 800 ˚C in at N_2 atmosphere for 5 hrs for obtaining the activated carbon.

3. Effect of pyrolysis parameter on pyrolysis yield

3.1. Microwave pyrolysis fuel yield

Figure 2 shows the effect of various temperature and different nitrogen supplied rate on the microwave pyrolysis fuel yield. Pyrolysis fuel yield improved with an increase in temperature to some extent. Due to the localized concentration of heat by microwave decrease pyrolysis oil yield at a higher temperature. The optimal temperature at which the supreme percentage of pyrolysis oil obtained was 350˚C. It is also observed that due to the increase in nitrogen flow rate beyond 2 L/min, the pyrolysis fuel yield was decreased. This is due to a reduction in the retention time of pyrolysis gases inside the condenser beyond specific nitrogen flow rate (i.e., 2 L/min).

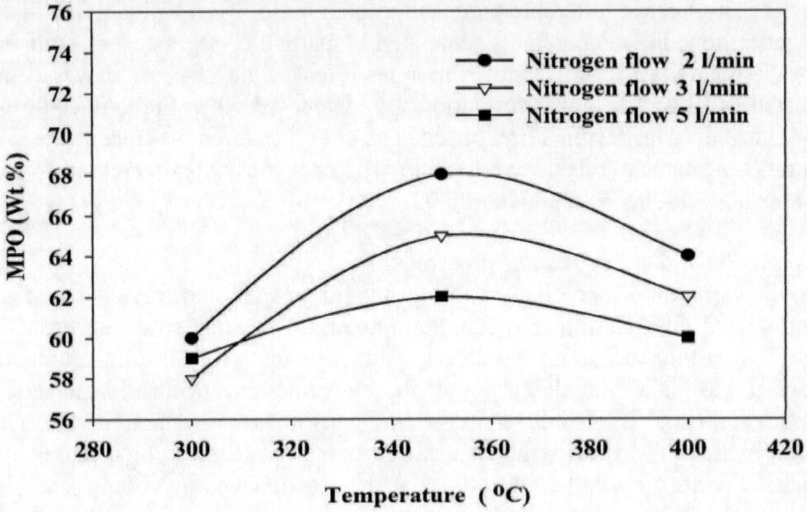

Figure 2. Comparison of pyrolysis oil yield with varying different parameters.

It was observed that when the nitrogen flow rate was increased from 2 L/min to 3 L/min, the pyrolysis fuel yield decreased by 4% at 350°C and increased to 5 L/min the pyrolysis fuel yield decreased by 8%. At 300 °C the pyrolysis fuel obtained from the entire nitrogen flow rate nearly the same. Compared to 350 °C pyrolysis temperature, the pyrolysis fuel yield obtained at 300 °C was less due to higher retention time enhances the secondary cracking of the pyrolysis gas. At 400 °C pyrolysis temperature, the kinetic energy of the gases is slightly high compared to 350 °C. The higher kinetic energy of the gases decreases the retention time of gases in the condenser and affect the pyrolysis fuel yield.

3.2. Microwave pyrolysis char yield

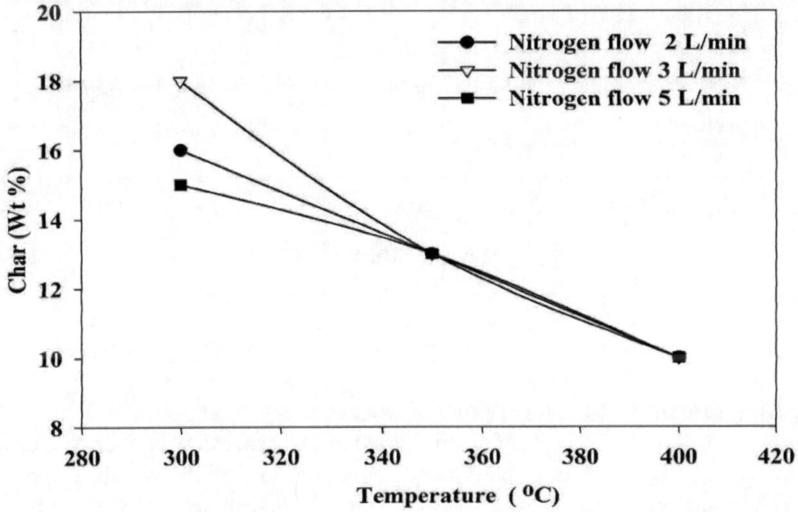

Figure 3. Comparison of char yield with varying different parameters.

The formation of char percentage in the microwave assisted pyrolysis was less. The heat generation on the susceptor leads to a higher localized temperature, which causes lower char formation. Figure 3 shows the effect of an increase in temperature on the char yield. Microwave pyrolysis improves the localized heat generation, the temperatures generated at particular region were high in which even carbon absorbs the microwave energy and evaporates. Hence, the char yield was decreased in the microwave assisted pyrolysis. The char formation was decreased when the temperature and nitrogen supplied rate was increased at 10 rpm stirrer speed. The char formation was nearly the same for the all nitrogen flow rate at the same pyrolysis temperature. The char formation decreased from 18% to 10% when increased the temperature from 300 to 400 ˚C.

3.3. Microwave pyrolysis non-condensable gas yield

Figure 4 shows the variation of non-condensable gas yield on the microwave-assisted pyrolysis with various temperature and different nitrogen supplied rate and a constant stirrer speed of 10 rpm. Non-condensable gas proportions were increased with increase in pyrolysis temperature and nitrogen supplied rate due to the higher kinetic energy of the molecule and localised heat generation, which decreased the retention time. The lower retention time causes for the reduction in condensation time and instantaneously carried away the pyrolysis gases from the condenser, which causes increased Non-condensable gases. Figure 4 it can be observed that the least NCG was obtained at 350˚C with the nitrogen flow rate of 2 L/min. when the nitrogen flow rate was increased from 2 L/min to 5 L/min, the NCG yield decreased by 6% at 350˚C. NCG at 300 ˚C remains same at the entire nitrogen flow rate. The higher NCG was noticed at the temperature reaches 400 ˚C with the nitrogen flow rate of 5 L/min because of its higher kinetic energy and higher internal heat generation compared to other temperature. Hence it can be concluded that microwave assisted pyrolysis improved the amount of non-condensable gases obtained at higher nitrogen flow rates.

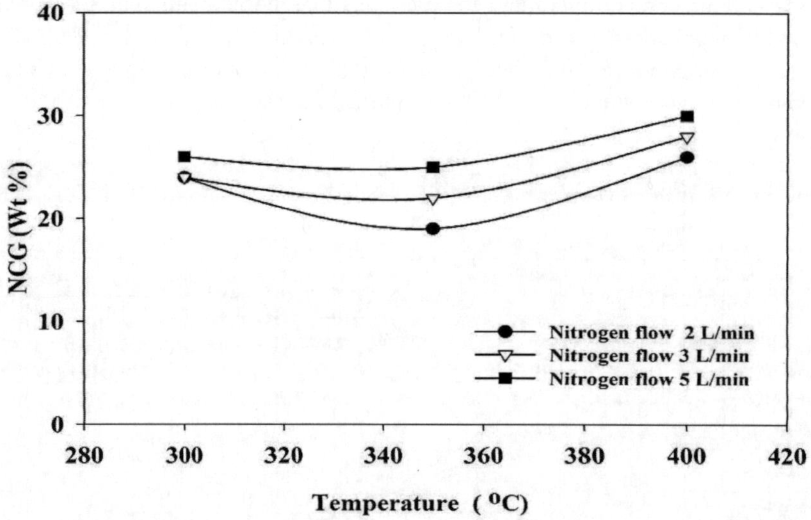

Figure 4. Comparison of non-condensable gas yield with varying different parameters.

4. Effect of heating rate and temperature on energy recovery

The essential parameter which can be studied in the pyrolysis process was energy recovery. It can be calculated from pyrolysis process time, the Energy consumption of the pyrolysis process, pyrolysis fuel yield and higher heating value of the pyrolysis fuel. The pyrolysis fuel yield, calorific value and process time varied depending upon the temperature and heating rate. Figure 5 depicts the energy recovery of the microwave assisted pyrolysis with varying temperature and power rate or heating rate.

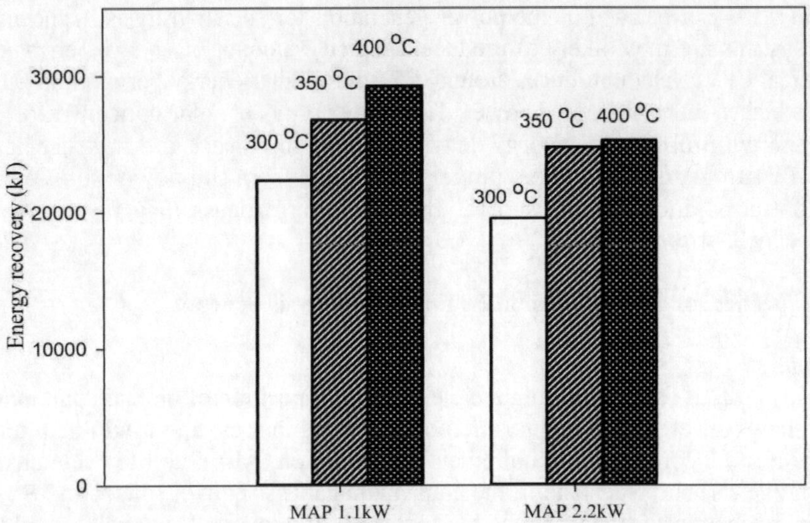

Figure 5. Comparison of energy recovery with various temperatures and heating rate.

In this process, the maximum energy recovery was attained at 1.1 kW power rating because of higher pyrolysis fuel yield, lesser power consumption and increased heating value of the obtained pyrolysis fuel. In microwave assisted heating, at 2.2 kW power rate, secondary cracking was decreased the amount of pyrolysis oil yield production. The amount of pyrolysis oil produced in 2.2 kW was less, and energy consumption was high compared with 1.1 kW heating is the reason for decreased energy recovery. The combined effect of fuel yield, energy supply and calorific value decreased the energy recovery in the microwave pyrolysis with a 2.2 kW power rating compared to 1.1 kW power rating. The higher energy recovery of 29274 kJ was attained in microwave assisted pyrolysis with 1.1 kW power rate at pyrolysis temperature 400°C. The maximum energy recovery is 15.75% higher than microwave pyrolysis with a 2.2 kW power rating at a similar temperature. Also, the energy recovery was 9% and 30% higher than 350 °C and 300 °C respectively at microwave assisted pyrolysis with 1.1 kW power rating. The energy recovery was improved with a rise in temperature due to a decrement in the processing time and hiked in the heating value of the pyrolysis fuel.

5. Solar assisted pyrolysis for better energy recovery

In solar thermal reactor for the pyrolysis process was classified into two types such as direct passionate and indirect passionate system. The direct, passionate pyrolysis reactor exposed directly in the solar radiation at the bright sun which fascinating the absorbed energy from the solar radiation. An indirect passionate system, the incident sunlight was transferred to the reactor from an opaque surface and generated the heat to the reactor surface [19]. Besides, the solar heating system further classified into the partial heating system and continues heating system. In the partial heating solar system, the pyrolysis reactor was passionate at a particular temperature using solar concentrator and further heated by the system like thermal, electrical or microwave. Moreover, the continuous heating solar system is passionately unremitting using solar radiation coupled with another system to complete the pyrolysis cycle.

Solar concentrators are classified into 4 types they are

1. parabolic-dish collector or concentrator (PDC);
2. parabolic-trough collector or concentrator (PTC);
3. double-concentration collector or concentrator (DCC);
4. heliostat-field collector or concentrator (HFC).

Furthermore, solar PV can be used as the power generation for solar pyrolysis. Concentrator Photo-Voltaic (CPV) systems are most likely to produce a massive amount of energy per rated power and module active area. CPV system produces around 2.5 times higher energy per module active area and around 1.3 times higher energy per rated power. These four types of solar concentrators are used to convert solar energy into thermal energy. It will be used for direct and indirect heating of the biomass. Solar PV can be used for direct power generation from the power source and it can be stored in the batteries and further used for the pyrolysis process may be the increasing the energy recovery by 2.5 times [20].

6. Essential properties for different biomass for the pyrolysis process

6.1. Ultimate analysis

The ultimate analysis used for determining the elemental composition of biomass and municipal solid waste. Table 1 shows the elemental composition of different biomass, and it will help to analysis the molecular formula and theoretical air requirement for oxidation. Also, the life cycle analysis will be carried out using the carbon percentage in the elemental analysis. Further, the N and S levels and its effect on the environment also studied in the biomass fuel. Moreover, C,H,O value used to determine the higher heating value of the biomass and pyrolysis fuels [21]. The organic contents, along with moisture ash, were analyzed using the final analysis. The elemental analysis was given the uniform system to analysis and compared the pyrolysis fuel. The chemical composition and generalized chemical formula were determined from the elemental analysis [22]. In addition to that, the ASTM standards are available for reassurance of elemental analysis of biomass pyrolysis fuels [23].

Table 1. Ultimate analysis of various biomass at dry basis (wt.%) [253, 254].

Biomass	C	H	O	N	S
Hazelnut shell	51	6	42	1	0
Tobacco stack	49	6	43	1	1
Almond shell	49	6	43	1	1
Rice husk	48.86	9	42	0.1	0.04
Lignite	66	8	21	1	4
Corn Stover	49	6.6	43.7	0.6	0.1
Corncob	49	6	44.2	0.8	0
Saw dust	53.9	7	37	2	0.1
Spruce wood	51.4	7.1	41.2	0.3	0

6.2. Proximate analysis

The essential characteristics of biomass and pyrolysis fuel are determined by ultimate and proximate analysis. The proximate analysis was used to determine the combustible and non- combustible constituents present in the biomass and pyrolysis fuel. In this analysis helped to predict the carbon life cycle assessment in the entire pyrolysis process. The commercial value of the pyrolysis fuel and biomass was determined from the proximate analysis. The proximate analysis investigation was critical to anticipate the burning phenomenon of pyrolysis fuel and biomass [22]. The gross value of composition in the biomass and biofuels also measure quickly. The biomass fuel E_870-06 standards are used to analyze the fixed carbon, volatile matters and Ash. Different ASTM standards used separately to find the individual components of biofuel and biomass. The fixed carbon value deviates from the elemental analysis of carbon content. The char yield was determined in the proximate analysis without including the volatile matters. Table 2 shows the proximate analysis of the different biomass sample [23].

IOP Conf. Series: Earth and Environmental Science **312** (2019) 012017 doi:10.1088/1755-1315/312/1/012017

Table 2. Proximate analysis of various biomass at dry basis (wt. %) [24].

Biomass	Fixed carbon	Volatile matter	Ash
Wheat straw	17.71	75.27	7.02
Switch grass	14.34	76.69	8.97
Willow wood	16.07	82.22	1.71
Rice hulls	16.22	63.52	20.26
Alfalfa stems	15.81	78.92	5.27
Sugar cane bagasse	11.95	85.61	2.44
Rice straw	15.86	65.47	18.67
Almond shells	20.71	76	3.29
Hybrid poplar	12.49	84.81	2.7

7. Pyrolysis fuel Upgradation

7.1. Ozone Treatment

Pyrolysis oil and tar comprise different chemical constituents and functional groups like sulphur, carbonyls; The phenolic carboxyls compounds which afford potentials benefits at the same time challenges for commercial usage. Duc to physical and chemical changes, the pyrolysis oil becomes more viscous at storage. This is because the reaction may continue during the period, and volatile components might be lost because of ageing. It would involve a decrease in the heating value of pyrolysis fuel and also increase the risks of the separation of phases. Today, these undesirable compounds are eliminated by several distillations. However, this method of purification increases the cost of fuel. At present, the techniques most widely used are the enrichment of the pyrolysis oil by hydrotreating, catalytic cracking and esterification. The hydrotreating forms the undesirable water at high hydrogen pressure and high temperature, under in the presence of catalysts. The hydrotreating was not economical due to the consumption of higher hydrogen consumption and finally increased the useless water content in the reaction. The disadvantage of catalytic cracking is the degradation of the catalyst, which constitutes an essential barrier to marketing. The aldehydes can be spontaneously involved in oligomer reactions to form the phenol composition during the storage. The resolve this issue, the ozone can be used as the oxidizing agent to convert the aldehydes into corresponding acids in an effective way. After that, the oxidized pyrolysis oil will be treated with alcohol in the presence of a heterogeneous acid catalyst. The aldehydes will be converted to the corresponding esters at the end of the process. Ozone has the potential of very high oxidation, just behind the free radical hydroxyl. This is derived from the oxidising activity of ozone and the oxidizing spaces produced by ozone, such as the radicals OH. Consequently, ozone can oxidise many organic compounds and inorganic. Moreover, because of its unstable composition, there breaks up with time, it is thus no residues after the treatment. For these reasons, ozone is used at many ends, such as disinfection, discolouration, and deodorisation. Finally, the pyrolysis oil and tar will be converted into corresponding hydrocarbons and esters.

7.2. Ultrasonic Treatment

The ultrasonic method is an effective method for upgrading the heavy pyrolysis oil mixed with a mutual solvent. However, low-cost ketones and long-chain hydrocarbons have been chosen for the mutual solvent for this study. Ultrasonic treatment widely used in the field of mixing of pyrolysis oil with mutual solvent improves stability due to high effectivity and short treated time. Different ultrasonic power, the different solvent ratio, different solvent type and different exposure time has been chosen for upgrading the pyrolysis oil. Initially, acetone, ethyl acetate, and n-octanol have been chosen for preliminary study. Further, the experiment will be carried out by the different solvent to find the maximum upgradation of the pyrolysis oil. Afterwards, the blend which had been treated with ultrasonic was put into the centrifuge tube. The sludge and heavy metals are separated at the bottom of the centrifuge the upgraded oil was collected from the top. The water dispersed in the pyrolysis fuel cab be removed when the ultrasonic wave passed in the pyrolysis fuel. The fuel-water interface destroyed during the ultrasonic sound propagation due to the cavitation effect. The cavitation effect decreases the surface tension of the fuel and mechanical vibration enhance the proper mixing and form the vortex to enhance the blending and stirring operation. The heavier hydrocarbon molecules break down into smaller molecules during the ultrasonic treatment, which causes increasing the calorific value of the fuel. Further, ultrasonic can lead to enhance then-octanol and acids reaction to synthesise the high energy dense esters. The small molecules constituents which are decamped in volatile material leaving the blend as gaseous form due to mechanical vibration and cavitation effect during the ultrasonic treatment.

8. Life Cycle Assessment

The direct and indirect environmental impacts of the pyrolysis process, biomass and waste feed and pyrolysis products were analysed using Life Cycle Assessment (LCA) models. The carbon footprint during the process and energy consumption and recovery along with source emissions also tabulated for accurate analysis of LCA. The different process flow diagram can be prepared, and a suitable model has been chosen for an exact prediction of LCA value. It includes accounting, assessing, manipulating and interpreting the environmental and carbon impacts generated from the LCA including source material, production time, energy consumption, waste utilization and emission reduction. The LCA for the biofuel analysis demonstrates the excess energy required to turn the energy availability in the raw material to convert the useful energy of the pyrolysis fuel. Also, The LCA provides evaluation and sustainability of pyrolysis fuel and its production. In this proposed project, the life cycle assessment of pyrolysis and gasification will be carried out and compare them for prediction of a suitable energy efficient, eco-friendly process for alcohol and biofuel synthesis.

9. Conclusion.

From this study, the important conclusions were mentioned below from the experimental results by different pyrolysis process.

- The energy consumption was decreased when activated carbon combined with waste by 40% of the weight. The energy recovery is 4 times higher than microwave assisted pyrolysis without susceptors.
- The optimal microwave pyrolysis temperature was attained at 350 °C process temperature with power supplied rating of 1.1 kW to attained maximum pyrolysis fuel yield and maximum energy recovery.
- The proper design of solar assisted pyrolysis with CPV will increase the energy recovery rate by 2.5 times.
- Ozone and ultra-sonication method used for fuel upgradation to decrease the energy supplied.
- Life Cycle analysis will be needed to find the carbon footprint on each and every process.

References

[1] Evangelopoulos, P., Kantarelis, E and Yang, W. Experimental Investigation of Pyrolysis of Printed Circuit Boards for Energy and Materials Recovery under Nitrogen and Steam Atmosphere. *Energy Procedia, **105**:986–991(2017).*

[2] Seltenrich, N.: Emerging waste-to-energy technologies: Solid waste solution or dead end? Environ. *Health Perspect. **124**, A106–A111 (2016).* doi:10.1289/ehp.124-A106

[3] Alston, S.M., Clark, A.D., Arnold, J.C., Stein, B.K.: Environmental impact of pyrolysis of mixed WEEE plastics part 1: Experimental pyrolysis data. *Environ. Sci. Technol. **45**, 9380–9385 (2011).* doi:10.1021/es201664h

[4] Lee, J., Song, H.T., Yoo, J.-M.: Present status of the recycling of waste electrical and electronic equipment in Korea. *Resour. Conserv. Recycl. **50**, 380–397 (2007).* doi:10.1016/j.resconrec.2007.01.010

[5] Ramasamy, K. K., & T-Raissi, A. Hydrogen production from used lubricating oils. *Catalysis Today, **129**(3–4): 365–37 (2007).*

[6] Dominguez, A., Fernandez, Y., Fidalgo, B., Pis, J.J., Menendez, J.A.: Bio-syngas production with low concentrations of CO2 and CH4 from microwave-induced pyrolysis of wet and dried sewage sludge. *Chemosphere. **70**, 397–403 (2008).* doi:10.1016/j.chemosphere.2007.06.075

[7] Balat M, Demirbas MF. Pyrolysis of waste engine oil in the presence of wood ash. Energy Sources, *Part A: Recovery, Utilization, and Environmental Effects.**31**:1494-9 (2009).*

[8] Kim YS, Jeong SU, Yoon WL, Yoon HK, Kim SH. Tar-formation kinetics and adsorption characteristics of pyrolyzed waste lubricating oil. *Journal of Analytical and Applied Pyrolysis;**70**:19-33 (2003).*

[9] Domínguez, A., Menéndez, J. A., Fernández, Y., Pis, J. J., Nabais, J. M. V., Carrott, P. J. M., & Carrott, M. M. L. R. Conventional and microwave induced pyrolysis of coffee hulls for the production of a hydrogen rich fuel gas. *Journal of Analytical and Applied Pyrolysis, **79**(1–2 SPEC. ISS.): 128–135 (2007).*

[10] . Jones, D. a., Lelyveld, T. P., Mavrofidis, S. D., Kingman, S. W., & Miles, N. J. Microwave heating applications in environmental engineering—a review. Resources, *Conservation and Recycling, **34**(2): 75–90 (2002).*

[11] Paradela, F., Pinto, F., Ramos, A. M., Gulyurtlu, I., & Cabrita, I. Study of the slow batch pyrolysis of mixtures of plastics, tyres and forestry biomass wastes. *Journal of Analytical and Applied Pyrolysis*, **85**(1–2): 392–398.(2009).

[12] Suriapparao, D. V., Vinu, R.: Bio-oil production via catalytic microwave pyrolysis of model municipal solid waste component mixtures. *RSC Adv.* **5**, *57619–57631 (2015).* doi:10.1039/c5ra08666c

[13] Minkova V, Razvigorova M, Bjornbom E, Zanzi R, Budinova T, Petrov N. Effect of water vapour and biomass nature on the yield and quality of the pyrolysis products from biomass. *Fuel Process Technol* 2001;**70**:53–61. doi:10.1016/S0378-3820(00)00153-3.

[14] K. Imazaki, Biomass Energy Conversion Apparatus, *Japanese Patent JP2010030870 (2010).*

[15] N. Ugolin, Method Using Solar Energy, Microwaves and Plasmas to Produce aLiquid Fuel and Hydrogen from Biomass or Fossil Coal, *US Patent. US 8388706(2010).*

[16] N. Limon-Vazquez, F. Chavez-Lara, Collector and Systems for Generating Steamfrom the Use of Solar Energy, *Mexican Patent MX 2010012078 (2012).*

[17] Nguyen TM, Jung J, Lee CW, Cho J. Effect of asphaltene dispersion on slurry-phase hydrocracking of heavy residual hydrocarbons. *Fuel. 2018 Feb 15*;**214**:174-86.

[18] Xu J, Jiang J, Dai W, Zhang T, Xu Y. Bio-oil upgrading by means of ozone oxidation and esterification to remove water and to improve fuel characteristics. *Energy & Fuels. 2011 Mar 11;**25**(4):*1798-801.

[19] Yadav D, Banerjee R. A review of solar thermochemical processes. *Renewable and Sustainable Energy Reviews 2016*;**54**:497e532.

[20] Sukhatme SP, Nayak JK. Solar energy. *McGraw-Hill Education; 2017* Sep 15.

[21] Sheet, R.F.: Characteristics of Biomass as a Heating Fuel. Renewable Fact Sheet. *Renew. Energy*. 1–3 (2008)

[22] Kirubakaran, V., Sivaramakrishnan, V., Nalini, R., Sekar, T., Premalatha, M., Subramanian, P.: A review on gasification of biomass, (2009)

[23] Basu, P., Basu, P.: Chapter 2 – Biomass Characteristics. In: Biomass Gasification and Pyrolysis. pp. 27–63 (2010)

[24] Jenkins, B.M., Baxter, L.L., Miles, T.R., Miles, T.R.: Combustion properties of biomass. *Fuel Process. Technol.* **54**, 17–46 (1998). doi:10.1016/S0378-3820(97)00059-3

IOP Conf. Series: Earth and Environmental Science **312** (2019) 012018 doi:10.1088/1755-1315/312/1/012018

Production of biodiesel from Aegle marmelos correa seed oil for fuel cell application

Vinoth Thangarasu, Anand Ramanathan*
Department of Mechanical Engineering, National Institute of Technology, Tiruchirappalli-620 015, India.
*Corresponding Author. Tel: (0431) 2503423, Fax: (0431) 2503423,
E-mail: anandachu@nitt.edu

Abstract. Energy acting an important role in the development of any country. In an emerging country like India, energy plays an important role to fulfil the energy demand by the industry as well as society. Each year there is an increase in demand for energy. Currently, most of the demand for energy is filled with a petrol-based product that creates the problem of sustainable supply of energy and increasing problems like, pollution, global warming, etc. The huge quantities of energy require a huge amount of fuel, spending this huge amount of fuel across the world have adverse effects on the environment which cannot be altered after damage being done. Energy production from fossil fuel is the main reason for environmental degradation which effects on human health, disturbs the ecological equilibrium and biotic diversity. There is a requirement of the alternate and renewable source to tackle the energy demand as well as keeping a safe environment. Bioenergy productions from biofuels have emerged as one of the most popular sources. The biofuels are a promising, renewable and eco-friendly source of energy that can replace fossil. To avoid the effects of growing energy demand on the environment, human life in near future and satisfy present energy demand, these biofuels have the potential to provide energy services with almost zero emissions of air pollutants and greenhouse gases. This paper aims at emphasizing the production of biodiesel from Aegle marmelos Correa seed oil. Maximum biodiesel yield of 95% was obtained using a potassium hydroxide catalyst. Physical and chemical properties of biodiesel were determined as per American Society for Testing Method D6751 standards. It can be concluded that biodiesel produced from Aegle marmelos Correa seed oil can be another energy source to substitute the petroleum diesel fuel. Further, AMC biodiesel can be a promising source to produce hydrogen for fuel cell application.

Keywords: Aegle Marmelos Correa Seed; Oil extraction; Acetone; Heptane; Design Expert; Biodiesel production; Fuel Cell.

1. Introduction

Because of the expanding vitality request and contamination issues created by the utilization of fossil fuels, it is essential to create elective energizes and additionally renewable vitality source. As an alternative fuel, biodiesel is turning out to be progressively imperative because of decreasing petroleum holds and antagonistic ecological results of fumes gasses from fossil fuel fuelled engines [1,2].

Our over-dependence on fossil fuel, increasing oil demand and monetary value, environmental problems caused by fossil fuels stimulated the search of a new alternative to fossil fuels. The emission of CO, unburned hydrocarbons, the particulate matter gets reduced by the biodiesel from edible oil, non-edible oil, animal fats algae, etc. compared to fossil fuels [3,4]. Biodiesel is an Eco-friendly and renewable fuel that can be utilized as an alternative to petroleum diesel due to the increasing energy demand [5].

IOP Conf. Series: Earth and Environmental Science **312** (2019) 012018 doi:10.1088/1755-1315/312/1/012018

Renewable energy is a source of energy which is not diminishing with human consumption, which is collected from natural resources like wind, sunlight, tide, rain, and biomass. In the areas of electricity generation, air and water heating and transportation, biomass, and geothermal energy played a significant role. Petroleum fuels are formed under the earth's surface due to entrapped biomass, which is a non-renewable source of energy.

A biofuel is a fuel derived from biomass, which included animal materials, industrial wastes, and algae's and plant materials. Due to the anaerobic digestion of this biomass, a chemical change takes place on the biomass leads to the production of chemicals which can be used as a fuel. The increase in pollution caused by fossil fuels leads to the searching of new alternative fuels like biofuels are an excellent alternative to petroleum fuels.

The principle feedstock's for biodiesel is edible oil, non-edible oil, waste cooking oil, animal fats and algae's, etc. The perspective of the limitation of satisfying oil for human use, the non-edible oils are the promising alternative for the production of biodiesel. In India, unflinching efforts are going on to find a new sustainable energy source. In the northeast part of India, lots of oil-bearing plants are growing, the seeds of these plants are not appropriately utilized [6]. The production of biodiesel from edible has been inconceivable in India since despite everything we depend on different nations for edible oil in order to meet the demand [7].

The cost of the biodiesel feedstocks, the cost of the methanol, by-products of biodiesel production assumes a noteworthy part in the overall economy of the biodiesel production process. At the time of biodiesel production, a considerable amount of biowastes like glycerol, seed cake, seed shell, etc., is generated. Utilization of these by-products is essential for improving the overall economy of the bio-diesel production process [8].

Jatropha, Karanja, Neem, Mahua, Rubber are some proven non-edible feedstocks. In order to meet the enormous energy requirement of a country like India, these proven non-edible feedstocks are not sufficient. Here comes the importance of finding new non-edible feedstock for producing biodiesel. Aegle Marmelos Correa is famous for medical uses. The seed is not yet appropriately utilized. The oil from the seed can be used for producing biodiesel.

Biodiesel can be produced by the chemical process such as transesterification of straight vegetable oils or animal fats with alcohol in the bearing of a suitable catalyst [9]. Transesterification is the process of reaction between triglycerides present in the oil with alcohol in the presence of a strong base to form a mixture of alkyl esters with glycerol. The transesterification reaction is a chain reaction in which triglyceride first converted into di-glyceride then into monoglyceride finally it forms a mixture of alkyl esters [10].

Biodiesel is a fuel derived from plant materials which consist of triglycerides. The triglycerides present in the plant material which compelled to react with methanol with a suitable catalyst produce biodiesel. Mainly biodiesel is mono-alkyl esters of free fatty acids. Laura et al., [11] used oil containing high free fatty acid for the synthesis of biodiesel. The properties of biodiesel are almost comparable to that of petroleum diesel, so it named as biodiesel Biodiesel can be used in diesel engine directly or with some engine modification or as a blend with petroleum diesel. The process of converting triglycerides into biodiesel called transesterification.

Edible oils are the oil, which is used for human consumption. The commonly used edible oils are coconut oil, palm oil, mustard oil, soybean oil, sunflower oil, etc. edible oils are used to produce biodiesel by transesterification method. Biodiesel is, for the most part, arranged from expectedly developed edible oils, for example, rapeseed, soybean, sunflower, and palm along these lines prompting to reduce nourishment versus fuel issue [12]. Around 7% of worldwide vegetable, oil supplies were utilised for biodiesel generation as a part of 2007. Full utilization of consumable oils may bring about other critical issues, for example, starvation in creating nations. The utilization of non-consumable plant oils when contrasted and eatable oils are incredibly critical in creating nations on account of the gigantic interest for eatable oils as sustenance, and they are costly to be utilised as fuel at present.

IOP Conf. Series: Earth and Environmental Science **312** (2019) 012018 doi:10.1088/1755-1315/312/1/012018

Non-edible oil is not used for human consumption. Hundreds of varieties of non-edible seeds are available in the world which is not properly utilizing. Presently, more than 95% of the world biodiesel is delivered from eatable oil, which is effortlessly accessible on an expansive scale from the rural business. Be that as it may, constant and substantial scale generation of biodiesel from eatable oil without appropriate, arranging may bring about adverse effects to the world, for example, consumption of nourishment supply prompting to financial irregularity. A conceivable answer for conquering this issue is to utilize non-eatable oil or waste cooking oil.

In this research, the effect of factors such as solid-liquid proportion and extraction time on Aegle Marmelos Correa seed oil yield were analyzed using Mixture of Simple lattice Design and Optimal Design to optimize the Maceration extraction method. Further, biodiesel was produced from extracted AMC crude oil for fuel cell applications.

2. Materials and methods

Aegle Marmelos Correa fruit is to be collected from Indian Council of Forestry Research and Education, Coimbatore, TamilNadu. Figure 1 shows the photograph view of the fruit and bi-section fruits. Methanol (99.9% purity), phosphoric acid (H_3PO_4 80%), Hydrochloric acid (HCl, 38%) Sulphuric acid (H_2SO_4, 98%), Sigma Aldrich, Bangalore.

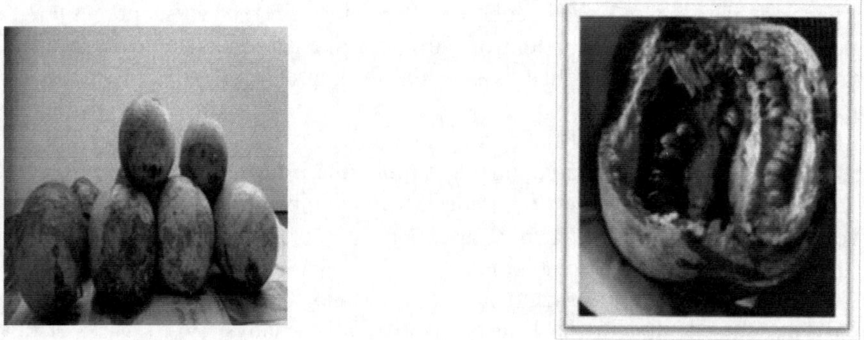

(a) (b)

Figure 1. A photographic view of (a) AMC fruit and (b) Cross-section of the fruit

2.1. Oil extraction

Aegle Marmelos Correa, generally known as bael, also vilvam, golden apple and wood apple, is a species of tree native to India, Srilanka, and Bangladesh. It is widely available in Southeast Asia countries like as Pakistan, Burma, Vietnam, and Thailand. Its fruits are utilized in Ayurvedic medicine and leaves of the tree used for Lord Shiva as part of worship. The bael fruit has a smooth wooden shell with green, grey or yellow skin. The maturation of trees takes about 11 months and can reach the size of a bigger grapefruit, and some are much bigger. The shell is very hard that its necessity to bisect with a pestle or hammer. The seed of the fruit is extracted and collected in a container. The preparation of the seeds before an oil extraction step depends on the kind of seed and the required quality of the oil. Initially; the seed is mixed with seed gel and fruit pulp. The seed is soaked in water for 2 hrs. for easy removal of gel, materials stick onto the surface of the seed. After that, the seed is cleaned using a metal strainer. The cleaned seed is dried in sunlight for 24 hours and stored in a plastic container for future use.

The dried seed is crushed into small particles using a household 750 W grinder. The crushed sample was sieved to 0.75 mm particle size and dried in sunlight before extraction. Heptane and acetone are used for the extraction of oil from the seed. Acetone (99%, Boiling range 55.5 – 56.5 °C, CDH) and Heptane (85%, Boiling range 90 – 100 °C, (Loba Chemie). The average particle size is estimated by the average size of the sieves between which the seed particles are caught:

IOP Conf. Series: Earth and Environmental Science **312** (2019) 012018 doi:10.1088/1755-1315/312/1/012018

$$\mathbf{d_p} = \frac{dp{,}max + dp{,}min}{2} \tag{1}$$

Maceration is carried out using 100 g seed in a 3 neck round bottom flask using 400 ml solvent (acetone: heptane). The flask is kept in a container ½ filled with water to keep the solvent temperature below the atmospheric temperature in order to reduce the solvent evaporation at the time of stirring. A stirrer rod was introduced into the flask through the top, and the speed of stirring is adjusted to 600 rpm by using a digital rpm indicator and the knob. The experimental setup used for oil extraction is given in figure 2.

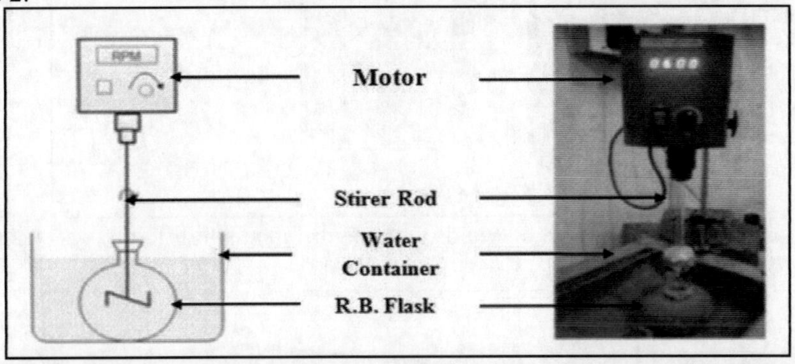

Figure 2. A schematic and photographic view of oil extraction setup

The extracted solvent seed mixture is filtered using a filter paper of Whatman Grade No.44 ashless filter paper by using vacuum filtration. The oil-solvent mixture is poured into a ceramic conical funnel; the filter paper was fitted in the bottom of the funnel. It took a few hours to filter 1 liter of solvent –oil mixture. The top of the funnel is closed at the time of filtration in order to prevent the evaporation of the solvent. Acetone and Heptane were evaporated at 50 °C and 94 °C respectively and collected in flask separately. Pure Aegle Marmelos Correa oil obtained after distillation.

2.2. Biodiesel production

The transesterification process was carried out in a hot plate system. The reactor system consists of 3-neck flat bottom glass type flask and reflux condenser. Initially, 50 g of AMC oil was preheated to 50 °C for 30 min to remove moisture content. Then methanol catalyst solution was prepared by taking a 1:6 molar ratio of oil to methanol and 1 wt% of potassium hydroxide catalyst and mixed thoroughly. After that reaction was carried out with 700 rpm stirring speed for 60 min at 60 °C in a 250 ml glass type flask fitted with a reflux condenser. After completion of the transesterification process, the resultant product was shifted into a separating funnel. After 6 hours of settling time, two distinct layers were formed. Top and bottom layer consist of biodiesel and glycerol respectively. The biodiesel was collected and was washed with hot water at 60 °C to remove the remaining catalyst, methanol, and glycerol

3. Results and discussion

3.1. Effect of various process parameters on oil extraction

The effect of solid to the liquid ratio on the extraction of AMC oil yield is shown in figure 3. The results indicated that no variation in the AMC oil yield was observed to increase in solid to liquid proportion from 1:4 to 1:6 for all 22 runs. In research work, a variation of solid-liquid ratios showed that insignificant role on the oil yield (figure 3). The result obtained from this design (1:4 solid to liquid ratios) was selected to be included in further Multilevel Categoric Design.

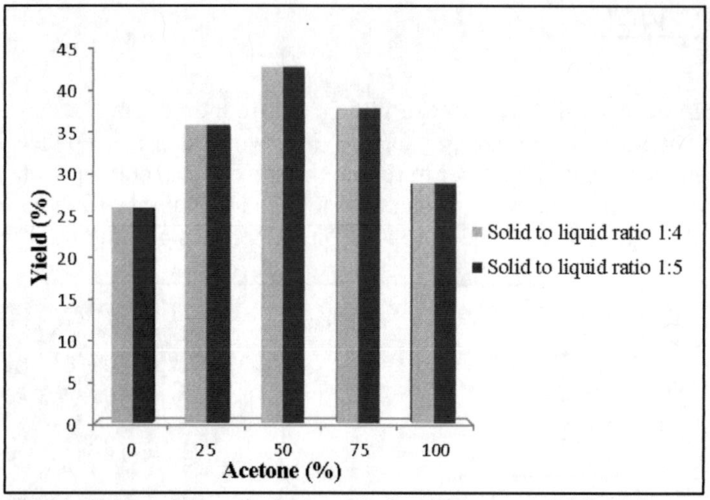

Figure 3. Variation of the oil yields with the different solid-liquid proportion

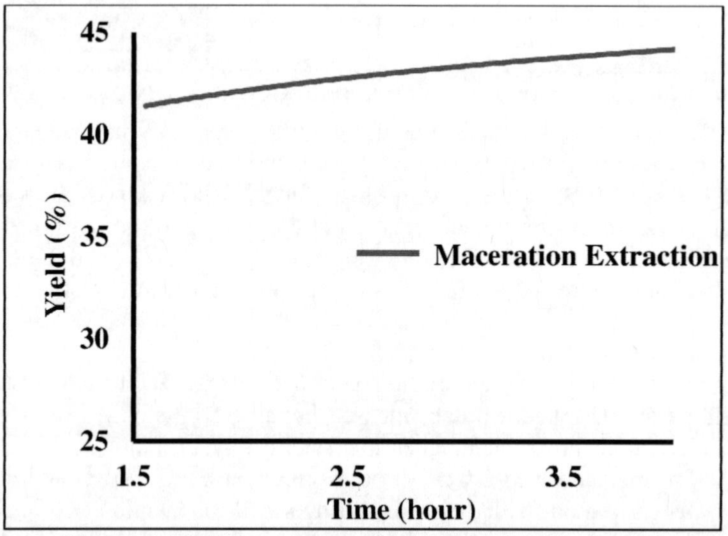

Figure 4. Variation of extraction time on AMC oil yield

The effect of extraction time on AMC oil yield is shown in figure. 4. The maximum yield of oil was achieved in 2.5 h (solid-liquid proportion (1:4), particle size (1.02), Acetone/Heptane (50/50 %, by volume)), but no significant difference was observed between 2.5 h and 4 h for the same solid-liquid proportion, particle size and Acetone/Heptane concentration. The graph showed that there was no substantial benefit to prolonging the duration (more than 2.5 h) of maceration since the solvent has already saturated.

4. Conclusions

Maceration assisted oil extraction from Aegle Marmelos Correa seeds by using Acetone and Heptane (50/50%,) solvent mixture exhibited improved oil yield than other solvents. The Acetone/Heptane solvent mixture 50/50 %, by volume, the solid-liquid proportion and of 1:4 and 0.75 mm particle size for 2.5 hours of extraction time gave a maximum oil yield percentage which is 43%. In this study, the results also exhibited the maximum oil yield than another soxhlet method of extraction. The oil gives an optimum conversion of biodiesel of 95% at a temperature of 56 °C, one wt% of KOH catalyst and 1:9 oil-methanol ratio. The fatty acid composition and physicochemical properties of Aegle Marmelos

Correa biodiesel show that can be used to produce biodiesel. It can conclude that, AMC biodiesel could be a promising alternative source for the hydrogen production through steam reforming process and which will be used for hydrogen fuel cell applications.

5. Scope for future work

Hydrogen production is a crucial development for the future alternative energy system to provide a clean and affordable system energy supply. Currently, interest is increasing in the development of fluid conversion technologies to produce hydrogen from hydrocarbons rich liquid fuels. This option does not involve any additional development cost infrastructure, which could help reduce costs introduction of hydrogen into energy systems. Also, hydrogen can be used in electrochemical fuel cell applications.

References

[1] Murugesan A, Umarani C, Chinnusamy T R, Krishnan M, Subramanian R and Neduzchezhain N 2009 Production and analysis of bio-diesel from non-edible oils-A review *Renew. Sustain. Energy Rev.* **13** 825–34

[2] Saba T, Estephane J, El Khoury B, El Khoury M, Khazma M, El Zakhem H and Aouad S 2016 Biodiesel production from refined sunflower vegetable oil over KOH/ZSM5 catalysts *Renew. Energy* **90** 301–6

[3] Karmakar A, Karmakar S and Mukherjee S 2010 Properties of various plants and animals feedstocks for biodiesel production *Bioresour. Technol.* **101** 7201–10

[4] Kasote D M, Badhe Y S and Hegde M V. 2013 Effect of mechanical press oil extraction processing on quality of linseed oil *Ind. Crops Prod.*

[5] Jesus A A, Almeida L C, Silva E A, Filho L C, Egues S M S, Franceschi E, Fortuny M, Santos A F, Araujo J, Sousa E M B D and Dariva C 2013 Extraction of palm oil using propane, ethanol and its mixtures as compressed solvent *J. Supercrit. Fluids*

[6] Canakci M and Van Gerpen J 2001 Biodiesel Production from Oils and Fats with High Free Fatty Acids *Trans. Am. Soc. Agric. Eng.*

[7] Pourzolfaghar H, Abnisa F, Daud W M A W and Aroua M K 2016 A review of the enzymatic hydroesterification process for biodiesel production *Renew. Sustain. Energy Rev.*

[8] Ayoola a a, Efeovbokhan V C, Bafuwa O T and David O T 2014 A Search for Alternative Solvent To Hexane During Neem Oil Extraction *Int. J. Sci. Technol.* **4** 66–70

[9] Gratuito M K B, Panyathanmaporn T, Chumnanklang R A, Sirinuntawittaya N and Dutta A 2008 Production of activated carbon from coconut shell: Optimization using response surface methodology *Bioresour. Technol.*

[10] Buasri A, Ksapabutr B, Panapoy M and Chaiyut N 2012 Biodiesel production from waste cooking palm oil using calcium oxide supported on activated carbon as catalyst in a fixed bed reactor *Korean J. Chem. Eng.*

[11] Naik M, Meher L C, Naik S N and Das L M 2008 Production of biodiesel from high free fatty acid Karanja (Pongamia pinnata) oil *Biomass and Bioenergy*

[12] Balat M 2011 Potential alternatives to edible oils for biodiesel production - A review of current work *Energy Convers. Manag.* **52** 1479–92

Acknowledgment

The authors would like to acknowledge the Department of Science and Technology, India for the financial support under Young Scientist Scheme. The authors also like to thank the Director and Head Department of Mechanical, National Institute of Technology, Tiruchirappalli for providing the experimental facility, valuable help, and support.

Thermal Performance of a Salt Gradient Non-Convective Solar Pond in Subtropical Region Climatic Conditions

Sendhil Kumar Natarajan*, Susant Kumar Sahu and Arjun Singh K
Department of Mechanical Engineering, National Institute of Technology Puducherry, Karaikal - 609 609, U.T. of Puducherry, INDIA.

*Corresponding Author: Assistant Professor and Head

Email: drsendhil1980iitmuk@gmail.com, sendhil80@nitpy.ac.in

Abstract. Renewable energy is becoming a significant source of energy due to the intensification in crude oil prices and the upsurge in greenhouse effects due to burning of fossil fuels. With only finite source of fossil fuel and exponential increase in the demand of power because of increase in human population, power generation from renewable energy promises a sustainable future for the mankind. Solar energy can be harnessed to meet our energy needs for a sustainable future. After the remarkable hike in oil prices numerous countries began to do an extensive research and development to utilize solar energy. There are many methods to trap solar energy, but an effective one was by using solar pond. Solar pond is an unnaturally constructed pond in which large temperature can be maintained in the lower convective zone by the way of higher salt concentration and by preventing natural convection so these ponds are termed as salt gradient solar pond. In this current research a salt gradient solar pond is designed and developed. Further the Thermal Analysis and Performance of the same model has been done in Vizag Climatic Condition. The same model can be utilized for generating sensible and latent heat for room heating and industrial heating applications. Moreover, electrical power can be generated by using suitable heat exchanger and working medium of low boiling point values such as ammonia.
Keywords:- Salt gradient solar pond, thermal efficiency, hot wire anemometer, solar power meter.

1. Introduction

Many researchers have been studied in both experimentally and analytically on thermal performances of solar pond based on different designs at numerous climatic conditions. Bezir Nalan et al [1] has created a salt gradient solar pond of surface area of 12.25m^2 area of the trapezoid cross-section pond was 1.7 m^2 and having a depth of 0.5 m. and a depth of 2 meter. The Thermal Efficiency was investigated for each zone theoretically and numerically. The pond was covered on the surface to minimize the thermal losses at the top during night time and also to increase the efficiency of the pond. Thermal analysis was carried out both theoretically and experimentally. P. Sarath Kumar et al [2] investigated the solar pond with PCM storage in which it was concluded that by adding Al_2O_3 with PCM improves the heat transfer coefficient along with heat storage capacity. Jimmy Leblanc et al [3] has fabricated a solar pond and introduced a novel system for extracting heat and thus the efficiency. Experimental studies of the system are studied and a close agreement with theoretical studies is found out. This study showed that an overall efficiency of about 55% can be increased with heat extraction from NCZ comparatively conventional way of extraction from lower convective zone. Jeffrey A Ruskowitz et al [4] has carried out an experimental investigation on the salt gradient solar pond and found out the suppression of evaporative losses and solar energy collection. The transparent material over the solar pond reduces the wind flow across the surface thus resulting in a decrease in UCZ-NCZ interface thereby increasing efficiency. A.A. Hill et al [5] in England has fabricated a solar pond porous material and investigated its linear instability in the lower convective zone (LCZ). It was found that the solar pond made of porous materials has more stability than the normal one. M.M. Ould Dah et al [6] has fabricated a mini solar pond of $0.64m^2$ and the performance and stability of it was carried out experimentally and numerically. In the experimental study it was observed to reach 54°C quickly after 20 days. G.S. Gunasegarane [7] has found out that for a monthly average solar radiation was 180 −270 W/m2 available, the optimum thickness of the three zones were determined from the study UCZ, NCZ and LCZ values were 0.2m, 1.3m and 0.9m respectively. These optimum values were determined for the maximum collection efficiency of the solar pond whose values are in the range of 0.22 to 0.17 when the pond is operated with and without heat extraction. Sathish Dhandapani et al. [8] in Tamilnadu has fabricated a portable solar pond with common salt (NaCl) and pebbles and investigated experimentally. D. Sathish et al [9] has built a portable solar pond with an area of the trapezoid cross-section pond about 1.7 m^2 and having a depth of 0.5 m. The Investigation has resulted that the maximum average storage temperature of NaCl salt was observed as 43°C and for coal cinder was 37°C. NaCl salt showed a better storage performance in comparison with the coal cinder one as more energy is stored due to presence of salinity profile. Jiansheng wang et al [10] constructed a mini trapezoidal shaped solar pond and analysed its energy efficiency at different zones. It was found that the energy efficiency is maximum at the lower convective zone compared to other zones. Franscisco Suarez et al [11] has fabricated and investigated a small scale solar pond and this has revealed that the results can be used to investigate the experiment performance of large scale solar pond and main issues that can decrease the thermal performance of the system . Ahmad Aizaz et al [12] has designed and fabricated a small scale solar pond. The experimental analysis was carried out and revealed that this technology can be used to meet the energy requirements of Pakistan. Jobanputra Deepak P et al [13] has designed and developed a solar pond and validated its performance as a laboratory model. Qi Wu et al [14] has investigated the solar pond and found that the heat storage capacity increases the heat extraction efficiency. With the experimental results obtained the heat extraction efficiency is observed to be 40.3% more. Gautam Saini et al [15] has experimentally investigated the effect of a translucent separator (glass) above LCZ which decreases the diffusion of salt from LCZ to NCZ and maintain the stability of salt gradient for a longer period of time. When there was no glass above LCZ then the gradient of salt start diminishing after one week and heat loss start increasing rapidly due to convection current. But if using a transparent 5 mm glass above LCZ than rate of diminishing salt gradient was very less and temperature maintain for a longer time period. The salt concentration in top of UCZ due to salt diffusion after 15 days was 8% without using glass and 4% with using glass. So this is a significant difference in salinity by using a glass above LCZ for maintaining the salt gradient for longer period of time. Ni Ni Aung et al [16] has investigated the research work numerically and experimentally with insulated solar pond throughout day and night time. The new setup was performed on a SGSP to be used for desalination. The data for daily temperature distributions, solar radiation and ambient temperature is collected from 16th of January until 16th of March 2017. The maximum storage temperature reached 305.55 K after two months of the filling process. Mohammad Reza assari et al [17] has fabricated two like solar ponds with circular cross sectional area $0.54m^2$ and two layers of pebbles and ball bearings as permeable medium are used. At the end of the experiment, the temperatures of LCZ of mixed medium and conventional are found out to be 75°C and 71°C. L.C. Ding et al [18] has investigated on a concept of passive electric power generation by using temperature gradient existing between LCZ and UCZ by using thermo electric devices which convert the heat in the lower convective zone to equivalent amount of electricity. Najam-ul-Hassan Shah et al [19] investigated the solar pond at lower convection zone for heat extraction. MATLAB code was used for numerical simulation. Finally, it was found to have more efficiency. Assad H. Sayer [20] has investigated a theoretical model of a solar pond considering upper, lower and non-convective zones. The temperatures at the upper and lower zones are considered. The result was validated against the already established ponds.

2. Development of Model

A salt gradient solar pond is a modified form of ordinary solar pond by creating gradient of salt to suppressing the convection current. It is a cheaper method for utilizing the solar energy without creating any pollution. There are three zones in a salt gradient solar pond as lower convection zone (LCZ), non-convection zone (NCZ), and upper convection zone (UCZ). LCZ is the lower zone of solar pond with highest concentration and used to store the heat energy from solar radiation. NCZ is the heart of solar pond because it suppresses the convection current due to the existence of the density gradient and helps out to store thermal energy. UCZ is consisting of very small salt concentration or fresh water. UCZ covers the NCZ and protect from external disturbance such as wind, dust, raindrop etc. Heat is stored in the LCZ and useful heat can be extracted by using the different types of heat exchanger. The working model of the solar pond of rectangular basin having dimension of (0.9×0.6) m^2 in cross section and 0.4m in deep has been developed by conventional carpentry process with appropriate design specification. The entire body structure of the model was used commercially used plywood except top surface. The interior of it was covered with galvanized sheet painted with black paint to enhance the absorptivity and proper attention was focused to make it leak proof. The inlet section of it was made of a poly vinyl pipe (PVC) of 12mm diameter to pour in the brine solution and covered a cap. Furthermore, a heat exchanger made up of copper tube of 5mm diameter in helical coil was rested at the lower zone of the solar pond which acted as an internal heat exchanger. The diameter of helical coil was of diameter 100mm. A 4mm Transparent Glass is used to place over the solar pond set up which decreases the diffusion of salt from LCZ to NCZ and maintain the stability of salt gradient for a longer period of time. To measure the incident solar radiation flux density a solar power meter was used. To observe the variation of solar radiation affected by air flow velocity a hot wire anemometer was adopted. The complete specifications of the experimental set up is given in table.1.

Table 1 Complete Specifications of investigational setup

S. No	Component	Items	Technical Specification
01	Basin	Plywood	(2000×1700×25) mm
		Galvanised Iron Sheet	1mm
		Glass	(900×600×4) mm
		Salt	14 kg
		paint	200 ml
		PVC pipe	12 mm
		Copper Pipe	5 mm
02	Solar Power meter	Sensor	High sensitivity silicon photodiode
		Spectral Response	400~1100nm
		Range	0~2000 W/m^2(0~634 $BTU/ft^2.h$)
		Accuracy (at 23°C, 60% RH)	±10W/m^2(±3$BTU/ft^2.h$) or ±5% whichever is greater
		Resolution	0.0~99.99 W/m^2:0.01 W/m^2
		Angular Accuracy	Cosine corrected < 7% (angle <60°)
		Tilt Angle Range	0°- 90°
		Sample Time	Approx. 0.4 second
		Operation Temp. & Relative Humidity	0°C~50°C (32°F~122°F), Less than 80% RH

03	Hot Wire Anemometer	Air Velocity Range	0.1-0.25 m/s
		Air Velocity Resolution	0.01 m/s
		Air Velocity Accuracy	±(5%+0.1m/s)
04	Temperature Sensor	Temperature	50~110°C
		Temperature measuremment accuracy	±1°C
		Brand	Ap Techdeals

3. Experimental Procedure

The primary objective of the research work is:

- To trap the thermal energy out of the incident solar irradiation.
- To extract the heat energy from the lower storage zone by using an internal heat exchanger
- To investigate the thermal performance of the system based on climatic changes

The real experimental set up is shown in figure 1. Similarly the solar power meter and the hot wire anemometer are shown in figure 2 and 3 respectively. The system consists of internal heat exchanger (IHE) embedded in the LCZ through which the fluid is circulated through pumping. Digital temperature sensors are used to measure the upper and lower convective zone temperatures.

Figure 1 Real Picture Captured during the preparation of Brine Solution

IOP Conf. Series: Earth and Environmental Science **312** (2019) 012019 doi:10.1088/1755-1315/312/1/012019

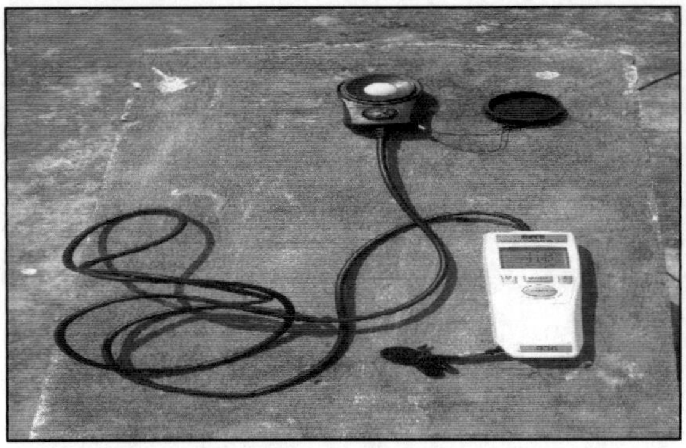

Figure 2 Real Photo of Solar Power Meter captured during experimentation

Figure 3 Real Photo of Hot Wire Anemometer captured during experimentation

The location and positioning of solar pond is very important for carrying out the experiments. The location with maximum exposure to sun radiation is selected. The experimental salt gradient solar pond set up is placed and experimentation is carried on the terrace at Mechanical Engineering Department in Raghu Engineering College, Visakhapatnam city which is situated on south east coastal part of India with a latitude and longitude of 17.68° N and 83.21° E where maximum sun exposure and water supply are easily available. The salt gradient solar pond is made up of rectangular shape with dimensions 0.9m by 0.6m with a depth of 0.4 m. The brine solution is prepared with 80 litres of water and 14 kg of common salt worked as thermal storage medium. The salt concentration was maintained 25% in lower zone consisted of 40 liters of water. A PVC pipe of 12mm diameter is used to add the brine solution of lesser concentration to the upper zones so that the brine solution in the lower storage zone doesn't get disturbed. The top of solar pond is covered by 4 mm thick transparent glass to prevent from the evaporation losses, wind disturbances and rainfalls. Experimental reading was collected daily from 9.30 am to 3 pm for 3 consecutive days starting from 21-02-2019 to 23-02-2019. A measurement system had been installed for recording the experimental data. Two digital temperature sensors with range of -50°C to +110°C temperature range and ±0.1°C accuracy were used to measure the upper zone air temperature and lower zone brine solution temperature for the required duration of pond operation. As we studied from literature review that more heat gets trapped in the lower zone, the temperature sensor is placed in the lower zone and only the mass of lower zone brine solution is considered during mathematical calculation and analysis. Two other temperature sensors are also used: Firstly, to measure the atmospheric air temperature and the other sensor placed on the pipe outside to measure the amount of heat extraction from the solar pond. Solar power meter was used to find the intensity of solar radiation of pond's surface. Hot wire anemometer is used to measure the wind velocity from the surrounding.

4. Result and Analysis

The day wise analysis and recorded data is given in this section.

IOP Conf. Series: Earth and Environmental Science **312** (2019) 012019 doi:10.1088/1755-1315/312/1/012019

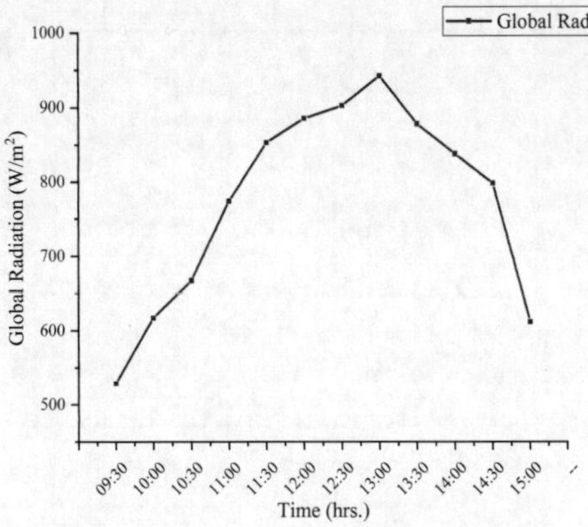

Figure 4(a) Variations of Global radiation with respect to time on 21.02.2019

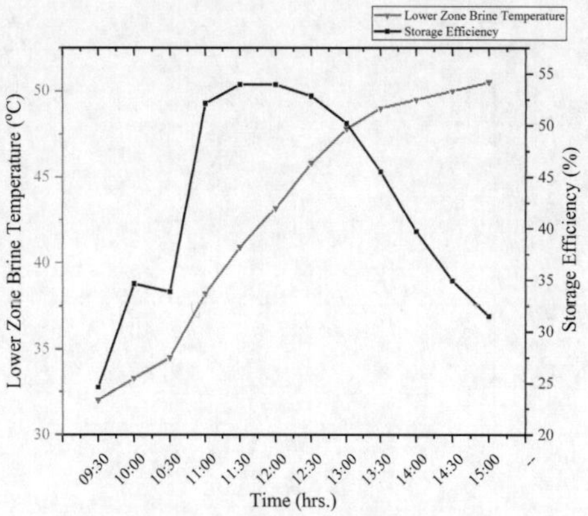

Figure 4(b) Variations of Lower Zone Temperature and Storage Efficiency with respect to time on 21.02.2019

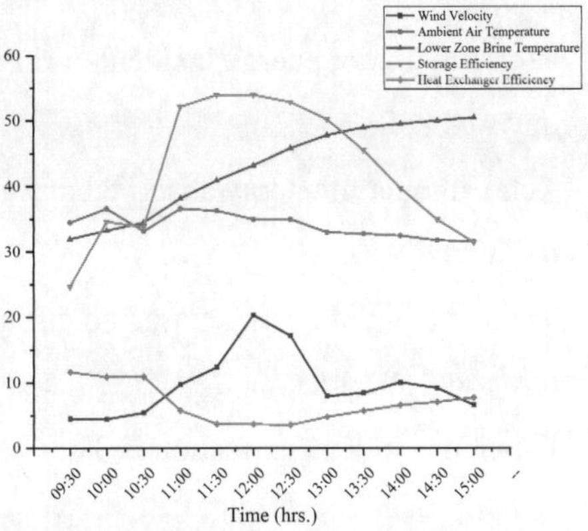

Figure 4(c) Performance analysis with respect to time on 21.02.2019

- Total amount of energy incident during the day was 9037.3 kJ
- Total amount of energy absorbed during the day was 4025.1 kJ
- Total amount of energy extracted during the day was 1363.0 kJ
- Efficiency of heat extraction is 33.86%

It has been observed from the figure 4(a), the Solar Radiation was increased from 9:30 am until it reached a maximum radiation of 942.5 W/m^2 at 1:00 pm, at which lower zone brine solution temperature is 47.8°C and then it started decreasing gradually.

From the above figure 4(b), we can get to a conclusion that the temperature of the lower zone brine solution kept on increasing till it reached a maximum temperature of 50.5°C at 3:00 pm because the energy is being incident on the system continuously.

As seen from the figure 4(c), the storage efficiency was maximum between 11:00 am and 12:00 pm and there after the storage efficiency goes on decreasing because the amount of heat incident decreased gradually after 12:00 pm. The ambient air temperature has started to decrease as soon as the wind velocity started to increase.

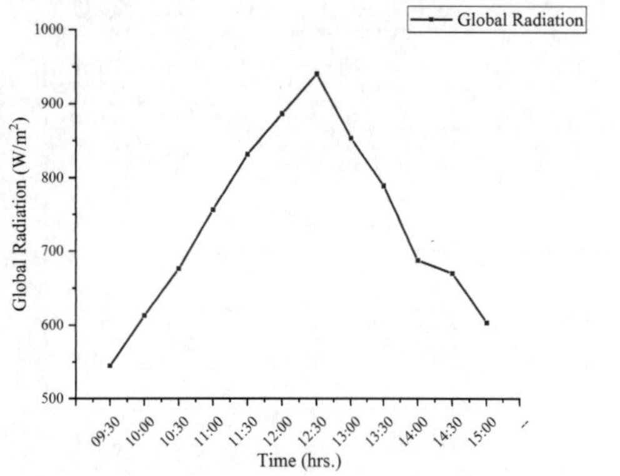

Figure 5(a) Variations of Global radiation with respect to time on 22.02.2019

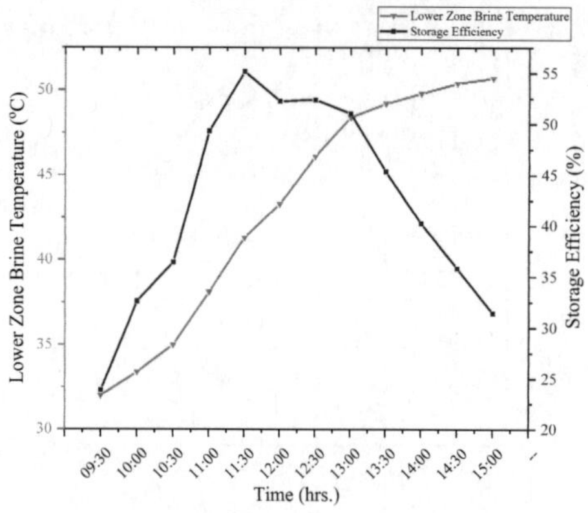

Figure 5(b) Variations of Lower Zone Temperature and Storage Efficiency with respect to time on 22.02.2019

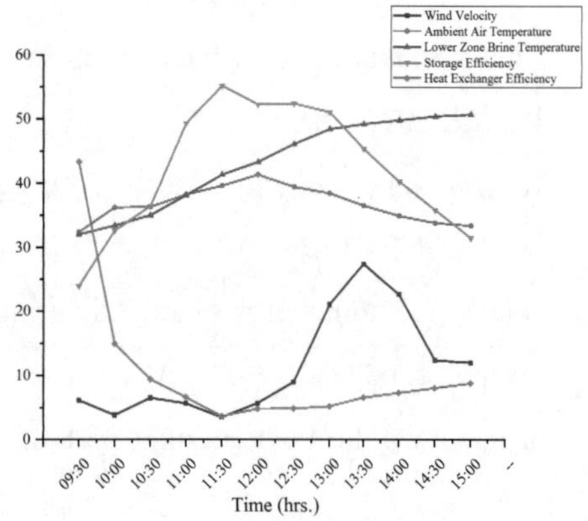

Figure 5(c) Performance analysis with respect to time on 22.02.2019

- Total amount of energy incident during the day was 8603.6 kJ
- Total amount of energy absorbed during the day was 4067.3 kJ
- Total amount of energy extracted during the day was 1601.1 kJ
- Efficiency of heat extraction is 39.36%

It has been observed from the figure 5(a), the Solar Radiation was increased from 9:30 am until it reached a maximum radiation of 940.1 W/m^2 at 12:30 pm, at which lower zone brine solution temperature is 46.1°C and then it started decreasing gradually.

From the above figure 5(b), we can get to a conclusion that the temperature of the lower zone brine solution kept on increasing till it reached a maximum temperature of 50.7°C at 3:00 pm because the energy is being incident on the system continuously.

As seen from the above figure 5(c), the storage efficiency was maximum between 11:00 am and 1:00 pm as the average maximum global radiation (940W/m^2) was observed during that time. The ambient air temperature was found to be maximum i.e. 41.3°C and then it started decreasing corresponding to increase in wind velocity.

IOP Conf. Series: Earth and Environmental Science **312** (2019) 012019 doi:10.1088/1755-1315/312/1/012019

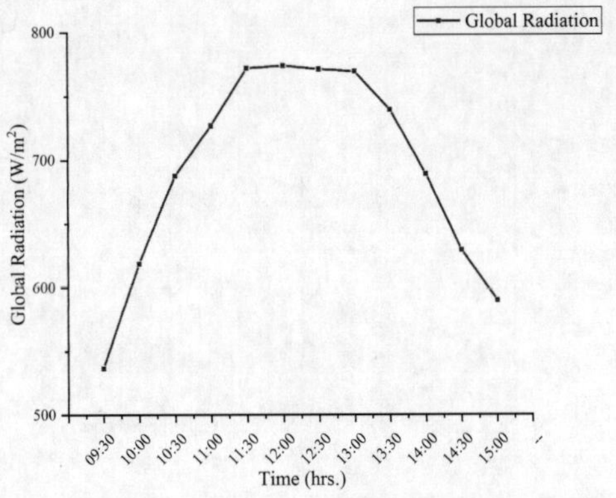

Figure 6(a) Variations of Global radiation with respect to time on 23.02.2019

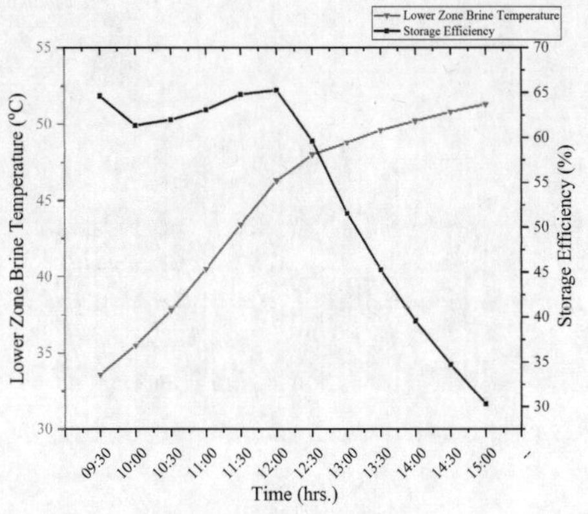

Figure 6(b) Variations of Lower Zone Temperature and Storage Efficiency with respect to time on 23.02.2019

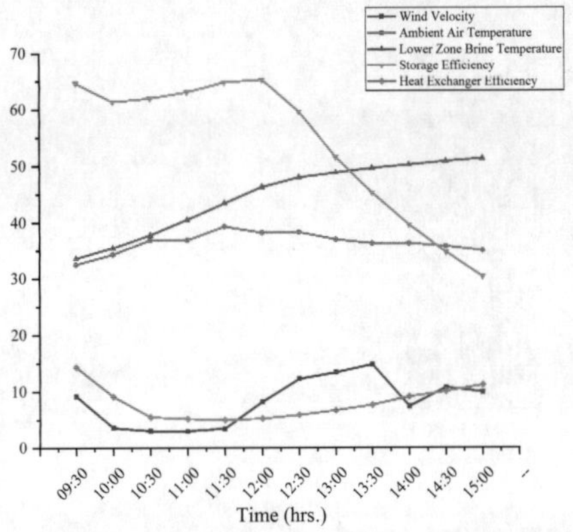

Figure 6(c) Performance analysis with respect to time on 23.02.2019

- Total amount of energy incident during the day was 8072.2 kJ
- Total amount of energy absorbed during the day was 4025.1 kJ
- Total amount of energy extracted during the day was 1851.8 kJ
- Efficiency of heat extraction is 46%

It has been observed from the figure 6(a), the Solar Radiation was increased from 9:30 am until it reached a maximum radiation of 773.9 W/m^2 at 12:00 pm, at which lower zone brine solution temperature is 46.3°C and then it started decreasing gradually.

From the above figure 6(b), we can get to a conclusion that the temperature of the lower zone brine solution kept on increasing till it reached a maximum temperature of 51.3°C at 3:00 pm because the energy is being incident on the system continuously.

As seen from the figure 6(c), the maximum storage efficiency was found to be maximum at 12:00 pm and a gradual decrease in storage efficiency is recorded since more amount of heat is extracted during the final hours of the experimentation and a corresponding decrease in global radiation was noted. The minimum wind velocity during the experimentation days (3.024 km/hr) was noted on the same day at 11:00 am.

IOP Conf. Series: Earth and Environmental Science **312** (2019) 012019 doi:10.1088/1755-1315/312/1/012019

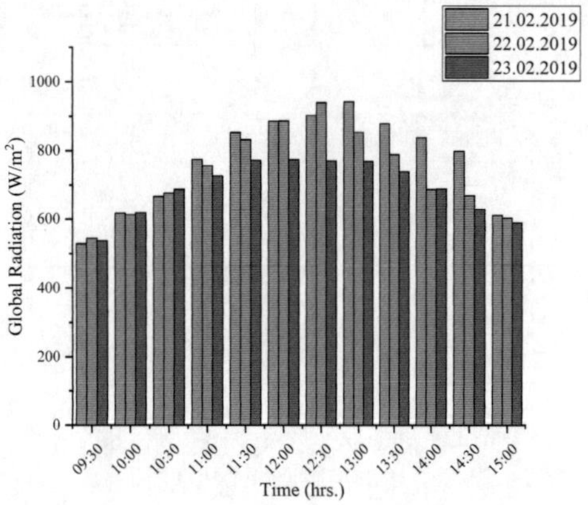

Figure 7(a) Variations of Global radiation

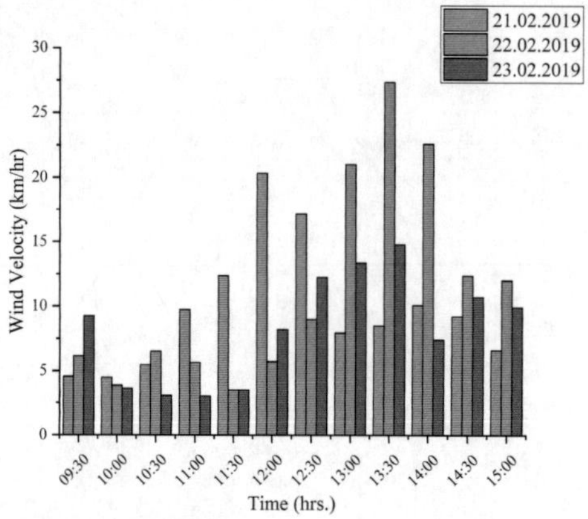

Figure 7(b) Variations Air Velocity

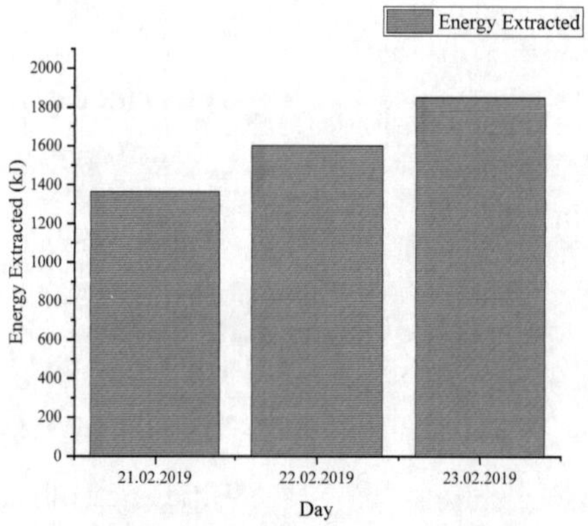

Figure 7(c) Variations of Energy Extracted

As seen from the above graph, the maximum average global radiation of 942.5 (W/m^2) was recorded on February 21, 2019 at 1:00 pm. The minimum average global radiation of 528.7(W/m^2) was recorded on February 21, 2019 at 9:30 pm. As seen from the figure 7(b), the maximum Wind Velocity was recorded on February 22, 2019 with 27.36 km/hr at 1:30 pm. The minimum wind velocity of 3.024 km/hr was recorded on February 23, 2019 at 11:00 am.

From the figure 7(c) it is clear that the amount of Heat Extraction is more on 23 February as compared to previous days because the temperature of the lower zone brine solution increased at a faster rate since there are no fluctuations in the climate on 23, February compared to day 1(i.e. 21 February), day 2(i.e. 22 February) where we had witnessed sudden changes like cloudy weather during the initial hours of the day.

5. Conclusion

From the Experimentation procedure, the results obtained are as follows:

- The maximum solar radiation 942.5 (W/m^2) was recorded on 21, February 2019 at 1:00 pm and the minimum solar radiation of 528.7(W/m^2) was recorded on the same day 21, February 2019 at 9:30 pm
- The maximum wind velocity was recorded on February 22, 2019 with 27.36 km/hr at 1:30 pm and The Minimum Wind Velocity of 3.024 km/hr was recorded on February 23, 2019 at 11:00 am.

- The maximum lower zone brine solution of was recorded on the last day of experimentation i.e. 23, February 2019 at 3:00 pm.
- The maximum storage efficiency was recorded on 23, February 2019 at 12:00 pm and minimum storage efficiency was recorded on 22, February 2019 at 9.30 am
- The amount of heat extraction was comparatively more on 23, February 2019 compared to previous days.
- The storage efficiency of the system after reaching its maximum value, undergoes a gradual decrease even though energy is being incident on it. This is due to the reason that the system here is an open system and cumulative value of heat incident and heat stored were taken into consideration during experimental calculation and analysis.
- The amount of heat extraction can be improved by increasing the number of heat exchangers in the set up.

References

[1] Nalan B C and Arzu Ş Ş 2011 Thermal efficiency for each zone of a solar pond *Chinese Phys. Lett.* **28** 1–3

[2] Sarathkumar P, Sivaram A R, Rajavel R, Praveen Kumar R and Krishnakumar S K 2017 Experimental Investigations on the Performance of A Solar Pond by using Encapsulated Pcm with Nanoparticles *Mater. Today Proc.* **4** 2314–22

[3] Leblanc J, Akbarzadeh A, Andrews J, Lu H and Golding P 2011 Heat extraction methods from salinity-gradient solar ponds and introduction of a novel system of heat extraction for improved efficiency *Sol. Energy* **85** 3103–42

[4] Ruskowitz J A, Suárez F, Tyler S W and Childress A E 2014 Evaporation suppression and solar energy collection in a salt-gradient solar pond *Sol. Energy* **99** 36–46

[5] Hill A A and Carr M 2013 Stabilising solar ponds by utilising porous materials *Adv. Water Resour.* **60** 1–6

[6] Ould Dah M M, Ouni M, Guizani A and Belghith A 2010 The influence of the heat extraction mode on the performance and stability of a mini solar pond *Appl. Energy* **87** 3005–10

[7] Gunasegarane G S 2015 Parametric-design-study-of-500m2-Salt-Gradient-Solar-Pond.doc *Int. J. Sci. Eng. Res.* **6** 1387–92

[8] Sathish D, Veeramanikandan M, Thirunavukkarasu R, Tamilselvan R and Karthickmunisamy T 2018 Thermal Performance on Portable Mini Solar Pond Using NaCl and Coal Cinder *Appl. Mech. Mater.* **877** 430–5

[9] Sathish D, Veeramanikandan M and Tamilselvan R 2017 Experimental Study on Portable Mini Solar Pond Using Salt and Pebbles *Int J Chem Sci* **15** 2–7

[10] Wu D, Liu H, Jiang L and Wang J 2017 Energy efficiency analysis of a trapezoidal solar pond *IOP Conf. Ser. Earth Environ. Sci.* **100**

[11] Suárez F, Ruskowitz J A, Childress A E and Tyler S W 2014 Understanding the expected performance of large-scale solar ponds from laboratory-scale observations and numerical modeling *Appl. Energy* **117** 1–10

[12] Aizaz A and Yousaf R 2013 Construction and Analysis of a Salt Gradient Solar Pond for Hot Water Supply *Eur. Sci. J.* **7881** 517–29

[13] P J D 2015 Designing of Portable Solar Pond for Improved Efficiency *Int. J. Sci. Technol. Eng.* **1** 426–30

[14] Wu Q, Wang H, Xie S, Zhang L, Wang J, Dong Z and Zhao T 2018 Effect of heat extraction on the thermal efficiency of salt gradient solar pond *Energy Explor. Exploit.* 1–14

[15] saini G, agrawal A, Varshney M and Rohatgi N 2017 Experimental analysis of Salt Gradient Solar Pond with and without using a transparent separator (Glass) above lower convective zone *Int. J. Mech. Eng.* **4** 5–10

[16] Aung N N, Soe M M and Mon Mon Aye 2017 Experimental and numerical study on

IOP Conf. Series: Earth and Environmental Science **312** (2019) 012019 doi:10.1088/1755-1315/312/1/012019

temperature distribution of a Salt gradient Solar Pond *Int. J. Mech. Prod. Eng.* **5** 30–5

[17] Assari M R, Basirat Tabrizi H, Parvar M, Kavoosi Nejad A and Jafar Gholi Beik A 2017 Experiment and optimization of mixed medium effect on small-scale salt gradient solar pond *Sol. Energy* **151** 102–9

[18] Ding L C, Akbarzadeh A, Date A and Frawley D J 2016 Passive small scale electric power generation using thermoelectric cells in solar pond *Energy* **117** 149–65

[19] Shah N, Arshad A, Khosa A A and Ali H 2017 Thermal analysis of a mini solar pond of small surface area while extracting heat from lower convective layer *Therm. Sci.* 1–10

[20] Sayer A H, Al-Hussaini H and Campbell A N 2016 New theoretical modelling of heat transfer in solar ponds *Sol. Energy* **125** 207–18

Numerical study of natural convection in flat receiver with and without secondary reflector for solar parabolic dish system

Sendhil kumar Natarajan*, B Raviteja, D Sri Harshavardhan, G Hari Prasath, Subbarama Kousik Suraparaju

Department of Mechanical Engineering,
National Institute of Technology – Puducherry, Karaikal 609609
U.T of Puducherry, INDIA.

Email: ***drsendhil1980iitmuk@gmail.com***

Abstract. The sustainability of our eco system is under threat due to frequent use of the fossil fuels. Due to the rapid depletion of the conventional energy sources and oil crisis, there is a great shift towards renewable energy sources. In the realm of renewable sources of energy, solar energy plays an impeccable role because of its abundance and dominance over the other sources. Therefore, efforts are being made to replace our existing conventional fuels with renewable energy sources. In this context, flow simulation and heat transfer are performed for the flat receiver with and without secondary reflector of solar parabolic dish collector is presented. A two-dimensional (2-D) model is used to investigate the approximate estimation of the natural convection heat loss from the actual geometry of the receiver for solar dish concentrator system. The Nusselt number correlations are proposed separately and compared for both receiver with secondary reflector (WSR) and without secondary reflector (WOSR) using power law and asymptotic computational fluid dynamics technique.

Keywords: Numerical analysis; Flat receiver; Secondary Reflector; Solar Parabolic Dish Collector; Nusselt number; Natural convection; ACFD.

1. Introduction

The issue of renewable energy is becoming significant due to increase in power demand instability of the oil crisis and environmental problems. Among all the available renewable energy sources, solar energy is the most abundant, dominant, powerful and efficient energy source. Solar energy can be utilized for the applications of heat as well as for the generation of electricity. Concentrating solar system is the current trending approach for co-generation of heat and electricity. In this regard, Reddy et al. [1] examined the thermal performance of $20m^2$ fuzzy solar dish collector and determined the solar dish focal image characteristics to propose the appropriate receiver design. Initially, the theoretical investigations on the dish with customized cavity receiver were performed at different operating conditions and estimated the total heat loss. Later, the real time investigation on the dish with customized cavity receiver was performed based on time constant, stagnation test and daily performance test. The investigation resulted that the efficiency was augmented with rise of volume flow rate. Sendhil Kumar et al. [2] numerically studied the natural convective heat loss for three different receivers of a solar parabolic dish concentrator system. Among those three receivers; the customized cavity receiver was the suggested

IOP Conf. Series: Earth and Environmental Science **312** (2019) 012020 doi:10.1088/1755-1315/312/1/012020

receiver for a solar dish collector system of fuzzy focal type. Ravi Kumar et al. [3] numerically analysed the 3-D porous line receiver for solar parabolic trough concentrator system and also investigated the effect of receiver design, thermal fluid properties and solar radiation on total heal collection by using renormalization-group (RNG) **k-ε** turbulent model with Therminol-VP1 as working fluid. The parametric investigation on the receiver resulted that the usage od porous medium in tubular receiver increases the system performance of system. Sendhil Kumar et al. [4] numerically studied the solar radiation and natural convention heat transfer in a cavity of trapezoidal shape absorber for Compact Linear Fresnel Reflector by using CFD FLUENT 6.3 and related with other cavity of closed model for validation. The simulation outcome was presented as correlation of Nusselt number to depict the consequence of the parameters on surface radiation and Natural convection. Reddy et al. [5,6] numerically studied about the 2-D surface radiation heat transfer and combined natural convection in a customized cavity receiver of parabolic dish concentrator system for investigating the consequence of the operating temperature, orientation, surface emissivity and the geometry on the overall heat loss originating at receiver and resulted that the convective heat loss was affected by receiver inclination whereas the radiation heat loss was affected by surface properties. The outcome of the current analysis was in good accord with models of open cavity. Sendhil Kumar et al. [7] developed and evaluated the performance of double axis tracking system for solar parabolic dish system. A 1-Watt photovoltaic panel was kept at focal point to evaluate the open circuit and short circuit voltage and resulted that the recommended system augmented short circuit current of 0.11A by 86.0%. Henceforth, it was concluded that the proposed system can be used only in parabolic system with concentrating photovoltaic module as a focal point for future researches. Reddy et al. [8] numerically studied the performance of the solar parabolic trough concentrator receiver by considering the various porous receiver geometries. The heat transfer and internal flow analysis was performed based on an RNG k-ε turbulent model, whereas foreign heat losses are considered as a laminar natural convection model and was solved using FLUENT. The results of the current analysis were in good accord with all other open cavity models and a comparative study was depicted for several porous geometries to get an optimum energy efficient receiver configuration. Sendhil Kumar et al. [9] numerically analyzed the 3-D solar dish customized cavity receiver with CPC, Trumpet and Cone reflectors for convective and radiative heat loss against the receiver for several operating temperatures and angles of inclination. The results depicted that trumpet shaped reflector on the receiver has given better performance when compared with the remaining designs. Sendhil Kumar et al. [10] numerically investigated the 2-D customized cavity receiver of solar parabolic dish concentrator system to evaluate the natural convection heat loss for both designs ''without insulation'' (WOI) and ''with insulation'' (WI) at the receiver and by varying inclination angles. The maximum heat loss through convection was observed during 0° inclination for both without and with insulation cases. The results of the current analysis were in good accord with other cavity models of open type. Sendhil Kumar et al. [11] numerically studied natural convection heat loss and radiation from a customized cavity receiver of solar parabolic dish of different inclination angles by using a trending approach i.e., Asymptotic Computational Fluid Dynamics (ACFD). individual Nusselt number correlations for heat loss through natural convection and combined heat loss through radiation and convection were given. It was noticed that the current heat loss model adheres to the same trend as of the other heat loss models. In this current article, the ACFD approach is utilized to advance a model with steady-state laminar natural convective correlation of Nusselt number for flat receiver without and with secondary reflector of a solar parabolic dish collector system.

2. Numerical modelling of flat receiver with and without secondary reflector

2.1. Mathematical Modelling

The receiver position is at the focal point of the parabolic dish collector. In real case the receiver is 3-dimensional in nature and it is approximated to 2-dimensional for simplicity. The receiver consists of water, which is considered to have identical temperature as of the surface. The simulations of heat transfer have been carried out depending on the concurrent outcome of system of the equations

IOP Conf. Series: Earth and Environmental Science **312** (2019) 012020 doi:10.1088/1755-1315/312/1/012020

portraying the conservation of energy, momentum and mass. The vector form of the energy equations momentum, and continuity.

1. Equation for Momentum:

$$q.\ \nabla q = X - \frac{\nabla P}{\rho} + \nu\nabla^2 q \tag{1}$$

2. Equation for Energy:

$$\nabla.\ (k_s\nabla T) = \rho\frac{DH}{Dt} - \frac{DP}{Dt} \tag{2}$$

3. Equation for Continuity:

$$\nabla \cdot q = 0 \tag{3}$$

The two-dimensional modelling and generation of grids have been carried out using ANSYS 19.1 R1 package. In real case, the flat receiver with and without secondary reflector are bounded by an infinite atmosphere. a rectangular enclosure (outer domain) is made such that the receiver with and without secondary reflector is placed centrally to attain such infinite atmosphere in numerical work. The dimensions of the rectangular arena should get enlarged until it has no significant outcome on the heat flow and operating fluid inside the receiver. The Grid independent analysis of the receiver is performed with and without secondary reflector on it, to acquire optimum cells. The region inside the cavity is given as fine grid and a coarse grid is maintained in the outer domain. At the end, overall number of optimum cells obtained are 47,816.

2.2. Boundary Conditions

The receiver's geometry had been set to several boundary conditions that prevails in the actual environment. The receiver's surface is constantly subjected to the solar insolation. Due to this, the surface area may achieve a constant isothermal state at the stagnation condition. Hence, the isothermal boundary condition has been applied to surfaces. The governing equations are based on energy conservation, momentum and continuity. The steady 2-Dimesional, state and laminar governing equations are elucidated using ANSYS FLUENT 19.1 by incorporating an implicit solver. For the velocity coupling and pressure, COUPLED algorithm was used. The discretization of energy and momentum had been performed using first order upwind scheme. The pressure was discretized by using body force weighted criteria. The criteria for minimum convergence were fixed at 10^{-6} for energy and 10^{-3} for velocity and continuity. To acquire the heat transfer and flow outcomes, the software commences the iteration till the principle of convergence is fulfilled. The working fluid properties are given based on the mean value of ambient air and receiver's surface temperature.

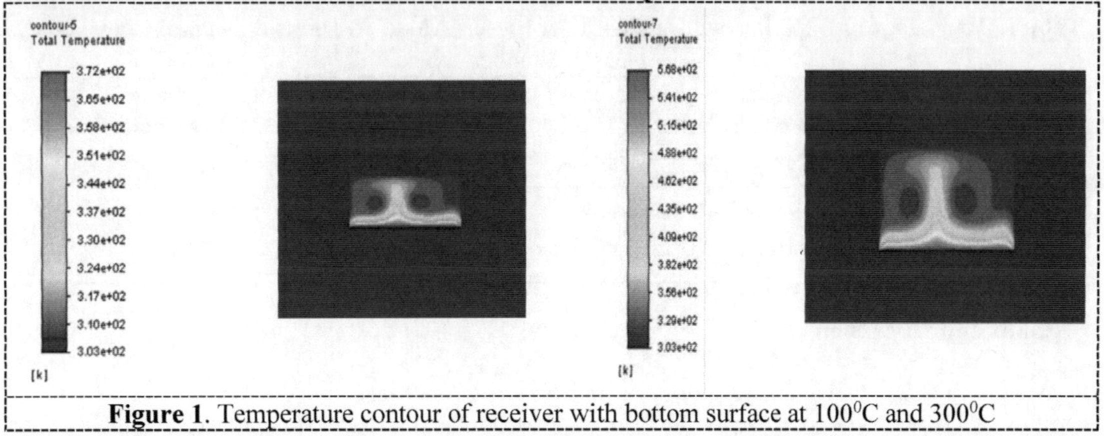

Figure 1. Temperature contour of receiver with bottom surface at 100^{0}C and 300^{0}C

For 100^{0}C, Grashof number (Gr)=1.5057×10^8, Rayleigh number (Ra)=1.5057×10^8, Nusselt number (Nu)=240.75, Heat transfer coefficient(h)=38.202 W/m^2-k and for 300^{0}C, Grashof number (Gr)=9.03×10^7, Rayleigh number (Ra)=6.15×10^7, Nusselt number (Nu)=217.65, Heat transfer coefficient(h)=44.015 W/m^2-k.

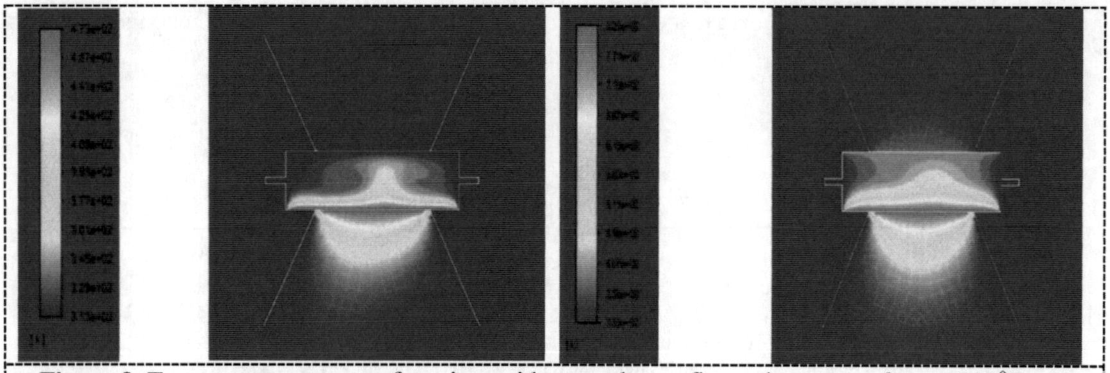

Figure 2. Temperature contour of receiver with secondary reflector bottom surface at 200^{0}C and 550^{0}C

For 220^{0}C, Grashof number (Gr)=1.262×10^8, Rayleigh number (Ra)=0.866×10^8, Nusselt number (Nu)=231.365, Heat transfer coefficient(h)=42.417 W/m^2-k and for 550^{0}C, Grashof number (Gr)=3.956×10^7, Rayleigh number (Ra)=2.66×10^7, Nusselt number (Nu)=139.931, Heat transfer coefficient(h)=35.216 W/m^2-k.

2.3. Validation of the natural convection model

The present geometry-Flat receiver with and without secondary reflector is relatively new and obscure. Therefore, the numerical simulation procedure was justified with natural convection model with a square cavity and also compared with the available literature. For validating the current numerical method of flat reciever, a classic case of natural convection in square cavity discussed by De Vahl Davis et al. [12] had been taken into consideration. The problem had been studied experimentally and numerically under laminar and steady flow with use of water. The outcomes of numerical and experimental investigation were expressed in terms of Rayleigh and Nusselt number correlations: Nu=2.56(Ra)$^{0.25}$ and Nu=2.83(Ra)$^{0.25}$ explicitly. With the help of this numerical method, the Nusselt number was estimated. The similarity between our current numerical study with De Vahl Davis's study is adorned in the table below. It is perceived that our numerical method is acceptable, when it is compared with available experimental data.

Table1. An analogy of Experimental Data with our Present Model in Terms of Natural Convective Nusselt Number

S.no	Rayleigh number (Ra)	Nusselt number (Nu) Nu_{Exp}	Nusselt number (Nu) Nu_{Num}	Our Present study Nu	Deviation in Percentage (%)
1	1.04×10^8	258.52	285.78	240.75	-6.87
2	4.57×10^7	210.48	232.68	197.98	-5.93

3. Results and Discussion

3.1. Nusselt number correlation developed by power law for flat receiver

By using the heat transfer coefficients values obtained from the numerical simulation Rayleigh number and Nusselt number are calculated for 22 data points. The 2-dimensional Natural convective Nusselt number relation for flat receiver is expressed as

$$Nu\ (WOSR)=2.0462(Ra)^{0.2586},\ 10^7\leq Ra\leq10^8$$

The above relation is obtained by fitting Nu vs Ra curve into power law in Axum software. The exactness of the fit is indicated by the R-square coefficient which is 0.986.

IOP Conf. Series: Earth and Environmental Science **312** (2019) 012020 doi:10.1088/1755-1315/312/1/012020

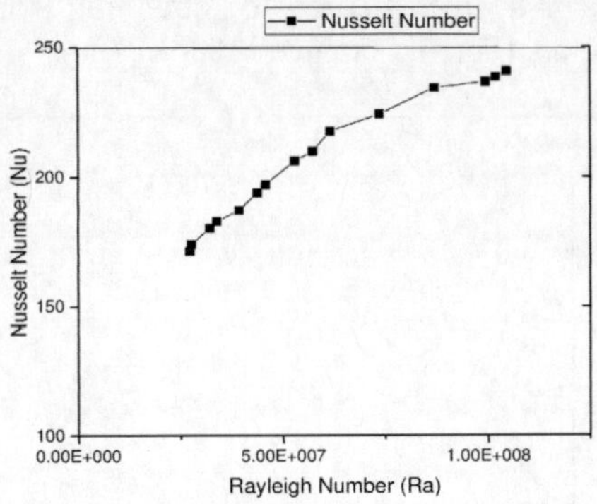

Figure 3. Nu vs Ra plot for flat receiver WOSR

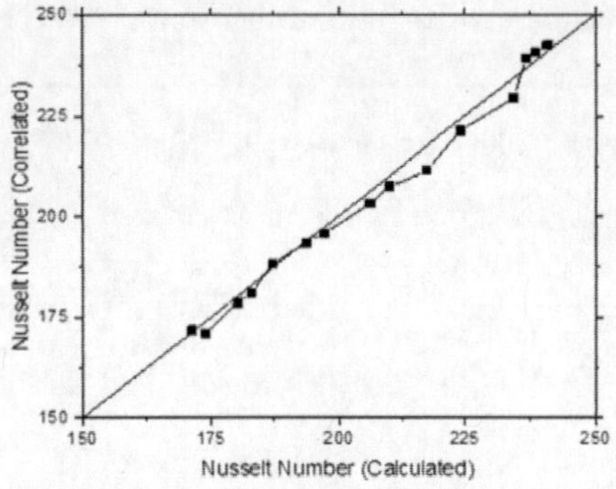

Figure 4. Uniformity plot of Nu for flat receiver WOSR

3.2. Nusselt number correlation developed by power law for flat receiver with secondary reflector

The 2-Dimensional Natural convective Nusselt number correlation for flat receiver with secondary reflector is expressed as

$$Nu\,(WSR)=0.1(Ra)^{0.42}\ ,10^{7}\leq Ra<10^{8} \qquad (4)$$

The above relation is obtained by fitting Nu vs Ra curve into power law in Axum software. The exactness of the fit is indicated by the R-square coefficient which is 0.984.

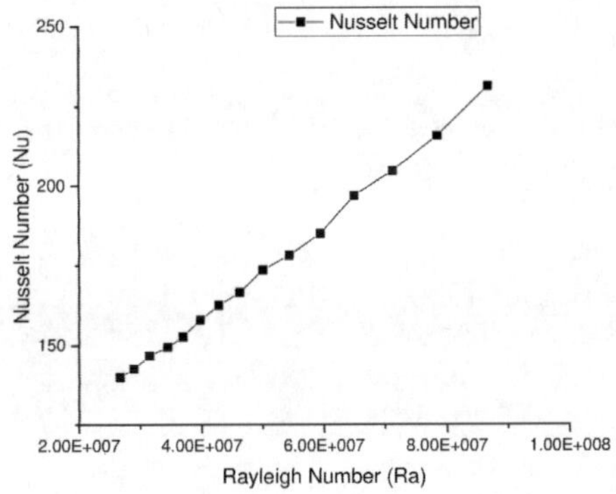

Figure 5. Nu vs Ra plot for flat receiver WSR

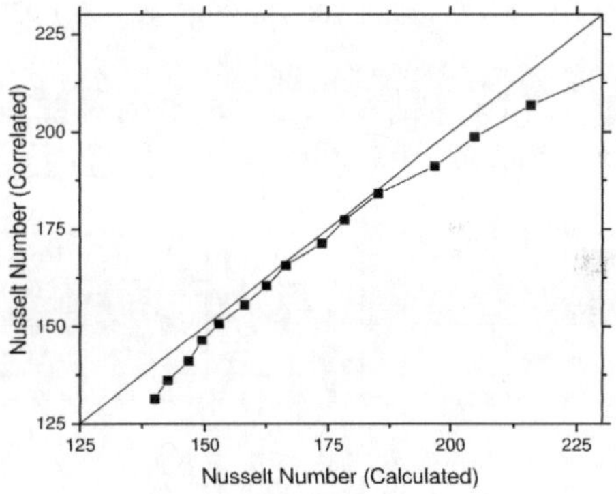

Figure 6. Uniformity plot of Nu for flat receiver WSR

3.3. Nusselt number correlation developed by Asymptotic Computational Fluid Dynamics Technique

In the problem with Natural Convection, the Nusselt number is a function of Rayleigh number.in accordance with the self-reliant variables, the conduct of natural convection in flat receiver can be designed by developing a Nusselt number correlation. The Natural Convective Nusselt number correlation can be designed by ACFD method. The usage of ACFD technique considerably decreases the number of data points required for attaining the authentic correlation. The effect of self-reliant variable is designed to figure out the correlation physical nature.

3.3.1. Nusselt number correlation for flat receiver: To evolve a correlation in ACFD technique, the median of a parameter was taken as reference to evolve correlations. Thus, $Ra_{ref}=5.92\times10^{7}$. in the present investigation, Rayleigh number is studied in a range of 10^{6} - 10^{7}. A linearized curve is achieved for the $(Ra/Ra_{ref})^{0.21}$ with respect to Rayleigh number. Obtaining the proper exponent of 0.21 is one of the eminent methods in the ACFD approach. It is a trial and error method for obtaining a linear relationship between ratio of the Ra.and Nu. The correlation coefficient obtained is 0.992.

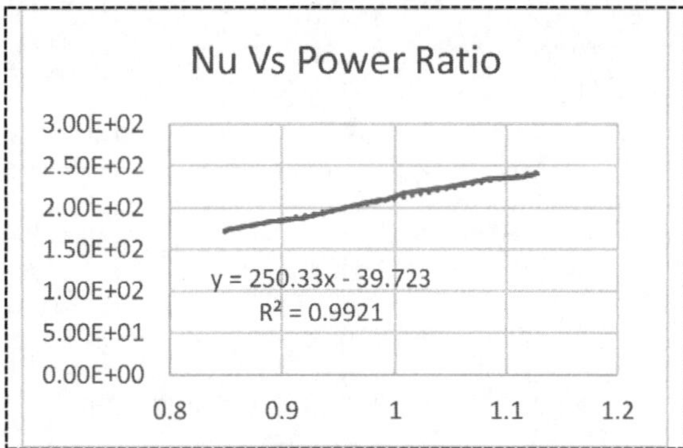

Figure 7. Nu vs Ra ratio raised to power 0.21 for flat receiver

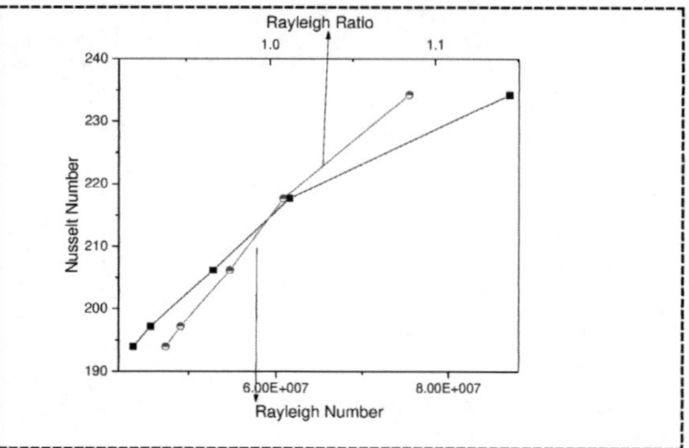

Figure 8. Deviation of Natural Convective Nusselt number with Rayleigh number of the flat receiver

The common form of Natural Convective Nusselt number correlation for flat receiver by ACFD technique is expressed as:

$$\mathbf{Nu=Nu_{ref}+(\emptyset_1-\emptyset_{1,\,ref})\frac{\partial Nu}{\partial \emptyset_1}} \tag{5}$$

Where $\emptyset_1=[Ra/Ra_{ref}]^{0.21}$

The coefficient $\frac{\partial Nu}{\partial \emptyset_1}$ is evaluated by achieving the result to the governing equations with all but one parameter fixed at the reference value.

The convective number correlation thus obtained for flat receiver is given by **Nu =214.5+($\emptyset_1$-1)247.** The correlation obtained gives us the acceptable results with minimum error.

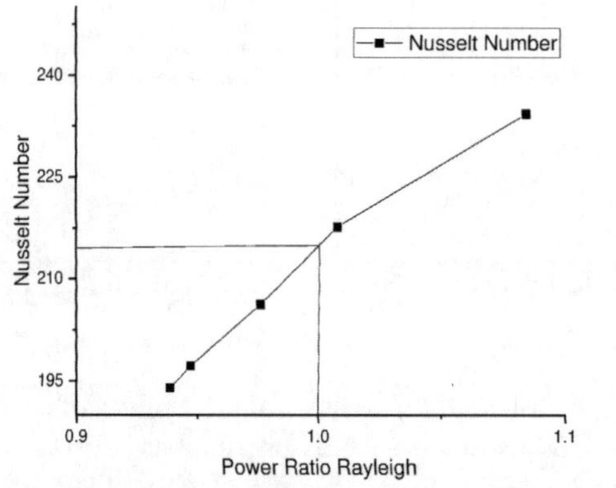

Figure 9. Deviation of Natural convective Nusselt number with self-reliant variable of the flat receiver

3.3.2 Nusselt number correlation by ACFD for flat receiver with secondary reflector: To evolve a correlation in ACFD technique, the median of a parameter was taken as reference to evolve correlations. Thus, Ra_{ref} =5.66×10^7. in the present investigation, Rayleigh number is studied in a range of 10^6 - 10^7. A linearized curve is achieved for the $(Ra/Ra_{ref})^{0.3}$ with respect to Rayleigh number. Obtaining the proper

IOP Conf. Series: Earth and Environmental Science **312** (2019) 012020 doi:10.1088/1755-1315/312/1/012020

exponent of 0.3 is one of the eminent methods in the ACFD approach. It is a trial and error method for obtaining a linear relationship between Nu and ratio of Ra. The correlation coefficient obtained is 0.977.

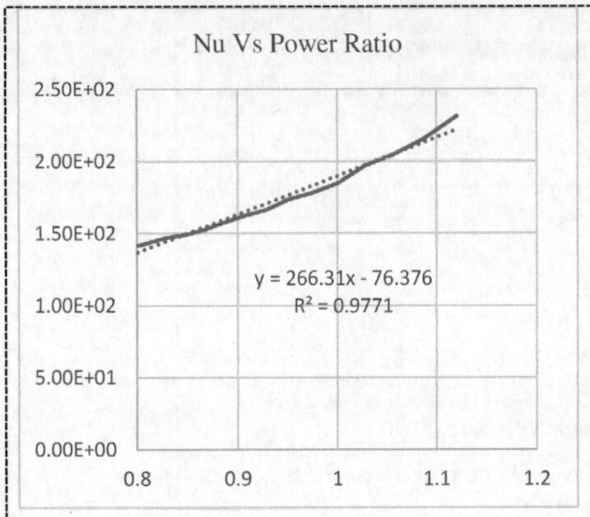

Figure 10. Nu vs Ra ratio raised to power 0.3 for flat receiver with secondary reflector

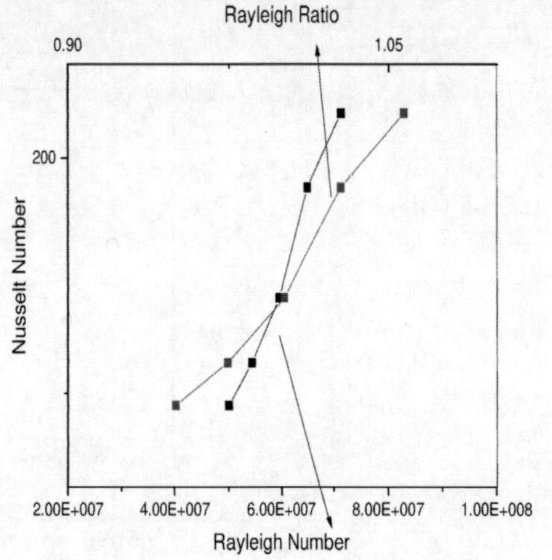

Figure 11. Deviation of Natural convective Nusselt number with Rayleigh number of flat receivers with secondary reflector

The general form of Natural Convective Nusselt number correlation for flat receiver is expressed as:

$$\mathbf{Nu=Nu_{ref}+(\emptyset_1-\emptyset_{1,\,ref})\frac{\partial Nu}{\partial \emptyset_1}} \qquad (6)$$

Where $\emptyset_1=[Ra/Ra_{ref}]^{0.3}$

The Natural convective number correlation thus obtained for flat receiver with secondary reflector is given by **Nu =184.5+(Ø₁-1)442.21.** The correlation obtained gives us the acceptable results with minimum error.

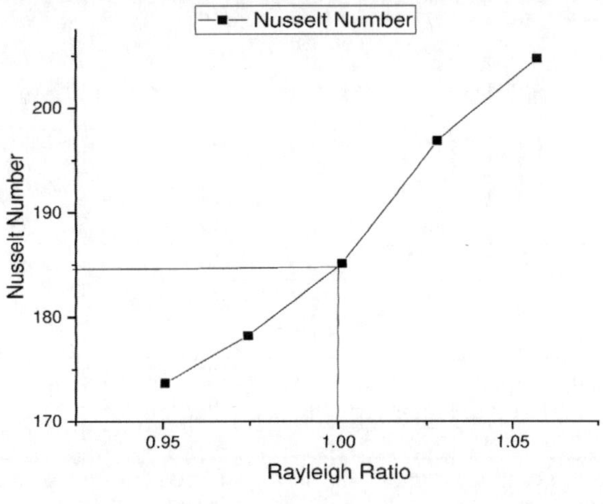

Figure 12. Deviation of Natural convective Nusselt number with self-reliant variable of flat receiver with secondary reflector

IOP Conf. Series: Earth and Environmental Science **312** (2019) 012020 doi:10.1088/1755-1315/312/1/012020

4. COMPARISON OF NUSSELT NUMBER CORRELATIONS

4.1. ACFD vs Power law Nusselt Number for flat plate receiver

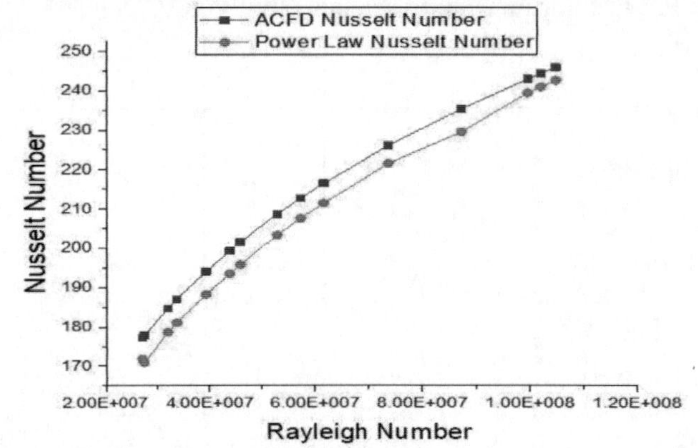

Figure 13. Comparison plot of ACFD and power law for flat plate receiver

In the above graph the Nusselt number varies from 171.21 to 240.75 for power law and from 127.08 to 240.75 for ACFD technique.

In the above graph the Rayleigh number varies from 2.72×10^7 to 1.046×10^8 for power law and ACFD technique.

It shows that convective Nusselt number obtained by adopting power law procedure and convective Nusselt number obtained by ACFD are comparable. The maximum difference between correlation developed by power law and ACFD Technique is obtained as 6.83% for both used and non-used data points.

4.2. ACFD vs Power law Nusselt Number for flat plate receiver with secondary reflector

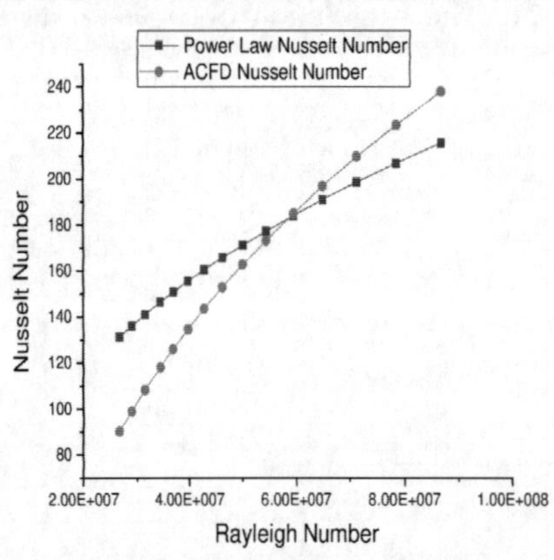

Figure 14. Comparison plots of ACFD and power law for flat plate receiver with secondary reflector

In the above graph the Nusselt number varies from 139.931 to 231.365 for power law and from 140 to 231 for ACFD technique. In the above graph the Rayleigh number varies from 2.67×10^7 to 8.66×10^7 for power law and ACFD technique. It shows that convective Nusselt number obtained by adopting power law procedure and convective Nusselt number obtained by ACFD are comparable. The maximum difference between correlation developed by power law and ACFD Technique is obtained as 9.71% for both non-used and used data points.

5. Conclusions

The results of 2-Dimensional numerical analysis were reported in terms of Nusselt number correlations for flat receiver without and with secondary reflector. The correlations are obtained by both power law and ACFD technique. It is found that the difference between the convective Nusselt numbers developed by power law and ACFD technique were found to be 6.83% for flat receiver and 9.71% for flat receiver with secondary reflector. It is also found that the number of data points needed were reduced considerably with the use of ACFD technique and also reasonably accurate Nusselt number correlation is obtained. The well-known benchmark model is compared with present numerical model. The correlation developed may effectively be used to measure the natural convective losses from the flat receiver with and without secondary reflector of a solar parabolic concentrator system.

Nomenclature

Ra	Rayleigh number
ø	independent variables
Gr	Grashof number
h	heat transfer coefficient (W/m^2-K)
T	Temperature (K)
t	Time (s)
ρ	density (Kg/m^3)
P	Pressure (N/m^2)
q	Energy flux (W/m^2)
Nu	Nusselt number

Subscripts

atm	Atmospheric pressure
Exp	Experimental
Num	Numerical
ref	Reference

References.

[1] Reddy K S, Natarajan S K and Veershetty G 2015 Experimental performance investigation of modified cavity receiver with fuzzy focal solar dish concentrator *Renew. Energy* **74** 148–57

[2] Sendhil Kumar N and Reddy K S 2008 Comparison of receivers for solar dish collector system *Energy Convers. Manag.* **49** 812–9

[3] Ravi Kumar K and Reddy K S 2009 Thermal analysis of solar parabolic trough with porous disc receiver *Appl. Energy* **86** 1804–12

[4] Natarajan S K, Reddy K S and Mallick T K 2012 Heat loss characteristics of trapezoidal cavity receiver for solar linear concentrating system *Appl. Energy* **93** 523–31

[5] Reddy K S and Sendhil Kumar N 2008 Combined laminar natural convection and surface radiation heat transfer in a modified cavity receiver of solar parabolic dish *Int. J. Therm. Sci.* **47** 1647–57

[6] Reddy K S and Sendhil Kumar N 2009 An improved model for natural convection heat loss from modified cavity receiver of solar dish concentrator *Sol. Energy* **83** 1884–92

[7] Natarajan S K, Thampi V, Shaw R, Kumar V S, Nandu R S, Jayan V, Rajagopalan N and Kandasamy R K 2019 Experimental analysis of a two-axis tracking system for solar parabolic dish collector *Int. J. Energy Res.* **43** 1012–8

[8] Reddy K S, Kumar K R and Satyanarayana G V. 2008 Numerical investigation of energy-

IOP Conf. Series: Earth and Environmental Science **312** (2019) 012020 doi:10.1088/1755-1315/312/1/012020

efficient receiver for solar parabolic trough concentrator *Heat Transf. Eng.* **29** 961–72
[9] Reddy K S and Sendhil Kumar N 2009 Convection and surface radiation heat losses from modified cavity receiver of solar parabolic dish collector with two-stage concentration *Heat Mass Transf.* **45** 363–73
[10] Sendhil Kumar N and Reddy K S 2007 Numerical investigation of natural convection heat loss in modified cavity receiver for fuzzy focal solar dish concentrator *Sol. Energy* **81** 846–55
[11] Kumar N S and Reddy K S 2010 Investigation of convection and radiation heat losses from modified cavity receiver of solar parabolic dish using asymptotic computational fluid dynamics *Heat Transf. Eng.* **31** 597–607
[12] De Vahl Davis G 1983 Natural convection of air in a square cavity: A benchmark solution *Int. J. Numer. Methods Fluids* **3** 249–64

Acknowledgement:
The authors would like to acknowledge the Department of Science and Technology, India for the financial support under Young Scientist Scheme DST-SERB PROJECT NO: YSS/2014/000748. Also, we would like to thank Director of NIT-Puducherry for providing the computational facilities for this project.

Experimental Investigation of Drying Potato for Karaikal Climatic Condition

Sendhil Kumar Natarajan* and Elavarasan E

Department of Mechanical Engineering, National Institute of Technology Puducherry, Karaikal – 609 609, U.T. of Puducherry, INDIA.
*Corresponding Author: Assistant Professor & Head

Email: sendhil80@nitpy.ac.in, drsendhil1980iitmuk@gmail.com

Abstract. Vegetables play an important and irreplaceable role in the human beings daily life and other living animals on the planet. Henceforth, wastage of vegetables should be kept away by developing techniques for conservation and preservation of vegetables to an extended period. In this article, a double slope solar dryer was designed and fabricated at the terrace of the science block, NIT Puducherry, Karaikal. The decrease in the wetness content of specimens in the double slope solar dryer was empirically inspected for Karaikal climatic conditions. The effect of double slope natural convective based solar dryer in drying of potato compared with open sun drying operation adjacent each other in similar condition. Experiments were carried out to determine the dryer surface temperatures, air temperature, and the percentages of the wet basis wetness content from 11.00 am to 6.00 pm for two days. In the investigation, it was clear that designed and fabricated double slope natural convective solar powered dryer removed the high wetness content about 52.29% and in open drying technique it removed wetness content of 42.99% from the potato slices on first day. On second day similar weighted samples were investigated wetness removed by solar dryer is 51% and 43.92% in open drying. So developed solar dryer removed high amount wetness content and quality of dried potato slices was high when compared to open sun dehydrating. Accordingly, the developed double slope natural convective based solar dryer can be effectively utilized to dry the vegetable samples.

Keywords: Experimental Studies, Double Slope Solar Dryer, Natural Convection, Open Sun Drying, Moisture Content.

1. Introduction

India secure fourth positions in area and generation of potato on the universe, In India out of the total potato delivered 17 percent is wasted, 0.03 percent is sent to other countries, 0.03 percent is prepared and 21.0 percent is utilized as seed. Out of the rest 61.94 percent is utilized for utilization as vegetable/other family activities [1]. Potato processing is a significant part of post-harvest the board and is a fast developing division on the world economy. In India, handling of potatoes constitutes less than 1.0 percent of total yearly generation and there is great scope for potato processing. Potato chip is the most widely recognized prepared potato item in India. Potato chip making includes washing, stripping, blanching, cutting and drying. Among these, drying is one of the unit activities consuming high amount of energy. The consequence of the techno- economic inspection of various drying choices demonstrates that sun powered drying is the least expensive [2].

Potato samples are generally dehydrated in simultaneous rotary dryers of 100 to 300 cm dia and 50 to 1500 cm length. The entry and exit air warmth are around 50 - 100° C. The crisp potato samples are set up by the slices like one utilized in sugar industry. The samples are dehydrated to a last wetness of twelve percent (w b). An important segment of the dehydrated potato sample is used as livestock feed since it is the most economical technique for protecting sample [3]. Potato slices are dried utilizing drum dryers. Before dehydrating, the samples experience cleaning, stripping, cutting, precooking, cooling, and mashing [4]. Mashed samples are then blended with a few added substances, to increase the texture and extend the shelf life of the samples; they additionally impact dehydrating procedure. Choy et al. [5] noticed that the general hue different in sample potato were least when exposed to a higher shifting dehydrated warmth. Other drying conditions were RH is 18.9%, 0.0087 kg/kg dehydrated air of absolute humidity, and air speed 2.4 m/s. Their rate decreases in general hue different for potato were 87 percent. Investigation recommended that period-different warmth dehydrating in a heat pump dryer operation has potential for dehydrating samples. Krokida et al. [6] examined the impact of microwave dehydrating on few quality characteristics of dried sample. The investigation demonstrated that microwave & microwave– vacuum dehydrating of sample resulted in decreased hue different contrasted with air-dehydrating. The hue difference of sample during dehydrating can be limited by different methods of pre-treatment. Sulfite treatment, whitening, osmotic and microwave pre-treatment have appeared to decrease the hue difference essentially during convectional dehydrating of sample [7]. McMinn and Magee [8] observed the air tunnel dehydrating (warmth 30 - 60°C, speed 1.5 m/s) of potato slices. The consequences of their examination showed that internal porosity of sample cylinders shape sample previously reduce with dehydrating period then it hiked to ten percent at the final dehydrating. The area of shrinkage displayed a linear correlation with respect wetness. The rate of shrinkage amid low- warmth dehydrating were more prominent than at high warmth and coefficient of rehydration higher about four percent with rising dehydrating warmth to 30 - 60°C.

Villareal and Griggs [9] recommended partial dehydrating of sweet potato cuts by dipped in circulate again concentrated sugar arrangement of 60 to 65 percent for a few hours before they are sun powered dried. Combo of microwave and warm air drying inspection has appeared to better the structure and bulk volume of dehydrated mushroom samples. Nonetheless, the geometry and dielectric characters of samples are such that potential for too heat of the middle hamper the utilization of these innovation [10]. Microwave– vacuum dehydrating of mushroom has been investigated by Pappas et al. [11]. The exploratory outcomes showed that microwave– vacuum dehydrating displays prevalent dehydrating execution just as better rehydration qualities. Shobhana et al [12] examined the evolution of generalized dehydrating characteristic curve. The outcomes demonstrates the approach suggested encourages to create single generalized characteristic curve represent sixteen dehydrating kinetics and dimensionless parameter called DPI characterizing the efficacy of potato's dryer setup.

Drying operation utilizing sun power carried out by conventional sun drying and different development sun power dryers. Normal open sun dehydrating is the economical techniques of sunlight power used that generally practiced in the countryside of the developing nations like India, where it is ordinarily utilized as a conventional food conservation method. A few foods items like vegetables samples [13], fish [14], Fruits samples [15], papads [16], pineapple slices and kachris can be solar dehydrated to preserve them for extended time. These items are delicious and need least cooking period. Also sun dehydrated Variety of cherries, carrots and cranberries can be utilized as an expansion cookies, leaves, breads, etc. Nonetheless, open dehydrating is a moderately slow procedure and in this procedure, extensive item losses takes place because of lacking dehydrating, insects, birds and rodents bringing about decrease of item quality. Subsequently, appropriately designed sun power dryers ought to be utilized for drying farming items in developing nations. A few researchers have created plan standards for different classes of sun powered drying arrangements like mixed-mode [17], greenhouse [18], direct [19] and indirect [20]. Among the different dryer development, the mixed mode dryers are noticed that best regarding item drying rate and drying cost [21], [22]. These mixed mode sun power dryers can better the quality of item, while decreasing the utilization of conventional fills.

IOP Conf. Series: Earth and Environmental Science **312** (2019) 012021 doi:10.1088/1755-1315/312/1/012021

Aghbashlo et al. [23] reported Page model to be the perfect fitted thin layer dehydrating behaviour of potato samples dehydrated in a semi-mechanical continuous band drier working three different air temperatures with three different air speeds and three different chain linear speeds. Demiray and Tulek [24] contemplated the impact of air three different temperatures on hue difference kinetics of carrot samples was inspected. Pareek and Kaushik [25] investigated the effect of drying techniques (solar dryer, Oven, Fluidized bed and Microwave) on the nature of Indian gooseberry powder amid storage. Surveyed characteristics include: Ascorbic acid, tannins, Total sugars, titrable acridity and reducing sugars. Chaouicha et al. [26] endeavoured to track the effect of different controlling variables of hybrid sun powered drying on the colour contrast as a quality parameter of dehydrated potato samples. It is apparent that, lot of research work is being conveyed out in the zone of sun powered dryer on natural and forced convective techniques for dehydrating vegetables. It is additionally apparent that, no literature review data is found for drying potato by double slope solar dryer. Subsequently, in the present work exploratory examination was carried out on the double slope sun powered dryer to dry the potato and compare the sun powered dryer results with open sun drying.

2. Experimental set-up

The fabricated exploratory arrangement operates on both natural convective double slope solar dryer and plate for open sun dehydrating. The dryer arrangement developed by GI sheet secured with plywood of 2.5 cm thickness all around (base, two sides and two front sides) the drying load. The GI sheet is secured on both inside and outside of the plywood. The top surface is shut with two plane glasses of 0.4 cm thickness with inclination of 10.9°, which is equivalent to the latitude of the area Karaikal. Two aluminium works of 241cm×74 cm are fabricated. One is fit inside the drying chamber for setting the vegetables items to be dried and another is fit outside the drying chamber for putting the vegetables items to be dried under open drying technique. The sun power dryer work is placed in the chamber to put the items to be dried. An arrangement is made so the work can be moved while stacking or emptying the items. Two hollow pipes are given on both right and left edges to let the atmospheric air into the sun power dryer and a vent gap at a higher height is also provided to let the hot air to go out. The whole setup is placed on four vertical mild steel L-edge plates of 2.5 cm. A same foldable L-point plate is given outwardly of the chamber for keeping an aluminium work for open sun drying. Both the components of the work are same. The photo of the double slope dryer is shows in Fig.1

Figure 1. Photograph of Double Slope Solar Dryer and open drying method for drying Potato

In order to notice the dissimilarity of the temperature inside the solar powered dehydrating chamber, eight different T-Type thermocouples are given. From the eight thermocouples, three are utilized to determine the temperature of the warm air, remaining five thermocouples are attached to determine the surface temperature of the dehydrating chamber. Plus, one J-Type thermocouple is utilized to measure the surrounding temperature. The global radiation is estimated by utilizing Hukseflux Pyranometer. All the nine thermocouples are associated with the Agilent 34972A Data acquisition system to consistently measure the readings. The photos of the Pyranometer and the Data acquisition system are appeared in Fig. 2.

IOP Conf. Series: Earth and Environmental Science **312** (2019) 012021 doi:10.1088/1755-1315/312/1/012021

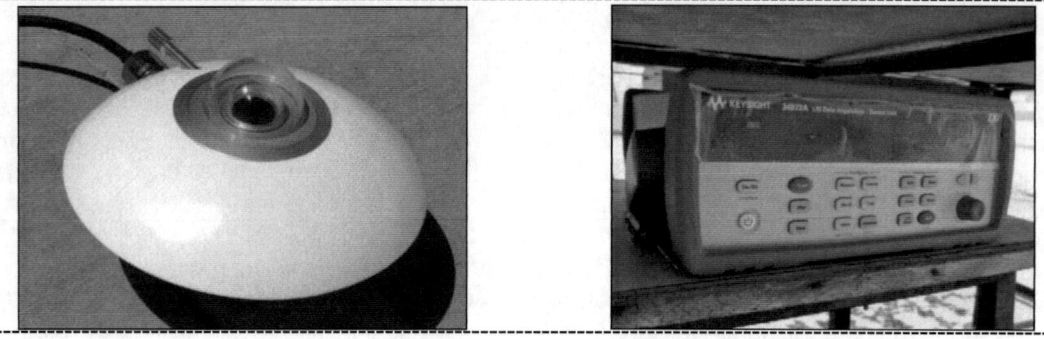

(a) **(b)**

Figure 2. Photograph of Pyranometer and Data Acquisition Unit

3. Results and Discussion

The investigation was done to explore the decrease of the wetness content of the potato in a fabricated sun powered dryer. The investigation was conducted by two days from11.00 am to 6.00 pm. On first day two equivalent weighted (1312 grams) slices of potato are taken; one lot of slices is put in the sun powered dryer and the second set of slices are placed in open sun drying process. The pyranometer readings and eight thermocouples readings including surrounding temperature are estimated in regular intervals. On day one the variation of the global radiation and mean air temperatures and surfaces temperatures of the sun powered dryer are appeared in Fig.3.

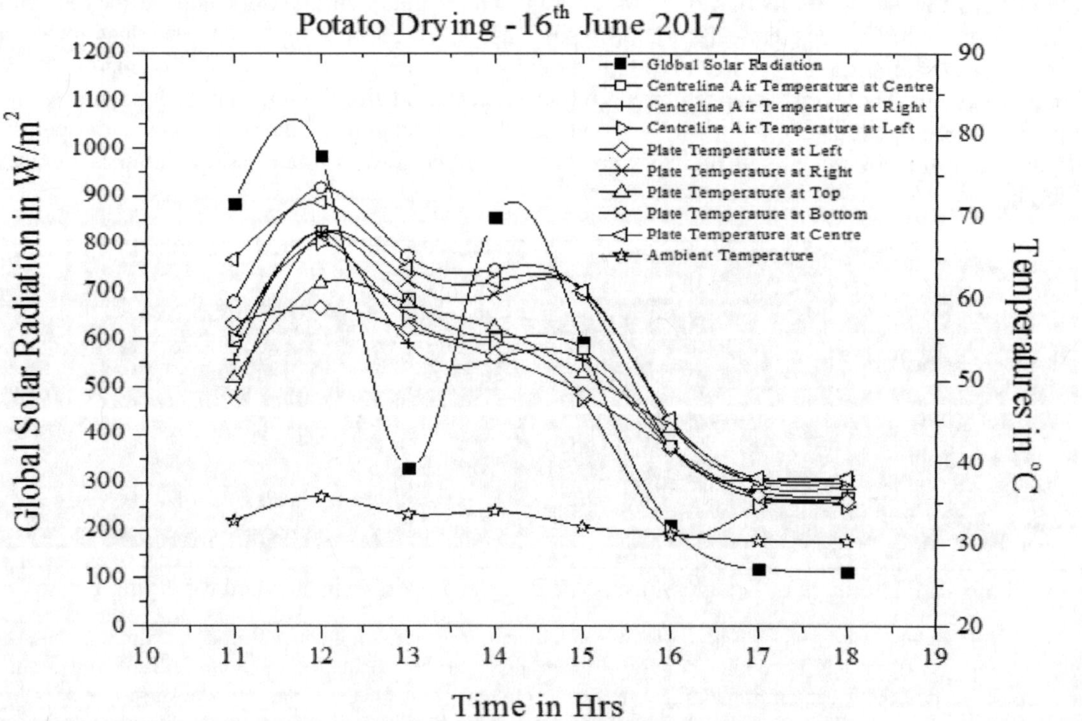

Figure 3. Variation of the global solar radiation, ambient temperature, average air temperatures and surface plate temperatures of the solar dryer on first day.

IOP Conf. Series: Earth and Environmental Science **312** (2019) 012021 doi:10.1088/1755-1315/312/1/012021

From the Fig.3, it is noticed that, the highest global radiation recorded was 1013.11 W/m^2 and the highest temperature got was 71.7 °c. The slices dried in open sun drying process lost 565 grams in the range of eight hours while the slices dried in sun powered dryer lost 686 grams. It is noticed that fabricated solar dryer eliminated the high wetness content of 52.29% from the potato slices, whereas, 42.99% the wetness content only eliminated by open sun drying technique.

On second day fresh two equivalent weighted (650 grams) slices of potato are taken; one lot of slices is put in the sun powered dryer and the second set of slices are placed in open sun drying process. On second day the variation of the global radiation and mean air temperatures and surfaces temperatures of the sun powered dryer are appeared in Fig.4. The pyranometer readings and eight thermocouples readings including surrounding temperature are estimated in regular intervals. It's observed that highest temperature got was 76.4 °c. The slices dried in open sun drying process lost 260 grams in the range of eight hours while the slices dried in sun powered dryer lost 349 grams. It is noticed that fabricated solar dryer eliminated the high wetness content of 54% from the potato slices, whereas, 40% the wetness content only eliminated by open sun dehydrating technique.

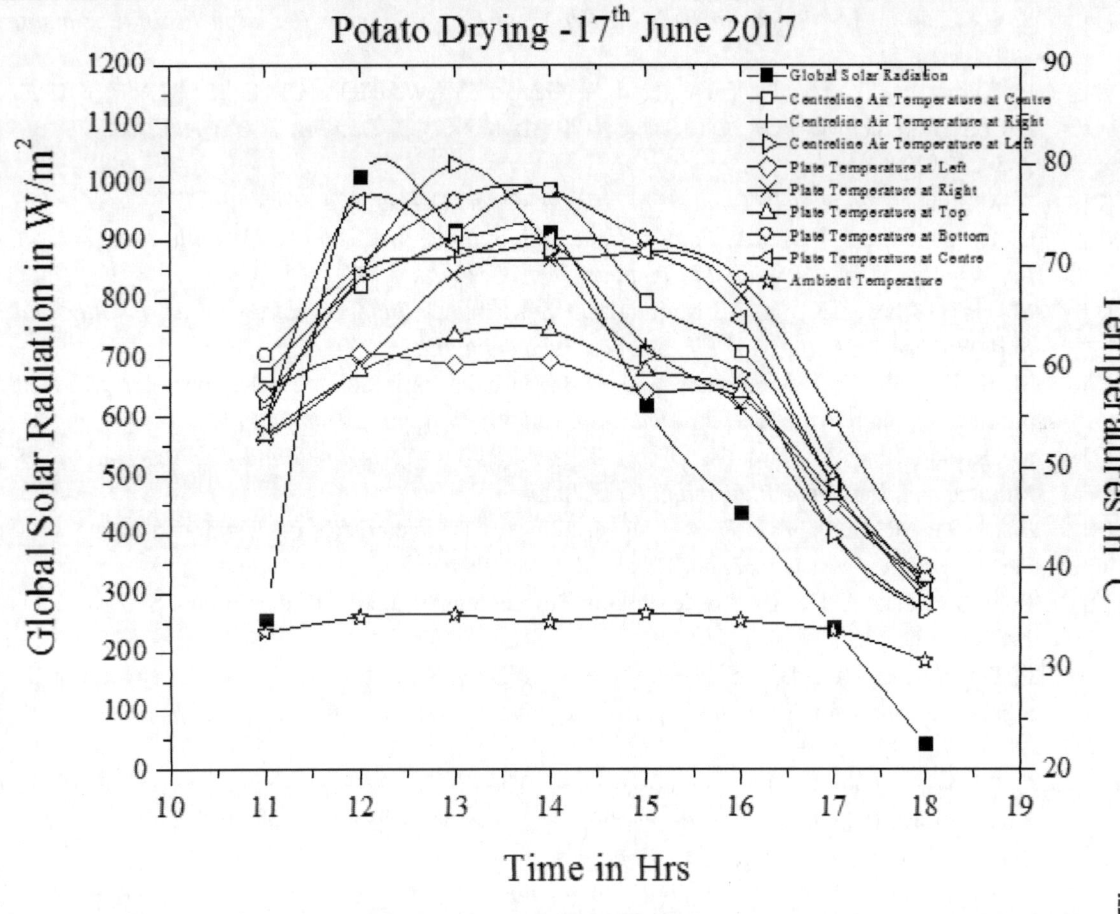

Figure 4. Variation of the global solar radiation, ambient temperature, average air temperatures and surface plate temperatures of the solar dryer on second day

The rate of dehydrating in sun powered dryer is more compare with the open sun dehydrating. The slices dehydrated in the sun based dryer took substantially less time when contrasted with the

IOP Conf. Series: Earth and Environmental Science **312** (2019) 012021 doi:10.1088/1755-1315/312/1/012021

open sun drying technique to accomplish similar wetness content. In this manner, the fabricated sun powered dryer setup can be adequately utilized to dry the vegetable items.

4. Conclusion

A double slope solar dryer appropriate has been developed and inspected have been directed to assess the rate decrease of wetness fraction for potato. For double slope sun powered dryer with natural convection, sun based dryer was 10.32% more effective when contrasted with open sun drying in case of potato on first day. During second day solar dryer was 14.21% more effective when compared with open drying. The items dried in the solar dryer took considerably less period when compared with the open sun dehydrating technique. Samples from the solar dryer were observed to better quality. It is reported that the drying time depends mainly on mass flow rate of dehydrating air, global solar radiation and air temperature. Drying rate in the natural convection sun powered dryer is higher than that of the open drying techniques.

References

[1] S. G.S. and G. J.S, *Potato in India. A Report of Central Potato Research Institute, Shimla.* 1992, pp. 6–11.

[2] P. Kumari, S. Yadav, L. Yadav, and R. Gael, "EVAWATION OF DIFFERENT SOLAR DRYING SYSTEMS FORPOTATO CHIPS MAKING," *J. Daiying, Foods H.S.*, vol. **25**, pp. 245–247, 2006.

[3] A. Mujumdar, *Handbook of Industrial Drying, Marcel Dekker, New York.* 1987.

[4] G. K. MJ Willard, VM Hix, *Potato Processing, 4th ed., WF Talburt, O Smith, Eds., an AVI Book by Van Nostrand Reinhold Company, Inc., New York.* 1987.

[5] M. H. HJ Choy, CK Jon, AS Mujumdar, CS Kiang, *Drying Technology on Agriculture and Food Science, AS Mujumdar, Ed., Science Publishing, Inc.,*. 2000.

[6] M. K. Krokida, C. T. Kiranoudis, and Z. B. Maroulis, "Viscoelastic behaviour of dehydrated products during rehydration," *J. Food Eng.*, vol. **40**, no. 4, pp. 269–277, 1999.

[7] M. Krokida and Z. Maroulis, *Drying Technology on Agriculture and Food Science , AS Mujumdar, Ed., Science Publishing, Inc.,*. 2000.

[8] W. A. M. McMinn and T. R. A. Magee, "Physical characteristics of dehydrated potatoes — Part I," *J. Food Eng.*, vol. **33**, no. 1–2, pp. 37–48, 1997.

[9] R. Villareal and T. Griggs, "Sweet Potato: Proceedings of the First International Symposium," 1982, pp. 82–172.

[10] P. Torringa, E., Esveld, E., Scheewe, I., van den Berg, R., & Bartels, "Osmotic dehydration as a pre-treatment before combined microwave-hot-air drying of mushrooms.," *J. Food Eng.*, vol. **49**, pp. 185–191, 2001.

[11] C. Pappas, E. Tsami, and D. Marinos-Kouris, "The Effect of Process Conditions on the Drying Kinetics and Rehydration Characteristics of Some Mw-Vacuum Dehydrated Fruits," *Dry. Technol.*, vol. **17**, no. 1–2, pp. 157–174, 1999.

[12] S. Singh and S. Kumar, "New approach for thermal testing of solar dryer: Development of generalized drying characteristic curve," *Sol. Energy*, vol. **86**, no. 7, pp. 1981–1991, 2012.

[13] D. Jain and G.N.Tiwari, "Thermal aspects of open sun drying of various crops," *Energy*, vol. **28**, pp. 37–54, 2003.

[14] J. Patterson and G. Ranjitha, "Qualities of commercially and experimentally sun dried fin fish , Scomberoides tol Qualities of commercially and experimentally sun dried fin fish , Scomberoides tol," *African J. Food Sci.*, vol. **3**, no. 10, pp. 299–302, 2009.

[15] I. T. Toğrul and D. Pehlivan, "Modelling of thin layer drying kinetics of some fruits under open-air sun drying process," *J. Food Eng.*, vol. **65**, no. 3, pp. 413–425, 2004.

[16] M. Kumar, P. Khatak, R. K. Sahdev, and O. Prakash, "The effect of open sun and indoor forced convection on heat transfer coefficients for the drying of papad," *J. Energy South. Africa*, vol. **22**, no. 2, pp. 40–46, 2011.

[17] F. K. Forson, M. A. A. Nazha, and H. Rajakaruna, "Modelling and experimental studies on a mixed-mode natural convection solar crop-dryer," *Sol. Energy*, vol. **81**, no. 3, pp. 346–357, 2007.

[18] E. A. Almuhanna, "Utilization of a Solar Greenhouse as a Solar Dryer for Drying Dates under the Climatic Conditions of the Eastern Province of Saudi Arabia," *J. Agric. Sci.*, vol. **4**, no. 3, pp. 237–246, 2011.

[19] M. S. Sodha, A. Dang, P. K. Bansal, and S. B. Sharman, "An analytical and experimental study of open sun drying and a cabinet tyre drier," *Energy Convers. Manag.*, vol. **25**, no. 3, pp. 263–271, 1985.

[20] E. K. Akpinar, "Drying of mint leaves in a solar dryer and under open sun: Modelling, performance analyses," *Energy Convers. Manag.*, vol. **51**, no. 12, pp. 2407–2418, 2010.

[21] I. N. Simate, "Optimization of mixed-mode and indirect-mode natural convection solar dryers," *Renew. Energy*, vol. **28**, no. 3, pp. 435–453, 2003.

[22] B. O. Bolaji and A. P. Olalusi, "Performance Evaluation of a Mixed-Mode Solar Dryer," *Au J.T*, vol. **11**, no. 4, pp. 225–231, 2008.

[23] M. Aghbashlo, M. H. Kianmehr, and A. Arabhosseini, "Modeling of thin-layer drying of potato slices in length of continuous band dryer," *Energy Convers. Manag.*, vol. **50**, no. 5, pp. 1348–1355, 2009.

[24] E. Demiray and Y. Tulek, "Color Degradation Kinetics of Carrot (Daucus carotaL.) Slices during Hot Air Drying," *J. Food Process. Preserv.*, vol. **39**, no. 6, pp. 800–805, 2015.

[25] S. Pareek and R. A. Kaushik, "Effect of drying methods on quality of Indian gooseberry (Emblica officinalis Gaertn.) powder during storage," *J. Sci. Ind. Res. (India).*, vol. **71**, no. 11, pp. 727–732, 2012.

[26] S. Chouicha, A. Boubekri, D. Mennouche, and M. H. Berrbeuh, "Solar drying of sliced potatoes. An experimental investigation," *Energy Procedia*, vol. **36**, pp. 1276–1285, 2013.

IOP Conf. Series: Earth and Environmental Science **312** (2019) 012022 doi:10.1088/1755-1315/312/1/012022

Production of Bio-Oil from Soapnut seed by Microwave Pyrolysis

A.Mathiarasu[1]*, M.Pugazhvadivu[2]

[1,2]Department of Mechanical Engineering, Pondicherry Engineering College

Pondicherry

mathianbu77@gmail.com

Abstract. Biomass is one of the potential source of renewable energy to meet the need of alternative fuels. Biomass can be converted into useful energy by thermochemical conversion processes such as gasification, combustion and pyrolysis. Microwave pyrolysis a non-conventional method of heating for the production of pyrolytic bio-oil, char and gas. In this work, soapnut seed is used for the bio-oil production by microwave pyrolysis. The experimental setup used in this work consists of a microwave oven of 2.45 GHz frequency with maximum input power of 800W. In this work, temperature distribution and pyrolytic products yield from the pyrolysis of soapnut with respect to microwave power input are studied. The microwave pyrolysis experiments are conducted at power densities of 8 W/gm, 7 W/gm & 6 W/gm for 20 minutes. The experimental shows that the bio-oil yield (50 wt %) is the highest at 7 W/gm. The flash point, viscosity and calorific value of the pyrolytic bio-oil are measured and presented. The results indicated that the bio-oil produced by the microwave pyrolysis can be upgraded and used as a potential fuel for IC engines

Keywords: microwave pyrolysis; soapnut; power density; bio-oil; bio-char.

1. Introduction

Renewable resources are considered as alternative energy sources for the production of bio-fuels and bio-energy. An extensive volume of researches is being carried out around the globe on biomass energy conversion systems. Biomass resources can be classified into edible and non-edible resources. The present need for food resources minimize the usage of edible resources for the production biofuels.

Pyrolysis is a process in which the biomass feedstock is decomposed in the absence of oxygen to liquid, solid and gaseous products. Considerable researches in conventional pyrolysis is available and the modernistic researches are on non- conventional type of pyrolysis. Microwave pyrolysis is a non-conventional method of pyrolysis in which electromagnetic waves that are produced at a specified frequency heats the feedstock (AV Bridgewater et al) [1] [2] .

Soapnut tree (*Sapindus mukorossi*) is a deciduous tree which is majorly found in India. The tree can grow to the height of 82 feet and width of about 5 feet. The tree is cylindrical and straight, the flowers are small and greenish white whereas the fruits are round shaped (Fig 1a). At initial stage of growth the outer shell is golden yellow and after dried it is brown in colour.(Fig 1b). It is well established that the saopnut seeds are rich in saponin content which makes it be a natural detergent that implies its application for cleaning purposes. However, there is no commercial exploitation of fruits of soap nut is made extensively. In chemical industry the pulp is considered as the raw material whereas the kernel is left out without any specified usage. A single soapnut tree produced about 125 kg of seeds annually [3] , which can be used as the source of bio-energy and biofuel production [4] [5] . Hence, the objective of the present work is to utilize soapnut seed as the feedstock in microwave pyrolysis. The specific objectives of the present work are i) to characterize the soapnut seed for its proximate and elemental compositions, ii) to use soapnut seed as the feedstock in microwave pyrolysis for producing bio-oil and iii) to determine fuel properties of soapnut bio-oil.

Fig 1 (a) Soapnut fruit

Fig 1 (b) Soapnut seed & shell

2. Materials and Methods

2.1 Soapnut seed

In this work, the Soapnuts are procured from the local market in Puducherry and it is manually broken using an iron mortar-pestle to separate the seed from the outer cover. The soapnut seeds are milled into a fine powder of approximately 2 mm.

2.2 Characterization of soapnut seed

2.2.1 Proximate and elemental analysis. The proximate composition (moisture, volatiles, fixed carbon and ash content) is found using a muffle furnace. The elemental composition, carbon, hydrogen, nitrogen, suphur and oxygen contents) is measured using Elementar CHNS analyser in Pondicherry University. The results of proximate analysis and ultimate analysis are shown in Table1. The high volatile matter (74.33%) and lower ash (2.81%) contents of soapnut seed shows that it is potential feedstock for pyrolysis.

2.2.2 Thermogavemetric Analysis soapnut seed, The thermogravimetric Analysis (TGA) of the soapnut seed is carried to determine its thermal decomposing behaviour in nitrogen atmosphere. TGA analysis is carried out in Simultaneous Thermal Analyser (STA) (Make: Netzsch - STA 449 F3 Jupiter) in

Pondicherry University in which temperature is regulated from room temperature to 800°C at 20°C/min as heating rate in nitrogen gas atmosphere using 50 mg of sample in Al_2O_3 crucible holder.

Table 1 proximate and ultimate analysis of Soapnut seed

	Soapnut seed
Proximate analysis (as received, wt. %)	
Moisture content	5.36
Ash Content	2.81
Volatiles	74.33
Fixed Carbon	17.5
Ultimate analysis (dry basis, wt. %)	
C	51.62
H	6.45
N	2.52
S	0.25
O	39.16

2.3 Microwave pyrolysis of soapnut seed

2.3.1 Microwave pyrolysis experimental setup. Microwave pyrolysis of soapnut seed is carried out in a domestic microwave oven made by LG microwave oven, India. Fig 2 shows the experimental microwave pyrolysis set-up. A regulator is used to vary the input power from 100 W to 800 W. The pyrolysis is carried out in a microwave resistant quartz three- neck round bottom flask of 250 ml capacity. Two allihn condensers are connected to the round bottom flask. A Nitrogen gas cylinder with pressure regulator is connected to the one neck of the round bottom flask. Holes are drilled at appropriate places to connect the condenser, thermocouples and nitrogen inlet. To find the reaction temperature a K-type thermocouple with temperature with AUDRINO microcontroller is used. Two conical flasks were used to collect the bio-oil [6] .

2.3.2 Microwave pyrolysis experimental procedure. The round bottom flask containing soap nut seed powder of 100 gm is placed inside the microwave cavity. Before the starting the experiment, the flask containing biomass is exerted with N_2 gas for about 7 min to make sure it is a no-oxygen zone.

The nitrogen flow is continued till the completion of pyrolysis process. The outlet of the three neck flask is connected to the condensers. Water at room temperature is circulated in the condensers at a constant flow rate. The temperature variation inside the microwave cavity is studied using AUDRINO UNO microcontroller connected to AUDRINO software for temperature data analysis. At the end of experiments, bio-oil produced is collected and weighed. The solid bio-char in the quartz flask is collected and then weighed. The uncondensed gaseous particles are left to in atmosphere. The percentage yields of bio-oil, bio-char and gas are calculated as shown in Eqn (1) to Eqn (3) respectively [7] .

$$bio - oil\ yield\ (wt\%) = \frac{mass\ of\ bio\text{-}oil}{massof\ biomass} \times 100 \quad (1)$$

$$bio - char\ yield\ (wt\%) = \frac{mass\ of\ bio\text{-}char}{massof\ biomass} \times 100 \quad (2)$$

$$gas\ yield\ (wt\%) = 100 - (bio - oil\ yield + bio - char\ yield) \quad (3)$$

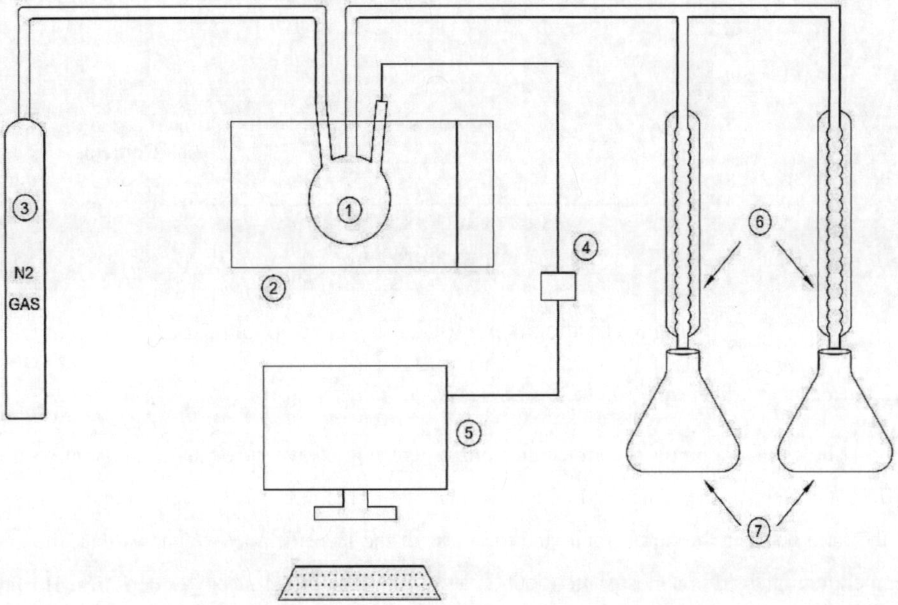

Fig 2 Experimental setup of Microwave pyrolysis and its components (1) Three-neck quartz flask, (2) LG microwave oven, (3)Nitrogen gas cylinder ,(4)Audrino Uno microcontroller, (5) Computer with Audrino software (6) Two Allihn Condensers, (7) Collecting Conical flask

2.4 Fuel properties of the soap nut seed bio-oil

The viscosity, flash point, calorific value and pH of the bio-oil produced by the microwave pyrolysis of soapnut seed are measured using Redwood viscometer, Cleveland open-cup apparatus, bomb calorimeter and pH meter respectively.

3. Result & Discussion

3.1 Thermogravimetric decomposition behaviour of soapnut seed

Fig 3 represents the thermal decomposition behaviour of soapnut seed from ambient temperature to 800°C. A drop in mass loss curve between room temperature and 100°C revealed the presence of moisture (4%) in the seed. It is seen that the volatiles release take place between 200°C- 500°C that

IOP Conf. Series: Earth and Environmental Science **312** (2019) 012022 doi:10.1088/1755-1315/312/1/012022

accounts to a mass loss of about 65%. The TGA analysis thus revealed that the soapnut has higher volatile matter content and it is a potential feedstock for microwave pyrolysis.

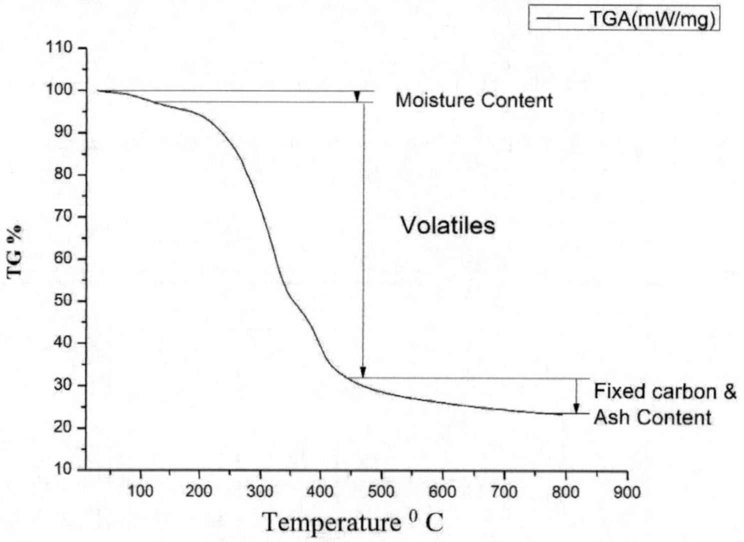

Fig 3 Thermogavemetric Analysis of Soapnut seed

3.2 Variation of temperature with respect to residence time

Fig 4 shows the variation of temperature during the microwave pyrolysis of soapnut seed at three different power densities. The maximum residence time of biomass is kept as 20 minutes. Power density is the ratio between the power input and the mass of the feedstock. Fig 6 shows that, the maximum temperature attained is almost same (550°C) at all the three different power densities. However, the time taken to attain the maximum temperature is varies with power density., that is 7min, 16 min and 11 min respectively when the power density is 6W/gm, 7 W/gm and 8 W/gm. This may due to the type of polarisation, specific area and surface morphology which is similar to the studies done by Suriapparao et al., 2015. Even though the maximum temperature is attained within the shortest time of 7 min at 6W/gm power density, there is drop in temperature after 10 minutes. This dissipation of heat during pyrolysis comparatively takes higher reaction time for complete decomposition. At 7 W/gm power density the maximum temperature was 528°C, but the temperature is not maintained throughout the reaction time. It is noticed that the temperature of 400°C and 500°C is maintained for 13 min at 7W/gm and 16 min at 8W/gm. On comparing the time taken for attaining maximum temperature, the pyrolysis which is conducted at 8 W/gm required less reaction time than pyrolysis conducted at 7 W/gm. But pyrolysis at 7 W/gm exhibits better sustainability on maintaining the temperature [8] . At the end of the reaction the temperature is 261.25°C. The maximum temperature attained at 6 W/gm is 550°C within 7 minutes. Similar type of temperature profile is studied by [9] [10] in which the maximum temperature for complete pyrolysis took 35 minutes.

IOP Conf. Series: Earth and Environmental Science **312** (2019) 012022 doi:10.1088/1755-1315/312/1/012022

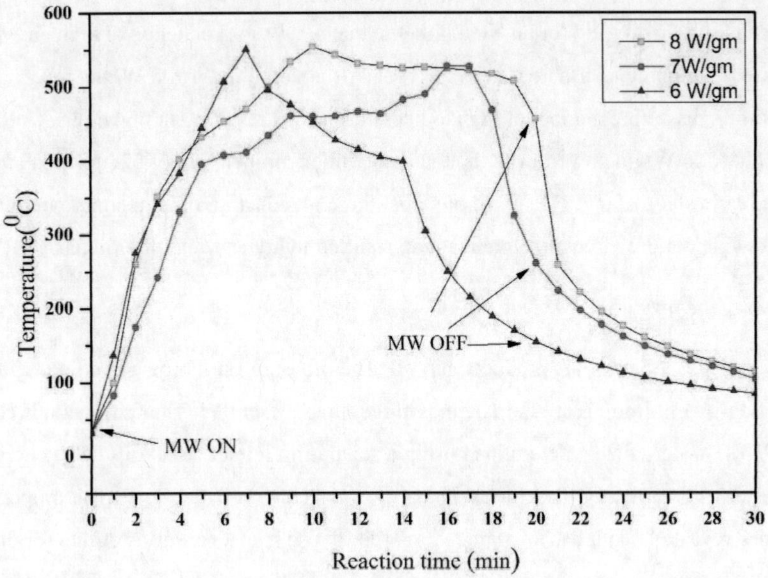

Fig 4 Temperature profile of microwave pyrolysis of soapnut seed

3.3 Product yields obtained by the microwave pyrolysis of soapnut seed

Fig 5 shows the bio-oil, bio-char and gas yields obtained by the microwave pyrolysis of soapnut seed at different power densities of 6 W/gm, 7 W/gm and 8 W/gm. The maximum bio-oil yield of 52% is obtained at a power density of 7W/gm. At this power density, the intensity of the electrons becomes stronger, this leads to a vigorous vibration of polar molecules.

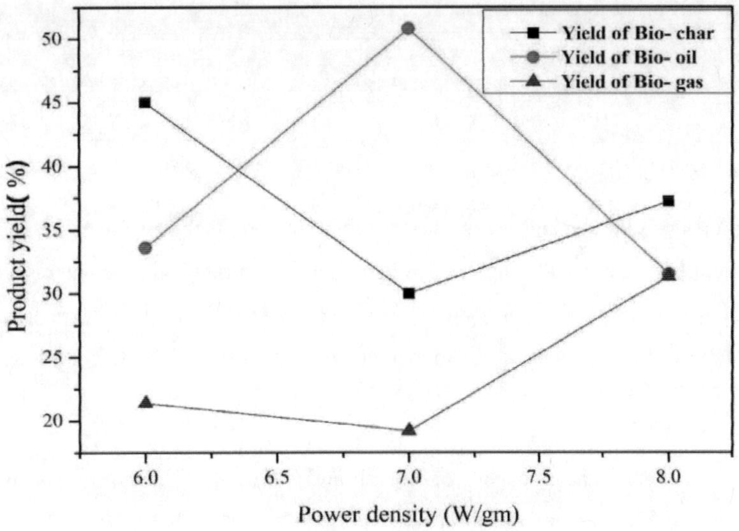

Fig 5 Product yield from Microwave pyrolysis of soapnut seed.

The field strength i.e the power intensity for rapid heating of the sample is due to the rotational or vibrational motion of the molecules. Therefore such kind of polarisation enables the increased production of bio-oil at this power density. The production of gas is high at 8 W/gm. It is been stated that, decreased pyrolytic oil yield for high temperature that results in production of high gas yield which might be caused by non condensable gases [11] [12] . It is noted that for 8 W/gm power density the temperature profile has more number of high temperature peaks resulting in higher gas yield. The bio-char yield is high at 6 W/gm power, even though a maximum temperature of 528 °C is attained by this reaction at the early stage of pyrolysis it is noted that there is reduction of temperature until the MW is off, thus lower temperature throughout the reaction resulted in higher yield of bio-char. [13] [14] .

3.4 Properties of the produced pyrolytic bio-oil

Table 2 shows the properties of the produced bio-oil. The bio-oil is dark brown in colour with a smoky odour. The viscosity of bio-oil is around 10 times higher than diesel fuel. The flash point is comparable to diesel fuel. The water content determined using a separating funnel is 10-15%. The energy content is about 50% lower than that of diesel fuel. The bio-oil is acidic in nature [15]. The properties of the bio-oil obtained revealed that it has to upgraded in order to use it as a partial substitute for diesel fuel

Table 2 Properties of the bio-oil

Properties	Produced Saopnut seed pyrolytic-oil
Appearance	Dark brownish colour
Kinematic viscosity at 40°C in cSt	36.7
Flash point (°C)	54
Fire point (°C)	74
Water content (%)	10-15
Gross calorific value (MJ/kg)	22
pH	5.2

4. Conclusion

Microwave pyrolysis of the soaput seed resulted with varying product yield fractions. The production of pyrolytic bio-oil is significantly high at 7 W/gm. The temperature profile for respective power densities enables to understand the feedstock's reaction to the incident microwave irradiation. The properties reveals that bio-oil needs upgradation to utilize it as a partial substitute for diesel fuel.

References

[1]Bridgwater, A.V., 2007. The production of biofuels and renewable chemicals by fast pyrolysis of biomass. *International Journal of Global Energy Issues*, *27*(2), pp.160-203.

[2]Czernik, S. and Bridgwater, A.V., 2004. Overview of applications of biomass fast pyrolysis oil. *Energy & fuels*, *18*(2), pp.590-598.

[3] http://www.sapindusmukorossi.com/;

[4]Suhagia, B.N., Rathod, I.S. and Sindhu, S., 2011. Sapindus mukorossi (Areetha): an overview. *International Journal of Pharmaceutical Sciences and Research*, *2*(8), p.1905.

[5] Chakraborty, M. and Baruah, D.C., 2013. Production and characterization of biodiesel obtained from Sapindus mukorossi kernel oil. *Energy*, *60*, pp.159-167.

[6] Meredith, R.J., 1998. *Engineers' handbook of industrial microwave heating* (No. 25). Iet.

[7]Wang, Y., Dai, L., Wang, R., Fan, L., Liu, Y., Xie, Q. and Ruan, R., 2016. Hydrocarbon fuel production from soapstock through fast microwave-assisted pyrolysis using microwave absorbent. *Journal of Analytical and Applied Pyrolysis*, *119*, pp.251-258.

[8]Lei, H., Ren, S. and Julson, J., 2009. The effects of reaction temperature and time and particle size of corn stover on microwave pyrolysis. *Energy & Fuels*, *23*(6), pp.3254-3261.

[9]Suriapparao, D.V. and Vinu, R., 2015. Bio-oil production via catalytic microwave pyrolysis of model municipal solid waste component mixtures. *RSC Advances*, *5*(71), pp.57619-57631.

[10] Suriapparao, D.V. and Vinu, R., 2015. Resource recovery from synthetic polymers via microwave pyrolysis using different susceptors. *Journal of analytical and applied pyrolysis*, *113*, pp.701-712.

[11] Borges, F.C., Du, Z., Xie, Q., Trierweiler, J.O., Cheng, Y., Wan, Y., Liu, Y., Zhu, R., Lin, X., Chen, P. and Ruan, R., 2014. Fast microwave assisted pyrolysis of biomass using microwave absorbent. *Bioresource Technology*, *156*, pp.267-274.

[12] Salema, A.A. and Ani, F.N., 2012. Microwave-assisted pyrolysis of oil palm shell biomass using an overhead stirrer. *Journal of Analytical and Applied Pyrolysis*, *96*, pp.162-172.

[13] Omar, R. and Robinson, J.P., 2014. Conventional and microwave-assisted pyrolysis of rapeseed oil for bio-fuel production. *Journal of analytical and applied pyrolysis*, *105*, pp.131-142.

[14] Mushtaq, F., Mat, R. and Ani, F.N., 2014. A review on microwave assisted pyrolysis of coal and biomass for fuel production. *Renewable and Sustainable Energy Reviews*, *39*, pp.555-574.

[15] Solikhah, M.D., Pratiwi, F.T., Heryana, Y., Wimada, A.R., Karuana, F., Raksodewanto, A.A. and Kismanto, A., 2018, April. Characterization of bio-oil from fast pyrolysis of palm frond and empty fruit bunch. In *IOP Conference Series: Materials Science and Engineering* (Vol. 349, No. 1, p. 012035). IOP Publishing.

Performance assessment of a sustainable Solar Air Heater (SAH) by using different profiles of absorber plates

Singuru Rajesh[1], Vandrasi Vinay Kumar[2] and Jana Sai Ram[2]

[1,2] Assistant Professor, Department of Mechanical Engineering,
Raghu Engineering College (A), Bhimunipatnam Mandal Visakhapatnam Dist., Andhra Pradesh, Pin-531162, India.

Corresponding Author: rajeshsinguru@gmail.com

Abstract. The main objective of solar air heater (SAH) is to make use of the solar potential for heating air and this is used for drying, crops, fishes, wet grains during post harvesting, space heating and in greenhouses. SAH is a black body which has less emissivity and high heat absorbing capacity. The air enters into the collector gets heated with addition of solar radiation. The collector is placed at an angle due to which the natural draught occurs. The main advantage of this type of air heater is compactness, maintenance, easy to assemble and dissemble. In this work a SAH is examined with the use of different materials, sizes and shapes (like plane, rugged) of absorber plates with optimum tilt angle for better thermal storage performance and greater efficiency.

Key Words: Air heater; collector; solar radiation; draught; absorber plate

1. Introduction

In the present situation, world is concerning on energy shortage and climate changes. Solution for these crises are implementation of clean and green alternating energy for energy production [1]. This results directed to think seriously for new alternate sources for sustainable energy production and products in all the fields. Concept of alternative energy sources relates with renewability, sustainability and low carbon emissions. Challenges such as high energy costs, fuel demand, global warming, unseasonal and acid rains are the targets of these products [2]. Sustainable products are defined as the design and development of products that have a minimal impact on the environment.

Sun is an alternative source of energy which releases abundant and sustainable energy to the environment, termed as solar energy. Various applications are there with this energy mainly used for thermal collection and storage. Solar energy is one of the best option to reach the demand of producing heat for various applications like drying agricultural products, space heating in buildings, textile, marine, dairy farms, desalination and regenerating dehumidification [3] without disturbing the ecosystem and environment. Solar Air Heaters (SAH) an option device in heating application, has been increasing in number day by day because of its simple construction, easy maintenance, feasible operation, made with locally available materials, eco-friendly, no requirement of fuel and cheap in cost [4]. On the other hand thermal efficiency of SAH was poor because of their inherently low convective heat transfer capability between the air flow and absorber material [8]. The SAH efficiency depends on surface geometry of collector and the solar radiation [5].

2. Literature Review

Karim et.al [6] led performance investigation of finned, v-corrugated and flat absorber collectors and found that air heater with v-corrugated was most efficient. Momin et al. [7] investigated on heat transfer effect of V-shaped ribs and fluid flow characteristics of rectangular duct SAH with V-shaped ribs. Esen [8] conducted a detail study to find the efficiencies of oval flat plate SAH with and without obstacles, found that higher yield efficiency values obtained for the collector with obstacles than without obstacles. Alvarez et al. [9] and Rajesh et al [10] developed an effective solar air collector with recyclable aluminum can absorber plate and compared the thermal efficiency. Koyunchu compared the performance of SAH and concluded that the shape of the absorbing surface effect the performance for crop drying [11].

2.1. *Concept Overview*

SAH mainly comprises of the following main components i) absorber plate ii) transparent cover, iii) heat transporter medium, iv) insulation, v) frame. The concept working of a SAH is, the ambient air enters into the collector from the bottom end and it gets heated with the addition of solar irradiation, falls over the collector face. This radiation penetrates into the collector through transparent cover and falls over the air heater absorber plate which results in increasing air temp. Because of the natural draught effect the hot air goes up and this will be collected from the outlet. With the understanding of air heater operation in figure 1, a model SAH was developed.

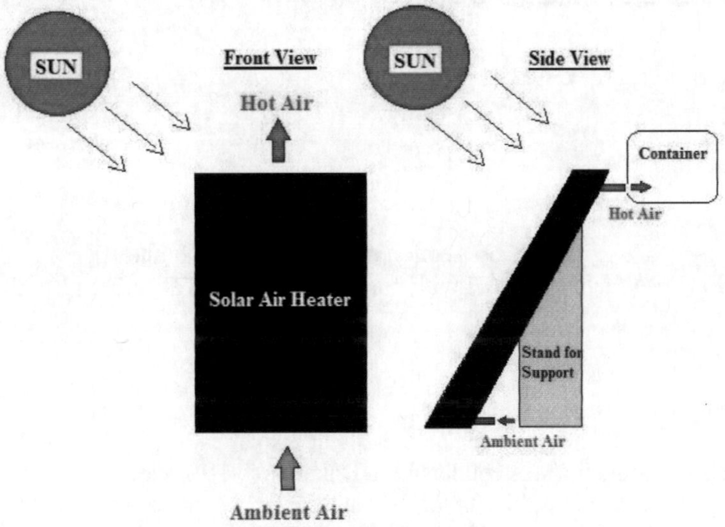

Figure 1 Concept working of sustainable SAH

2.2. *Methodology*

The objective of this study is to compare four varied types of SAH with the main fabricated model by considering different types of absorber plates such as **Type I**: *SAH without Al sheet*, **Type II**: *SAH with Al Sheet*, **Type III**: *SAH with black painted flat Al Sheet and* **Type IV**: *SAH with black painted rugged Sheet Al Sheet.* A model with all the SAH components was developed, to conduct experimentation shown in figure 2 with design considerations and components for fabrication. In this model we are inserting the above types of absorber plates and the values are considered for the performance analysis. Figure 3(a) depicts the cross section of SAH which will be for Type II and III. Figure 3(b) depicts the cross section of Type IV.

IOP Conf. Series: Earth and Environmental Science **312** (2019) 012023 doi:10.1088/1755-1315/312/1/012023

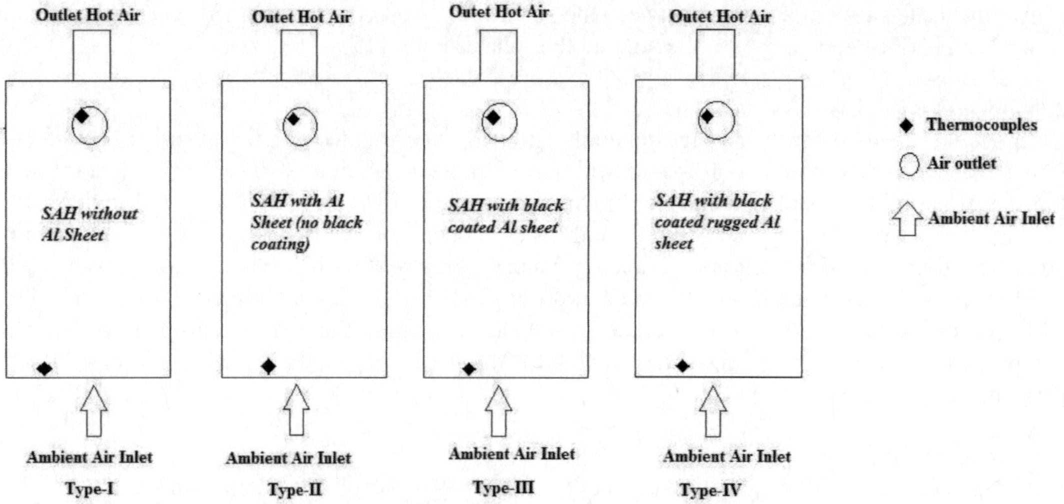

Figure.2 Experimental arrangement of all types of Air Heaters

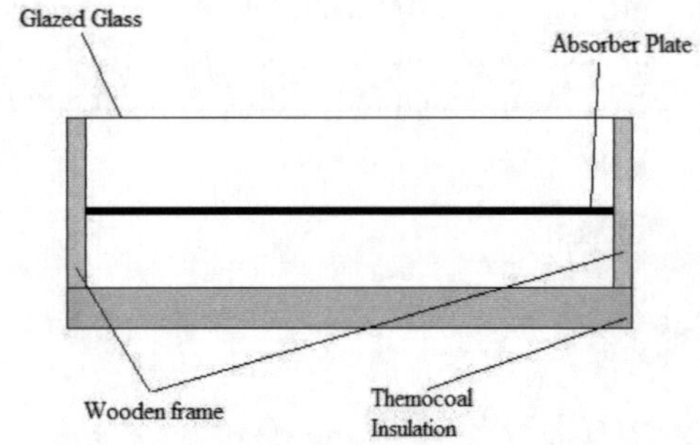

Figure 3. (a)

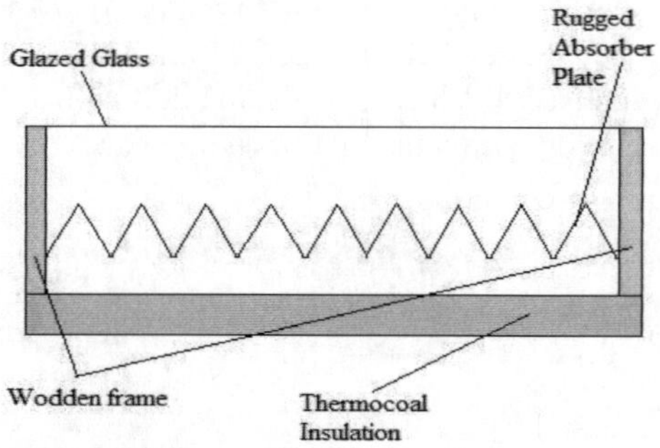

Figure 3. (b)

3(a) Cross section of SAH for Type II and III;
3(b) Cross section of SAH with black coated rugged absorber plate (Type IV)

3. Experimental Conditions

3.1. Geographical Conditions

For experimenting on a SAH the geographical details and solar radiation data of the testing location are required. The experimentation was conducted in Dakamarri (Latitude of 17.9885°N and Longitude of 83.4010°E), Visakhapatnam, Andhra Pradesh. Satellite image of experimental testing location is shown in the figure 4 and which shows the availability of the location. The experiment was conducted for 4 hr per day. The experimental values were taken at the below Table 1 conditions.

Figure 4. Satellite view of experimental testing location of SAH

Table 1. Geographical details of experimental setup.

Geography	Details
Location	Raghu Engineering College, Dakamarri, Visakhapatnam, Andhra Pradesh, India
Local Time	10:00 AM to 2:00 PM
Date(s)	12th - 15th March 2019
Latitude	17.9885°N
Longitude	83.4010°E

3.2. *Solar Irradiation*

A monthly average radiation data for the experimental location (Dakamarri) are presented in Table 2. The Daily Irradiance of this area as is calculated from the Synergy Environmental Engineers solar irradiance calculator with support of National Renewable Energy Laboratories (NREL), India. Experimental testing are conducted in the month of March, 2019 for 4 days with the different types of air heaters.

Table 2. Daily Irradiance of testing location Dakamarri (Synergy Environmental Engineers) and Annual Average Irradiance is 5.5 kWh/m^2 per day

S.No	Month	Irradiance kWh/m^2 per day
1.	January	5.09
2.	February	5.84
3.	March	6.45
4.	April	6.78
5.	May	6.50
6.	June	5.41

7.	July	5.11
8.	August	4.88
9.	September	5.09
10.	October	5.03
11.	November	5.07
12.	December	4.79

In order to reach the maximum output of SAH, it should be with the optimum tilt. The Figure 5 depicts the optimum tilt of solar collector for every month in the region of Visakhapatnam, Andhra Pradesh. The optimum inclination of present tested SAH in the month of March is 72°. Tilt angle is calculated with the help of Solar Calculator for Collectors optimum inclination.

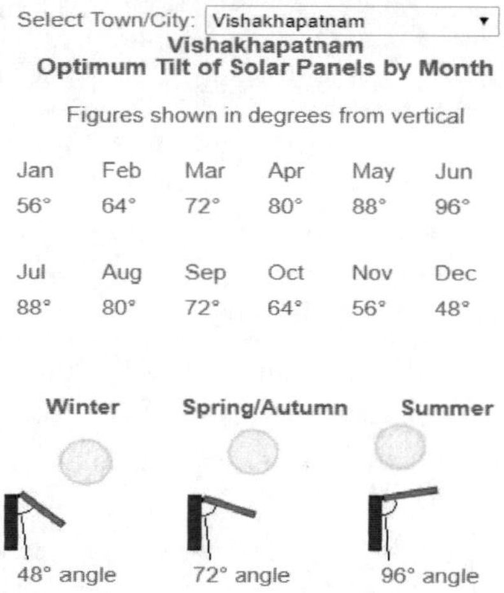

Figure 5. Optimum Tilt of Solar Panel for each month in Visakhapatnam, A.P.

(Courtesy: http://solarelectricityhandbook.com/solar-angle-calculator.html)

4. Experimental Setup

The experimental setup of SAH mainly comprises of black coated aluminum sheet (absorber plate), tempered glass (transparent cover), ambient air (heat transporter medium), thermo-coal sealing (back insulation), induced fan, wooden frame, solar panel for power supply, thermocouples (temperature measurement) and supporting stand.

The fabrication of SAH is started with the selection of country wood for wooden capsule of 1.2 x 0.6 m^2 at a low cost. Wood is cut into the dimensions (1.2m x 0.05m) of the outer frame of the collector and the thickness of frame is 5cm. A Plywood Sheet of 6mm is fixed thickness and 0.72m^2 area at the back of frame. A square hole of 120mm x 120mm of at the top of the collector is made for the exhaust of hot air from collector and a 6mm guide way on the side of the wooden piece is given for inserting glass. Back and Sides of wooden frame is sealed with thermo-coal for thermal insulation. An exhaust fan is fixed on the top of wooden frame to create artificial draught in SAH for the purpose of constant air flow during experiment.

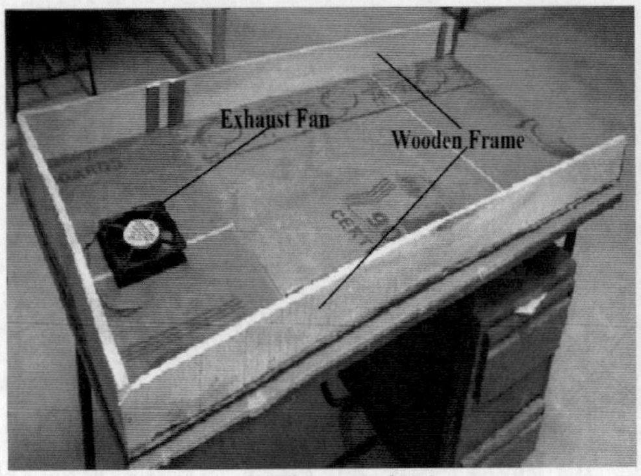

Figure 6 (a)

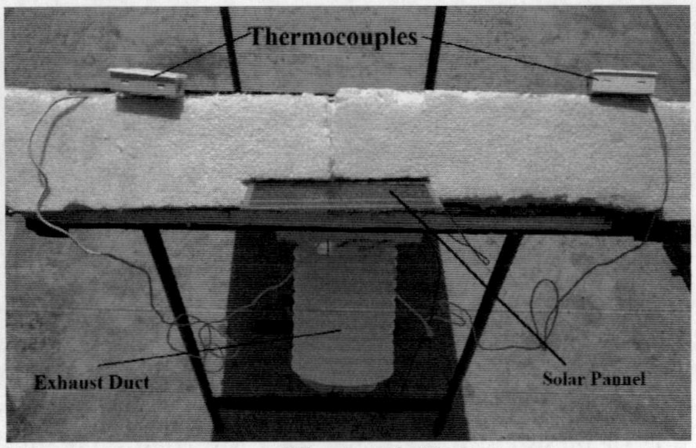

Figure 6 (b)

6(a) wooden frame with exhaust fan; **6(b)** Insulation and fixing of accessories

Tempered glass is fixed on the upper part of the wooden frame with the help of Silicon sealant to guide the inlet air from the lower end to the upper end of the air heater and leaves it into the duct provided with the exhaust fan, which was run with the help of solar panel (I=0.028A & V=30V). Measurement of the airflow at outlet and inlet of the collector was measured by Anemometer. T-type thermocouples were installed at outlet and inlet of the air heater to measure temperatures. Temp range of thermocouple is 120-150°C. Insulation is provided to the duct with foam for maintaining constant temperature shown in figure 6. An iron frame is fixed to the back part of the wooden frame for supporting entire setup with a tilt angle so that the sun rays gets incident on the air heater. Optimum Tilt angle of collector is 72° obtained for testing location is from solar-angle-calculator. Table 3 gives the conditions of the considered solar air heater.

Table 3. Specifications of SAH

S.No	Description(s)	Specification(s)
1.	Solar irradiation	$1000 W/m^2$
2.	Collector area	$1.2 \times 0.6\ m^2$
3.	Absorber plate temperature	Constant
4.	Absorber material	Black painted Al
5.	Double glazing	Applied
6.	Wind velocity	1.8 - 2.1 m/s
7.	Solar flux over collector face (I_T)	$950\ W/m^2\ °C$
8.	Transmissivity $(\tau_\alpha)_{avg}$	0.85
9.	Coefficient of top loss (U_t)	$6.2\ W/m^2\ °C$
10.	Coefficient of bottom loss (U_b)	$0.8\ W/m^2\ °C$
11.	Emissivity of top and bottom surfaces $\varepsilon_p = \varepsilon_b$	0.95
12.	Tempered glass emissivity	0.88
13.	Tempered glass transmissivity	0.85
14.	Tilt angle of collector	72°

IOP Conf. Series: Earth and Environmental Science **312** (2019) 012023 doi:10.1088/1755-1315/312/1/012023

SAH thermal efficiency is calculated from the ratio of the total useful specific enthalpy to the solar radiation flux incident on the transparent glass [12].

$$\eta=\frac{\int_{t_1}^{t_2}\dot{m}C_p}{A_c}\frac{(T_o-T_i)\,dt}{\int_{t_1}^{t_2}G.dt} \qquad (1)$$

Aluminum (Al) sheet of 30 gauge as baseplate (absorber plate) to SAH to ensure more heat absorption and reflection. Al Sheet is cut into different sizes for the different types for SAH shown in figure 7. For the Type *I* the no absorber sheet and the bottom of the frame is coated with black paint. In *Type II* the absorber plate as Al sheet without black paint of dimensions 1.2 x 0.6m^2 and a square hole is made at the top is used. In *Type III* the Al sheet with black paint is used as absorber plate. In *Type IV* Al sheet with black coated rugged sheet is used as absorber plate. All the above types are inserted into the main wooden frame during testing time. After the completion of main frame SAH fabrication, it is tested in sunny condition by using different profiles of al sheets for the performance assessment.

7(a) 7(b) 7(c) 7(d)

Figure 7(a) **Type I:** SAH with black paint; 7(b) **Type II:** SAH without black painted Al sheet; 7(c) **Type III:** SAH with black paint Al sheet; 7(d) **Type IV:** SAH without black painted rugged sheet

5. Results and Discussions:

The air heater tests were taken on clear sky circumstances. The ambient air entered in the collector was heated through the black painted absorber plate with minimum solar irradiation of 1000W/m^2 applied. Due to the air flow given by the induced fan, constant mass flow rate results in uniform temp rise, which occurs at inside the SAH of different flat and rugged Al absorber plates. With help of Aluminum (Al) as heat absorber material, noble heat transfer coefficient is obtained, leads to temp increase of absorber plate in SAH. The temperature distributions along the collector at different places were recorded by using T- Type thermocouples shown in figure 8 and relation between Time vs Temperature of collector represented graphically for all the types of SAH in figures 9 and 10 shown below.

Figure 8. Measuring and recording results of air heater

In Figure 9(a) indicates the Type I: SAH without Al sheet the min inlet temp is 33°C and the final out put temperature reaches to max of 45°C. Figure 9(b) shows the Type II: SAH with Al sheet but not black coated, the minimum inlet min temp is 36°C and the final out put temperature reaches to max of 47°C. The Figure 9(c) shows the Type III: SAH with black coated Al sheet, the minimum inlet min temp is 36°C and the final out put temperature reaches to max of 58.7°C. Finally the Figure 9(c) shows the Type IV: SAH with black coated rugged Al sheet, the minimum inlet min temp is 35°C and the final out put temperature reaches to max of 64.3°C.

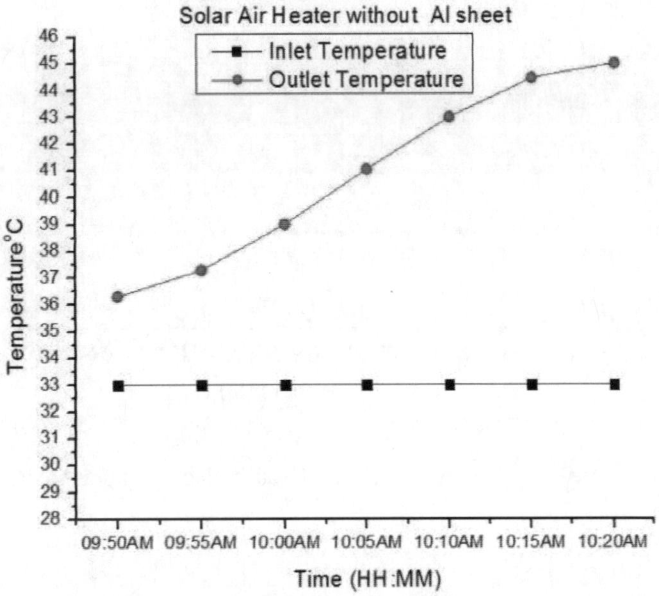

Figure 9(a) Temp vs Time SAH without Al sheet

IOP Conf. Series: Earth and Environmental Science **312** (2019) 012023 doi:10.1088/1755-1315/312/1/012023

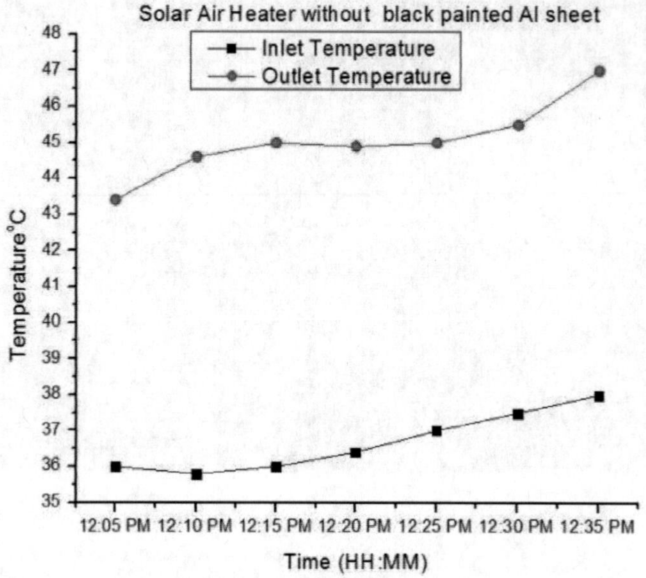

Figure 9(b) Temp vs Time SAH with Al sheet (no black coat)

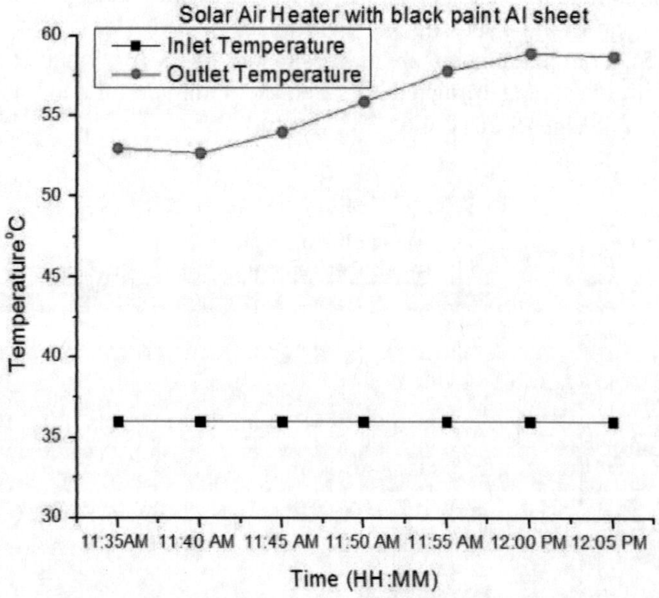

Figure 9(c) Temp vs Time SAH with black painted Al sheet

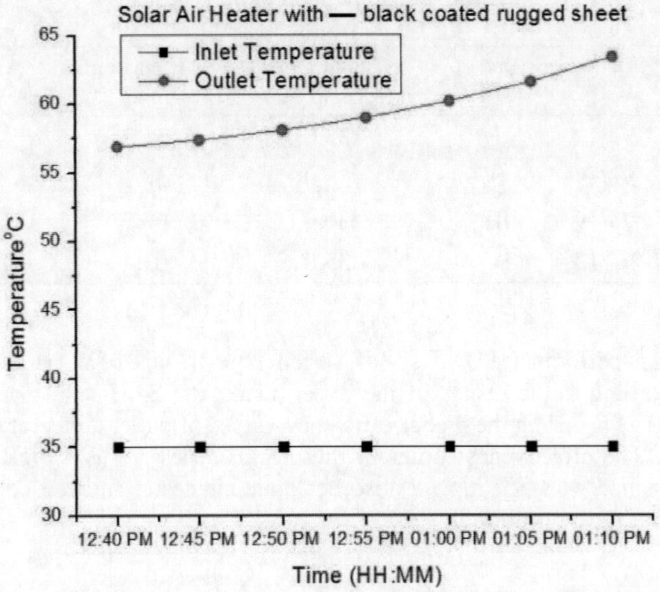

Figure 9(d) Temp vs Time SAH with black coated rugged Al sheet

Graphical data illustrates, when time changes, collector temperature increased in between 10:00 AM to 2 PM. Thermal efficiency also plays vital role in performance of SAH. Hence, depending on solar radiation with AM to PM, the thermal efficiency increasing gradually. Figure 10 is the comparison graph of all the 4 types of solar air heaters, in which Type IV is having the higher in efficiency by obtaining outlet temp because of the double pass of air.

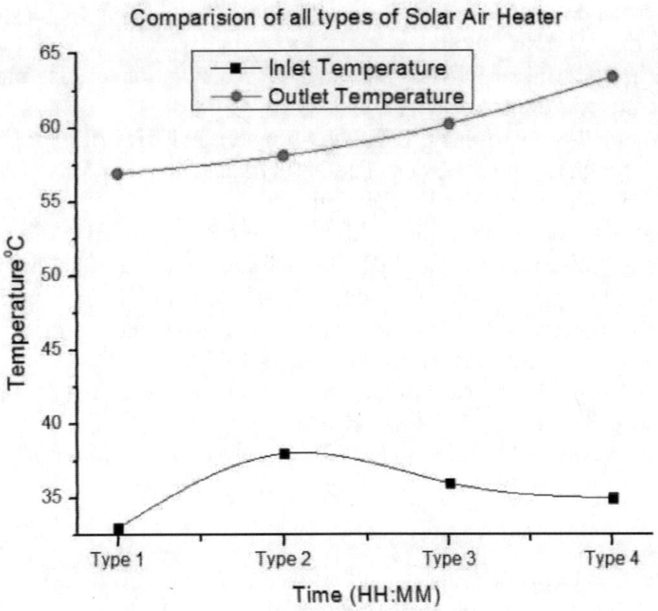

Figure 10. Temp vs Time graph in comparison of 4 types

Table 4. Comparison of all types SAH

S.No	Date	Type of SAH	SAH min inlet Temp (°C)	SAH max outlet Temp (°C)	Air flow Avg (m/s)
1	12/3/19	I	33.0	56.9	1.8
2	12/3/19	II	38.0	58.1	1.7
3	12/3/19	III	36.0	60.2	1.7
4	12/3/19	IV	35.0	63.4	1.8

6. Conclusion:

This paper deliver the performance analysis of varied types (four) of SAH by changing different absorber plates. According to the results of the experiments, the SAH with black coated Al rugged absorber plate (Type IV) is having the higher efficiency values of outlet temperatures, when compared with the other types. The effectiveness order of these SAH types are as Type IV, III, II and I in a descending manner. Since v-shaped rugged absorber plate air heater showed better values, this type was considered for heating application.

References

[1] Sreekumar A Development of solar air heaters & thermal energy storage system for drying application in food processing 2007 Thesis on Doctor of Philosophy, Department of Physics Cochin University of Science and Technology Kochi - 22 India

[2] Franco A, Shake M, Kalubi D and Hostettler S A Review of Sustainable Energy Access and Health care facilities 2017 *J. Sustainable Energy Technologies and Assessment* **22** 92-105

[3] Rajesh S and Choudary RB A cost effective desalination plant using a solar chimney with recycled aluminum can collector 2016 *J. Solar Energy* **2016** 1-10

[4] Abene A, Dubois V, Le Ray M and Ouagued A Study of a solar air flat plate collector use of obstacles and application for the drying of grape 2004 *J. Food Eng* **65** 15-22

[5] Karsli S Performance analysis of new-design solar air collector for drying applications 2007 *Renewable Energy* **32** 1645-60

[6] Karim Md A and Hawaldar MNA Performance investigation of flat plate, v-corrugated and finned air collectors 2006 *Energy* **31** 450-70

[7] Momin M E, Saini J S and Solanki S C Heat transfer and friction in solar air heater duct with Vshaped rib roughness on absorber plate 2002 *Int. J. Heat and Mass Transfer* **45** 3383-96.

[8] Esen H Experimental energy and exergy analysis of a double-flow solar air heater having different obstacles on absorber plates 2008 *Build Environ* **43** 1046–54

[9] Alvarez G, Arce J, Lira L and Heras MR Thermal performance of an air solar collector with an absorber plate made of recyclable aluminum cans 2004 *Solar Energy* **77** 107-13

[10] Rajesh S and Choudary RB A cost effective solar air heater with recycled al can (RAC) collector 2017 *J. Mat Sci and Mechanical Engg* **4** 45-50

[11] Koyunchu T Performance of various design of solar air heaters for crop drying applications 2006 *Renewable Energy* **31** 1073-88

[12] Sukhatme SP and Nayak JK 2008 *Solar Energy- principles of thermal collection and storage* (New Delhi: McGraw Hill Education) chapter 4 pp 173-199

Acknowledgements

Authors wishing to acknowledge assistance and cooperation from Project Associates Mr.R.Haibabu, Mr.Y.Muralikrishna, Mr.P.V.S.Venumadhav Reddy and Mr.Sk.Jilani from Department of Mechanical Engineering in Raghu Engineering College, Visakhapatnam, Andhra Pradesh, India.

IOP Conf. Series: Earth and Environmental Science **312** (2019) 012024 doi:10.1088/1755-1315/312/1/012024

Design and fabricate a solar torrefaction unit to improve the sugarcane bagasse quality for pyrolysis reaction

R Nagarajan[1], S Murugavelh[2]

[1]School of Mechanical engineering, Vellore Institute of Technology, Vellore, India.
[2]CO_2 Research and Green Technologies Centre, Vellore Institute of Technology, Vellore, India.

Corresponding Author: murugavelh.s@vit.ac.in

Abstract. High fertile soil and optimum sunlight lead to steady increasing sugarcane production in India. It directs second place in sugarcane production. At the same time, the sugarcane-based industries are facing a huge hurdle of disposal of solid residual called biowaste. It is a major issue in a fast-developing country like India. Pyro-oil from biowaste is one of the favourable resources for alternate energy. In this paper, a solar concentrator was designed and fabricated for torrefying the biowaste to improve the bio-oil yield from pyrolysis reaction. The biowaste form sugarcane was characteristic by the thermal-degradation analysis and internal composition analysis. The biowaste was torrefied at various temperature 160 °C, 170 °C, 180 °C, 190 °C. The optimum temperature for the torrefaction was 180 °C at 1hour. The pyro-oil yield was 25.8% and 22.1% for torrefied biowaste and wet biowaste respectively.

1. Introduction

Relic-based carbon fuels fulfil 80% of energy needs for the world. Report from the department of economic and social affairs from UN states that every year 83 million people were newly added in the inhabitants [1]. Due to the exponential growth in inhabitants directs to high utilization of energy. It has been reported that a surge of 24% in energy demand by 2040 results depletion and up-market in crude oil [2]. Another reports from BP energy economic states that the estimated lifetime for fossil fuel was 40-60 years and coal will sustain for another 100 years [3]. Therefore, the upcoming trend in the energy sector was using economic fuel such as coal instead of crude oil. Our current technology to create electricity was burning coal or fossil fuels in power plant and it was the major contributor for greenhouse gas like carbon-di-oxide, methane and Sulphur-di-oxide [4]. These gases are a major contributor to the chance of climate variation and surface temperature. The most of the country's in the world face the global warming effect such as irregular weather pattern, typhoons and el-Nino [5]. The situation guides to avoid relic-based fuels and recommends to find a suitable environmentally viable fuel.

Bio-waste from sugarcane is one of the most capable substitute resources for fuels [6]. Due to its extravagant production in nature and high-level carbon content. India is the leading cultivator of sugarcane in the world. India cultivates almost 290 million tons of sugarcane per year. It is equal to 25% of sugar for the entire world. From one ton of sugarcane, nearly 305 kgs of bio-waste can be collected. In India, around 435 sugar mills have a steam boiler. Bio-waste is used as a fuel for the boiler due to its cheap availability. The direct burning of bio-waste from sugarcane will produce roughly about 2.5 kgs of steam for one kilo of biowaste. The evidence from the sugar mill association from India reports that 35.7 million tons of sugar are manufactured at 2018-2019 years [7]. It is 4 million tons greater than in

IOP Conf. Series: Earth and Environmental Science **312** (2019) 012024 doi:10.1088/1755-1315/312/1/012024

previous years. Due to the instant progress in the sugarcane sector will generate a large amount of bio-waste. Therefore, special consideration is necessary to manage the bio-waste of sugarcane. Traditionally, the bio-waste from sugarcane has been an open fire in the fields. It causes air contamination and heads to asthma, eye irritation and even lunges cancer. To overcome these hindrances, people are using the bio-waste as an additive for green brick, natural acoustic damper, producing ceramic, glass material and biofuel like ethanol [8] & bio-oil.

Pyrolysis is the most suitable process to converted biowaste into bio-fuels. In pyrolysis, bio-waste is warm up to 350 °C to 550 °C in the passive atmosphere [9]. Due to that hemicellulose, lignin and cellulose will rupture apart into small molecular. Its results, discharge of flammable gases such as H2 and CH4. In additional pyro-gas can convert into bio-oil by reducing the entropy. The left-over biochar is promptly used to water distillation and soil amendment [10]. To improve the pyrolysis efficiency of biowaste, moisture content should be diminished. It has been reported that biowaste from sugarcane has 45% - 51% of water content [11]. Torrefaction process is required to eradicate the bound moister. To minimize the power consumption for the torrefaction process. The solar collector is used to power the device for the desired amount. Effect of torrefaction in biowaste from sugarcane conducted in temperature 150 to 170 °C for residence time one hour. It's led to improvements in properties homogeneous configuration of biomass, higher bulk energy capacity, high porous structure, water repulsion capacity and eradicate the rotting small from the biowaste [12].

In this study, a solar collector was designed and fabricated, the optimum time and temperature were tested for torrefaction, Composition analysis and TGA/DTG analysis was tested for biowaste. Pyrolysis was performed for wet and dry biowaste from sugarcane.

2. Material and Experimental methods

Biowaste of sugarcane was collected from juice shop inside the VIT, Vellore campus. The wet and dry biomass was chopped and ground into small pieces, then it fed into a multistage sieve machine to segregate the desired particle size. The wet and dry biowaste were subjected to ultimate and proximate analysis. On the other hand, wet biowaste was dried using the solar collector. The solar dryer consists of a frame with parabolic shape, reflect material, a small reactor made up of copper tube (figure 1).

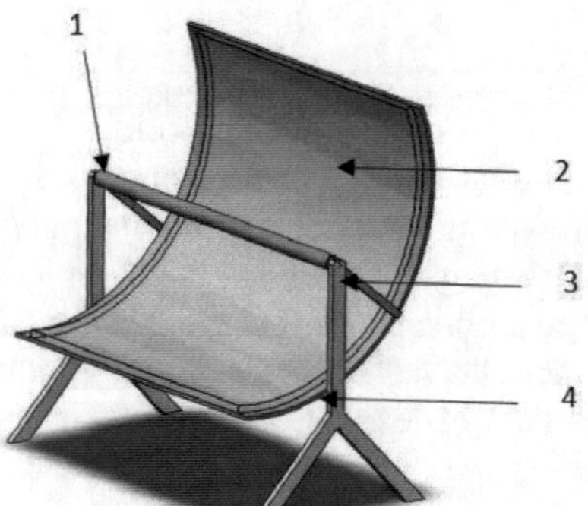

1. Copper Tube
2. Reflecting surface
3. Stand for solar collector
4. Parabolic frame

Figure 1. Solar collector for torrefaction

The space occupied by the solar collector was 1m × 1m × 1m. A steel frame is used to make an outer skeleton for the parabolic shape, where all the glass mirrors can fit in as a reflector. All reflector was focused on a single point to increase the temperature of the copper tube. All biowaste is kept inside the copper tube and kept on the focal point of the collector. The bio-oil were extracted from wet and dry biowaste by pyrolysis reaction. The reactor consists of a furnace, biomass chamber, condenser unit, vacuum pump (figure 2).

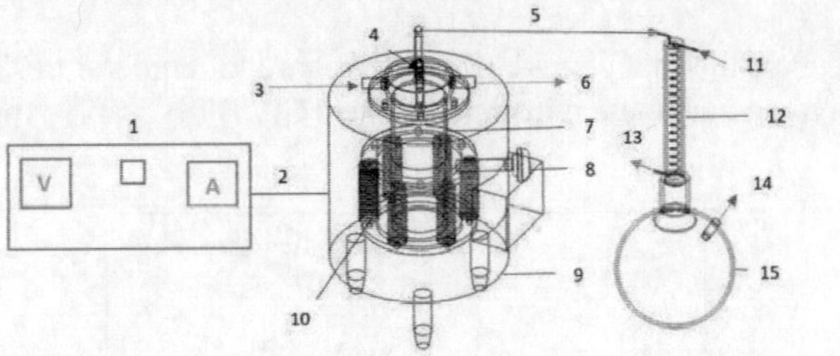

1.	Control Unit.	5.	Pyrolysis gas outlet.	9.	Outer body of reactor	13.	Water outlet from condenser
2.	Control line.	6.	Water outlet for silicon O-ring.	10.	Heating coil.	14.	Excess gas outlet.
3.	Water inlet for silicon O-ring.	7.	High-pressure reaction chamber.	11.	Water inlet to condenser	15.	Round bottom flask.
4.	Nitrogen inlet.	8.	Thermocouple.	12.	Condenser		

Figure 2. Fixed bed pyrolysis reactor for Sugarcane waste

Three-dimension measurement of the reactor is 0.25 m × 0.095 m × 0.005 m. The biomass chamber was constructed by 309 (S) stainless steel to resist high temperature. The furnace was covered by ceramic wool to prevent the heat loss. The total condenser unit was designed by the glass for easy visibility of oil flow. The vacuum pump is to suck the pyro-gas through the condenser column. Initially, wet biowaste of sugarcane was fed into the copper chamber and closed both the ends. The copper chamber was placed in the focal point of the parabolic solar collector. Due to the increases in temperature all unbound moisture will remove. Then the dried biowaste was tested ultimate and proximate analysis. On other hand, the dried biowaste was directly fed into the pyrolysis chamber. The temperature of 450 °C was set based on the thermogravimetric analysis and derivative gravimetric analysis. Finally, the pyro-oil yield was measured separately for wet and dry biowaste.

3. Theoretical Calculation

The volume of the solar power collector and the energy desired for torrefaction was computed. The equation is exploited to determine the complete energy for torrefaction [13].

$$Q_{total} = Q_{drying} + Q_{torr} \tag{1}$$

Total energy essential for dry out was computed by the equation.

$$Q_{drying} = [(1\text{-}m)/100\ C_{pb}\ (T_{finial} - T_{initial})] + [m_w C_{pw}\ (T_{finial} - T_{initial})] + L \tag{2}$$

The maximum and minimum energy required for torrefaction was 600 - 1000 kJ/kg. Assumed the torrefaction required 800 kJ/kg for 1 kg od biowaste.

$$Q_{torr} = [(1\text{-}m)\ C_p\ (T_{finial} - T_{initial}) + 800] \tag{3}$$

Solar energy should be equal to the total energy for the drying and torrefaction. The total energy is the minimum energy for drying and torrefaction process. To obtain the desired total energy, area and length of the solar collector play the crucial role. The area of the solar collector was evaluated by the equation.

$$\text{Average of the collector} = (Q_{total} * 1000)/\text{Average solar radiation} * \text{Collector efficiency} \tag{4}$$

The diameter and the focal length were denoted as d and f respectively. Length of the solar collector was evaluated by the equation.

$$\text{Length of the collector} = \text{Area} / [(d/2) + (1 + (d^2/16f^2))^2 + 2f \ln (d/4f) + (1 + (d^2/16f^2))^2] \quad (5)$$

4. Results and analysis

4.1. Feedstock analysis

The TGA/DTG graph denotes thermal degradation with regard to temperature. The TGA/DTG analysis was handled at 5 °C/min in an inactive environment. The TGA/DTG graph exposed in figure 3.

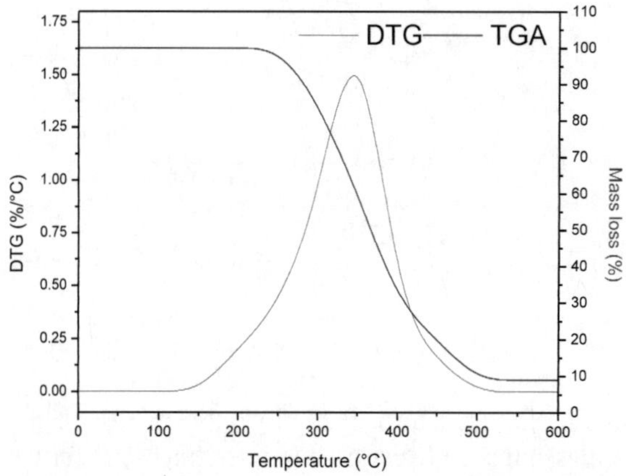

Figure 3. DSC/TGA analysis of sugarcane waste

The total degradation occurs between 0 °C - 600 °C. The TGA curve can separate into four areas. The first area denotes that removing of moisture and it occurs at 0 °C - 180 °C. In the second area, all cellulose and hemicellulose will rupture at 200 °C - 450 °C and it emit the combustible and non-combustible gases like H2, CO, CO2, CH4 and C_xY_x. The final area also called char formation area. It occurs at 450 °C – 600 °C. After 600 °C are called inactive area in that no mass loss occurs.
The proximate analysis shows 45.1% of moisture content, 69.5% of volatile content, 2.2% of ash content and 11.2% of fixed carbon content. Biowaste is more suitable for pyrolysis process due to its high volatile content and low ash content. These two parameters were more significant for high yield at low temperature. The proximate, ultimate analysis was revealed in Table 1.

Table 1. Psychochemical analysis of sugarcane biowaste

Proximate Analysis		Ultimate Analysis		Chemical Composition	
Moisture (%)	45.1	C (%)	42.98	Cellulose	43.3
Volatile (%)	69.5	H (%)	4.6	Hemi-Cellulose	31.5
Ash (%)	2.2	O (%)	22.18	Lignin	12.9
Fixed carbond (%)	11.2	N (%)	39.12	-	-
Q_{net} (KJ/kg)	1875	S (%)	-	-	-

The ultimate analysis shows 42.98%, 4.6%, 22.18% and 39.12% for carbon, hydrogen, oxygen and nitrogen respectively. The oxygen content was considerably lesser than wood biomass (40% to 45%). The quality and oxidational stability of pyro-oil will boost by the lesser oxygen content. Biowaste contains zero percentage of sulphur content. It means fuel taking out from biowaste is suitable for BS-VI norms. The energy value of biowaste was 18.75 MJ/kg It is roughly equivalent to brown coal. The chemical composition indicates 43.3% of cellulose content, 31.5% of hemicellulose content and 12.9% lignin content. Cellulose and hemicellulose content were much higher than the typical wood. High lignin was unfavourable substance for incineration process but biowaste from sugarcane shows much lower than the other biomass. It is a more positive approach to the pyrolysis method.

4.2. Experiment
From equation (1) to (5) is used to calculate the basic parameters for the collector (table 2).

Table 2. Calculate value for the solar collector

Parameters	Value
Q_{total}	1.59 MJ/kg
Area	1.052 m^2
Length	0.91 m
Diameter	1 m
Focal	0.25 m
Rim angle	90°

Torrefaction is a pre-processing method used to shrink the unbound moisture and intensify the energy value of the biowaste. It further diminishes the unsaturated group and promotes the hydrogen bonds. In torrefaction, mass loss occurs and volatile gas rises by converting the cellulose component into hemicellulose. One kilogram of biowaste from sugarcane was packed into the copper tube. To avoid heat environmental contact both side of the tube was sealed by a copper cap. The copper tube was placed at the focal point of the parabolic collector. The parabolic collector guides the sun radiation to the focal point to increases the temperature (Figure 4).

Figure 4. Parabolic solar collector

The experimentation was carried out at four different temperature (160 °C, 170 °C, 180 °C and 190 °C) for one hour to estimate the unbound moisture (table 3). The moisture content was degraded from 45% to 7.56% at 190 °C but few biomasses started the carbonization process. It clearly shows 190 °C was not suitable for the pyrolysis process. The next least moisture content was 190 °C. It reduces 45% moisture to 8.01%. The least moisture without carbonization process is better for the pyrolysis. After torrified, the biowaste was sent to the grinding machine to convert in the form of powder.

Table 3. Moisture content after torrefaction of sugarcane biowaste

Temperature (°C)	Moisture (%)

160	9.56
170	8.72
180	8.01
190	7.56

Then, the powder was filled in the biomass chamber and sealed to prevent the contact from the atmosphere. The experiment was performed in three different temperature 450 °C, 500 °C, 550 °C. The highest yield was attained from 500 °C. On the other hand, an identical experiment was conducted for wet biowaste (Figure 4.).

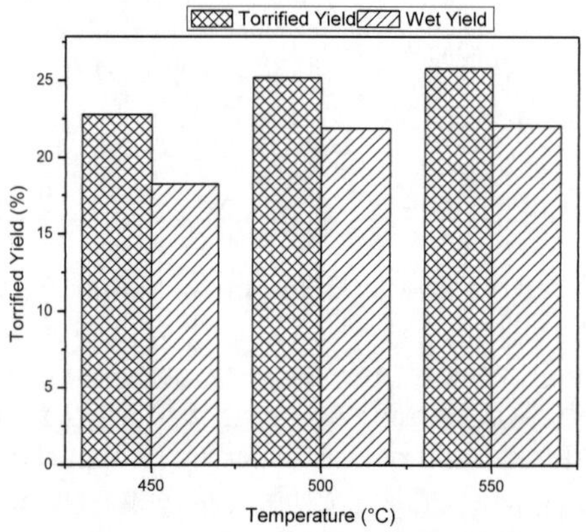

Figure 5. Pyro-oil yield for torrified and wet biomass

The graph shows that the pyro-oil yield for torrefied biomass was slightly higher than the wet biomass. It is due to the conversion of hemicellulose to cellulose component by limiting the OH group. It makes the biowaste into a high energy source. Torrefaction process leads the biowaste to a brittle material. Due to that, biowaste can convert into a fine particle which enhances the heat conduction property for the biowaste. This property plays a vital role in the pyrolysis yield.

5. Conclusion

Biowaste from sugarcane has high unbound moisture and it is the major hindrance for the pyrolysis process. To rectify this drawback a solar collector was designed and manufactured for the torrefaction. The research was conducted to find the optimum temperature for the torrefaction process and a higher yield on bio-oil.

1. Four different temperature (160 °C, 170 °C, 180 °C and 190 °C) were studied on torrefaction process. From that, 180 °C showed low unbound moisture without starting of carbonization process.
2. Pyrolysis was conducted for wet and torrefied biowaste. Wet biowaste produced less pyro-oil and high-water content. On the other hand, torrefied biowaste produces less water content and more pyro-oil. The highest yield for pyro-oil was 25.8% and 22.1% for torrefied biowaste and wet biomass respectively.

The inferences of this project confirm that the torrefied process diminishes the unbound moisture with the help of free solar energy and indicates that torrified biowaste at 500 °C produces extreme pyro-oil yield.

6. Reference

[1] Yanto D, lianto S D and Budimanb A 2015 *Ener. Procedia* **68** 157
[2] Exxon Mobil 2013 *The Outlook for Energy*: A View to 2040.
[3] Chen W H, Chen C J, Hung C I, Shen C H and Hsu H W 2013 *Appl. Energy* **112** 421.

IOP Conf. Series: Earth and Environmental Science **312** (2019) 012024 doi:10.1088/1755-1315/312/1/012024

[4] Patel B, Gami B and Bhimani H 2011 *En. Sust. Dev* **15** 372.
[5] Chen W H and Sheen HK 2012 *Bioreso. Techl* **118** 195.
[6] Al Arni S, Bosio B and Arato E 2010 *Renew. Ener* **35** 29.
[7] Pellegrini L F, Junior S O and Burbano J C 2010 *Energy* **35** 1172.
[8] Swarnakar A R, Choubey S 2016 *Int. Res. Jou. of Eng. and Tech* **03** 249.
[9] Kothari K, Shah A and Murugavelh S 2015 *Int.Jou. of Chem. Tech. Research* **8** 349.
[10] Cunha J A, Pereira M M, Valente L M M, Piscina P R, Homs N and Santos M R L 2011 *Biomass Bioene* **35** 2106.
[11] Wang Q, Liu S, Yang G and Chen J 2013 *Bioresour. Technol* **129** 676.
[12] Sarkar M, Kumar A, Tumuluru J S, Patil K N and Bellmer D D 2014 *Appl. Energy* **127** 194.
[13] Swaminathan R and Nandjembo F N P 2016 *Jou. of Sus. Bioe. Sys* **6** 66.

Acknowledgement
This project was sponsored by Science and Engineering Research Board, SERB, Science and Engineering Research, Department of Science & Technology, Government of India. Under ECR scheme.

Three dimensional numerical validation and investigation on air cooling system of Li-ion battery used in hybrid electric vehicles

Bhargav Ganesh Chekuri[1,3], Priyanka Satya Subhasree Vaskuri[1,4], S Shiva Kumar Chary[1,5], Thundil Karuppa Raj R[2,6]

[1]Student, Department of Automotive Engineering, Vellore Institute of Technology Vellore, Tamil Nadu, India
[2]Head of the Department of Automotive Engineering, Vellore Institute of Technology Vellore, Tamil Nadu, India

Email: [3]chekuribhargav.ganesh2018@vitstudent.ac.in
[4]priyanka.subhasree2018@vitstudent.ac.in [5]sshiva.kumarchary2018@vitstudent.ac.in
[6]thundil.rajagopal@vit.ac.in

Abstract The lithium ion batteries are most commonly used in hybrid electric vehicles for obtaining high energy density among the condemnatory issues that have been in automotive industry. As the thermal management system is very prominent in automotive batteries, an effective design consideration is to be carefully opted for cooling the battery. Especially, a forced cooling is followed as an empirical choice rather than natural cooling in the automotive industry. In this paper, a three dimensional battery pack with multiple cells and cooling passages is modelled and validated to maximize the durability and battery performance. An enumerated design of air-cooled system is validated from the literature and then numerically investigated the overall distribution of the temperature with different cases by changing the positions of the inlets and outlets of the battery to satisfy the thermal specification requirements of the battery. Since, a distinctive battery system is used in hybrid electric vehicle which contains multiple cells in parallel, cooling efficiency is accomplished by a uniform diffusion of air flow in the coolant passages which dissipates the heat evolved from the individual lithium-ion battery cell. The above four cases are compared and demonstrated that the efficient cooling performance is attained with the one inlet and one outlet which are placed at the opposite sides of the battery. This is acquired because, the faster the air flow rate, the more the system cools due to the high temperature dissipation.

Keywords: Temperature, cooling efficiency, fluid air,air flow rate,air cooling

IOP Conf. Series: Earth and Environmental Science **312** (2019) 012025 doi:10.1088/1755-1315/312/1/012025

1. Introduction

The common expectation from the hybrid electrical vehicles is to provide wide driving range and sensible refueling time for the modern hybrid passenger vehicles. The greatest advantage of these vehicles is the usage of smaller gasoline engine and the motor compared to the conventional and electric vehicles. In the hybrid vehicles the output power is obtained by the combination of powers of the gasoline engine and the motor. So these vehicles yield higher efficiency which is equal to the efficiency of the electric vehicles with the small sized motors. The dual purpose motor/generator promotes the regenerative braking in which the kinetic energy is recovered and used for charging the battery. Therefore, the batteries are needed to put up electrical currents as high as possible [2] for many charging and discharging cycles. The greatest power and energy is provided by lithium batteries in hybrid electric vehicle applications. For achieving the high voltage of the battery, the cells in the battery are chained in series to effectively run the hybrid electric vehicles [3].

Generally, in the lithium ion batteries the flow is happened in between the electrodes. During charging, the lithium ions shift from anode to cathode and in discharging, the lithium ions directly shift from cathode to anode. In the typical lithium ion battery lithium is used as anode and nickel oxide, cobalt, or manganese is used as cathode. In order to acquire more energy density of the battery the various electrolyte materials are explored [4]. In general there are two types of lithium ion batteries, one which uses the liquid or gel type of electrolyte and the other uses the solid polymer which acts as separator as well as electrolyte. The organic solution of electrolyte contains various solvents which show volatility and flammability [5] characters. The battery life depends on the calendar life in which the permanent loss of the capacity is observed. Not only high density but also safety and long durability issues commercialize the hybrid electric vehicles [6] .

Because of potential overheating [7] and exposure to more temperature [8-10] the battery undergoes very fast degradation. So, an appropriate cooling system is designed to maintain optimum battery temperature in various operating conditions [9] on which the life span of the battery is depended. Every 1^0C rise in temperature leads in reducing the life span of the battery by two months [11]. The battery temperature should be kept below 40^0 C and the difference in the temperature should be below 5^0 C [12]. Various cooling strategies are used in thermal management systems such as air cooling (natural air cooling, forced air cooling and indirect cooling), liquid cooling(direct and indirect liquid cooling) using water, acetone, refrigerants etc [14,15], air conditioning cooling and phase change phenomena cooling by using heat pipe [16]and PCM systems [18]are examined to enhance the cooling efficiency. A new technique is introduced by changing the phase of the material [17].In natural and forced convection circumstances the simulations have been processed for obtaining the heat dissipation and generation rates [8].

Wu et al [19] proposed the most effective cooling system in which forced air convection with heat pipe is used for controlling the high temperature of the battery. *Mahamud* and *park* [11] suggested a reciprocating air flow cooling system for reducing maximum temperature in cell and for enhancing the temperature uniformity. *Karimi* and *Li* [13] introduced another effective cooling system in which the distribution of forced convection is effective, cost effective and also maintains the uniform temperature and voltage across the battery. Air ventilation in some batteries is important for sending out the dangerous gases [20,21]. As the cooling is essential for battery thermal management system, various improved cooling techniques have been proposed effectively. Since air is the surrounding medium of the passenger vehicle, a better cooling system is recommended which reduces the cost and weight of the battery. Separate ventilation is provided in the front area of the passenger vehicle as entrance of the air which provides better cooling. In this paper, a forced air cooling system is used for the lithium-ion battery in hybrid electric vehicles for the various boundary parameters. And numerical simulation is carried out to attain the overall temperature distribution of the battery cells and coolant passages in the system.

2. Methodology

The validated battery system in this paper consists of 72 battery cells and 74 coolant passages. The coolant passages are placed in middle of the battery cells for dispelling the heat flux. A rectangular manifold is considered for both the inlet and outlet of the battery system. The battery cells considered in the system are prismatic in nature. The complete dimensions of the model has clearly mentioned in the below table [1].

Table 1. The dimensions of the battery system

	Length(mm)	Breadth(mm)	Height(mm)
Battery system	225	191	787
Coolant passages	3	65	151
Manifolds	--	--	20

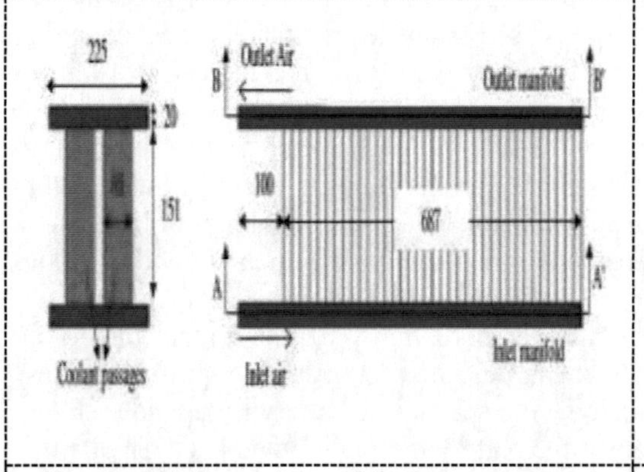

Figure 1. Dimensions of the battery system

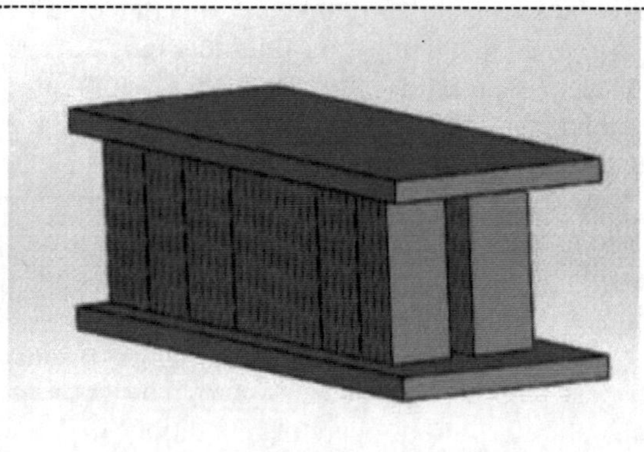

Figure 2. The solid works model

The above geometry taken from the literature [1] is modelled in the Solid woks and further taken to the numerical simulations. The geometry modelled in the solid works is then imported to cfd modelling tool ICEM CFD. In this, the whole geometry is specified and named accordingly. The blocking and the tetrahedral mesh of the considered design have obtained. The following figures show the modelled and meshed geometry in ICEM CFD.

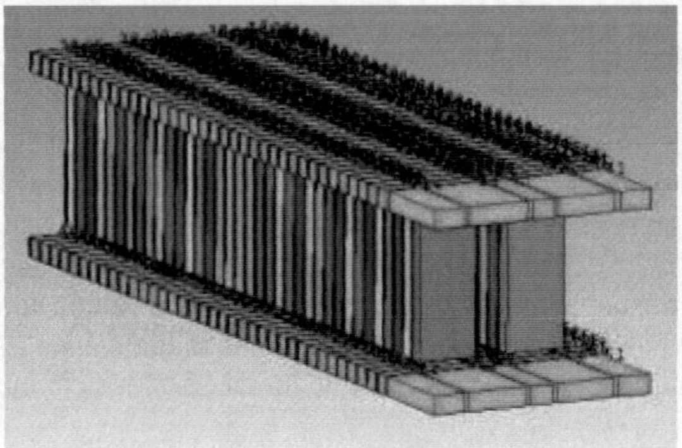

Figure 3. ICEM CFD geometry

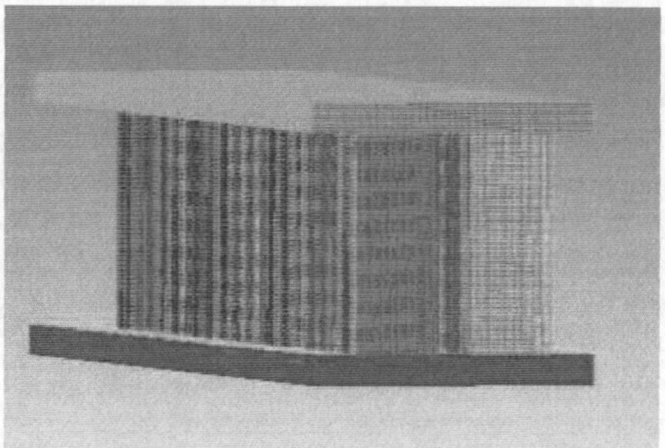

Figure 4. Meshed geometry

IOP Conf. Series: Earth and Environmental Science **312** (2019) 012025 doi:10.1088/1755-1315/312/1/012025

The final meshed model of the battery system is imported in to the ANSYS CFX where the considered model is simulated and numerically investigated. In the pre -processor of CFX the material properties and the boundary conditions are provided and simulated. The inlet temperature of the battery system is considered as 313 K. Air with the mass flow rate 0.045 kg sec^{-1} [1] is blown from the inlet of the battery system. The material of the battery is taken as aluminium from the literature. The following table shows the boundary conditions and the material properties of the model. In this paper, four cases are considered by changing the direction of the fluid air in the inlets and the outlets. The first case (validation) consists of single inlet and outlet which are placed on the identical sides of the battery. In the second case two inlets and two outlets are taken on the opposite sides of the battery. In the third case only one inlet and one outlet is taken on the opposite sides of the battery. The above four cases are simulated in the pre-processor of CFX and numerically investigated.

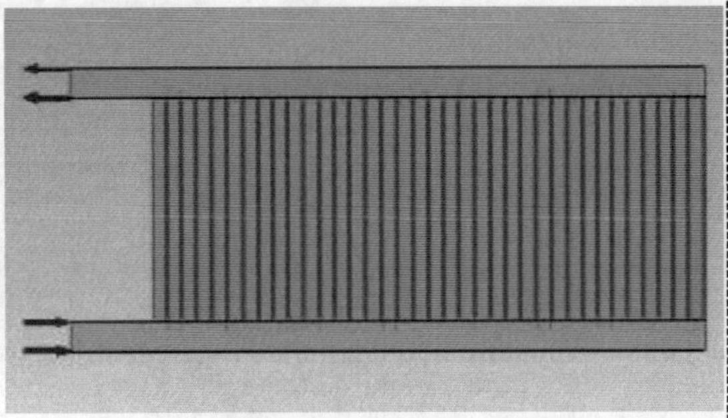

Figure 5. Case 1

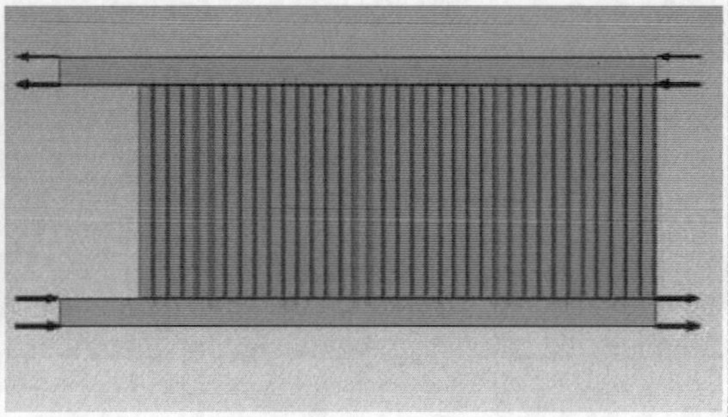

Figure 6. Case 2

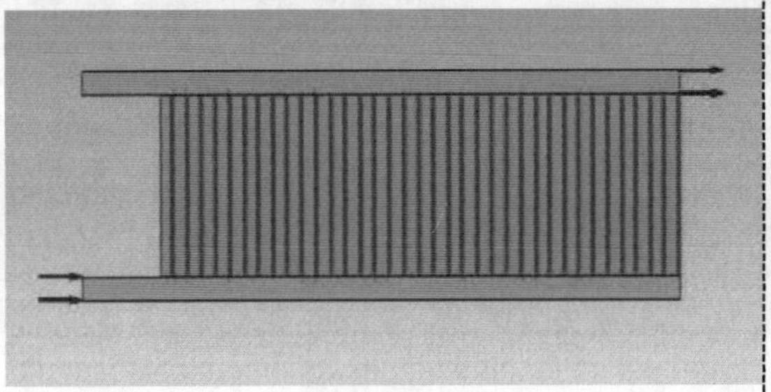

Figure 7. Case 3

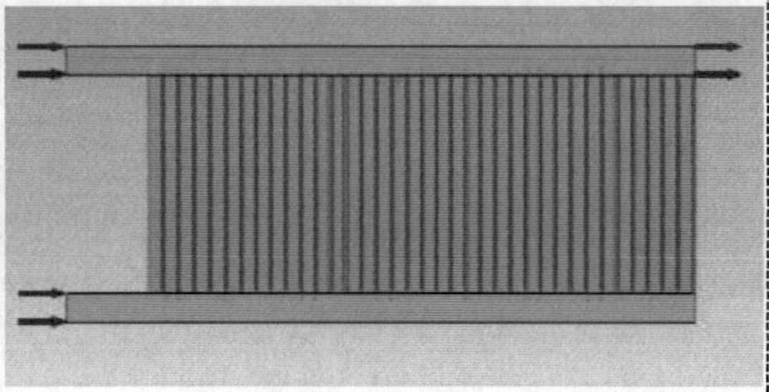

Figure 8. Case 4

Table 2. The material properties of the battery and air

	Fluid air	Battery cell
Density	1.184	2700
Dynamic viscosity	$1.98*10^{-5}$	---
Specific heat	1003	900
Thermal conductivity	435	240
Heat flux	--	254

IOP Conf. Series: Earth and Environmental Science **312** (2019) 012025 doi:10.1088/1755-1315/312/1/012025

3. Results and Discussions

The overall temperature distribution is obtained by conducting the numerical simulations for the four cases and the optimized design for the fluid flow is predicted. The required results can be extracted in the post-processor of the CFX.

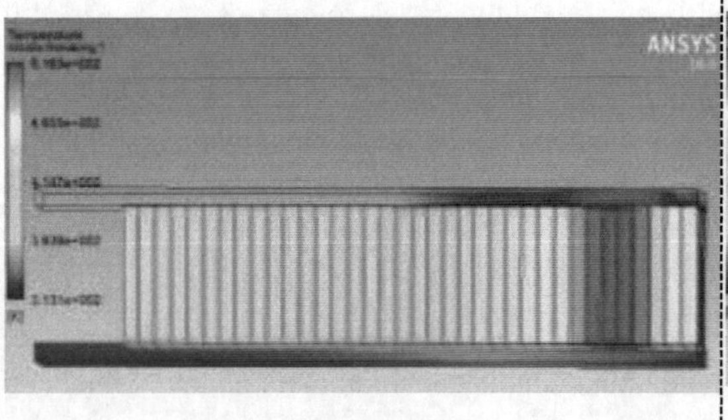

Figure 9. Case 1

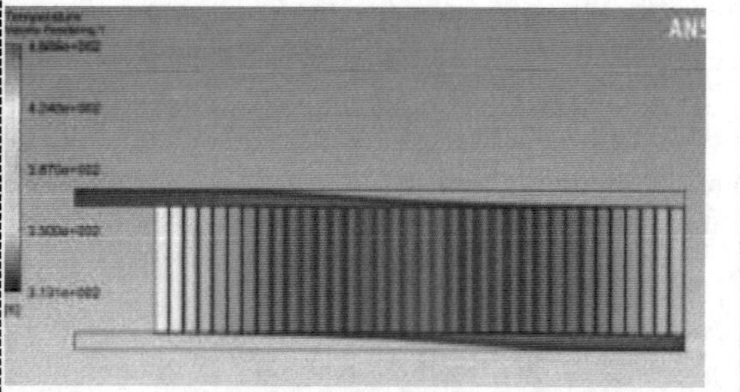

Figure 10. Case 2

Figure 11. Case 3

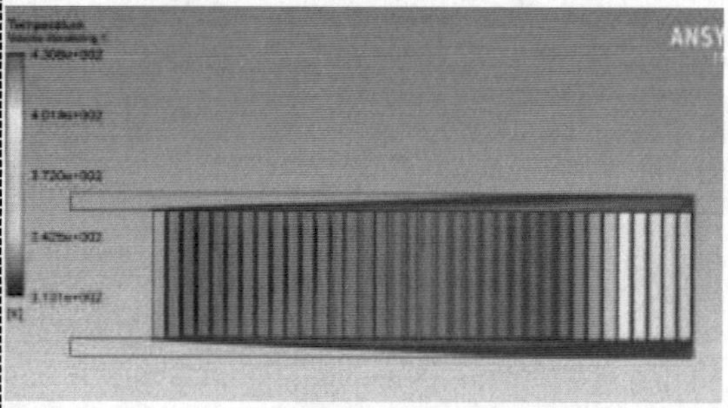

Figure 12. Case 4

The numbering of the cells and the coolant passages in the battery is given from the inlet manifold. The effective cooling is required for the lithium ion battery to obtain the optimum parameters. In the validation the air flows across the channels of the battery and reaches the top portion by cooling the cells of the battery .The hot air comes out from the top manifold which is in the opposite direction of the inlet flow. The heat transfer takes place in the whole battery and thereafter the temperature of the cells gets decreased. The amount of heat transfer directly depends on the air flow rate in the corresponding coolant passages. The outermost cells temperature is very critical due to its inefficient flow rates in the passages. So, the temperature of these cells increases.

The fluid air temperature in the first case is noted as 322.28 K and the highest temperature of the battery is 516 K observed at the 31st and the 67th cells of the battery. The lowest temperature is 372 K observed at the 1st and the 37th cells of the battery. The temperature distribution directly affects the battery SOC and the discharge rates. The temperature variation between the cells is higher than the desired threshold values of the battery which does not yield the optimum parameters.

In the second case the recorded fluid air temperature is 319.04 K. The highest and the lowest temperatures are 460 K and 420 K observed at the 24th, 60th and the 1st, 37th cells.

In the fourth case the noted fluid air temperature is 319.93 K. The highest and the lowest temperatures are 430 K and 375 K observed at the 4th, 40th and the 36th, 72nd cells of the battery. For the sake of obtaining the utmost cooling ability, there should be the uniform temperature distribution in between the cells. The above requirements are obtained in the third case where the fluid air temperature is 321.71 K. And the highest and the lowest temperatures are 420 K and 363 K observed at the 3rd, 39th and the

36th, 72nd cells of the battery. The temperature diffusion graph is plotted only for 1-36 cells because the same trend is followed by the 37-72 cells. The battery cells temperatures are plotted and compared for the four cases in the following figure.

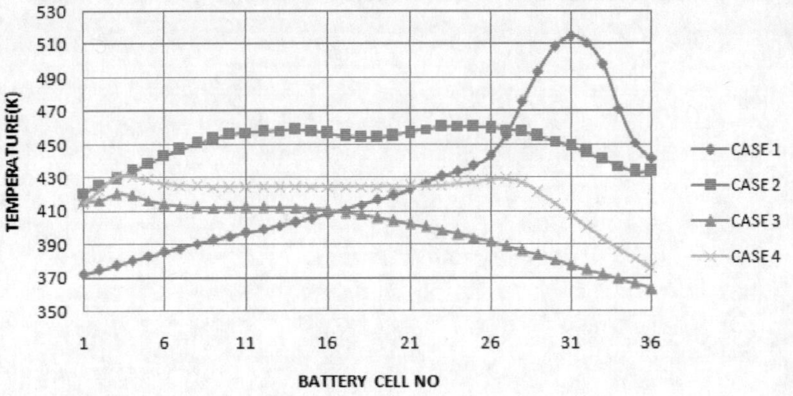

Figure 13. Overall temperature distribution in the battery cells.

4. Conclusions

The forced air cooling in the hybrid electric vehicles for the battery system is mostly suggested because of its low cost, compactness and the adaptable design. The different configurations of the air flow are developed and numerically calculated in this study. The recommended cooling performance is attained by providing one inlet and one outlet at the opposite sides of the battery. In the third case temperature consistency is obtained. This investigation can be further carried out by using forced convection with heat pipe which decreases the temperature and also obtain the temperature consistency. The simulations that have been done in this paper can further carried by introducing the tapered inlet and outlet manifolds which even more decrease the temperatures of the battery. It is anticipated that the above approach will be the standard one not only for the hybrid electric vehicles but also for the electric vehicles and fuel cell vehicles in which the battery cells are used.

References.

[1] H.park/*Journal of power resources* **239**(2013)
[2] J. Xun, R. Liu, K. Jiao, *J. Power Sources* **223** (2013) 47e61.
[3] M.R. Giuliano, S.G. Advani, A.K. Prasad, *J. Power Sources* **196** (2011) 6517e 6524.
[4] M.R. Giuliano, A.K. Prasad, S.G. Advani, *J. Power Sources* **216** (2012) 345e352.
[5] T. Tamura, T. Hachida, K. Yoshida, N. Tachikawa, *J. Power Sources* **195** (2010)6095e6100
[6] Q. Wang, P. Ping, X. Zhao, G. Chu, J. Sun, C. Chen, *J. Power Sources* **208** (2012)210e224.
[7] R. Kizilel, R. Sabbah, J.R. Selman, S. Al-Hallaj, *J. Power Sources* **194** (2009)1105e1112.
[8] C. Zhu, X. Li, L. Song, L. Xiang, *J. Power Sources* **223** (2013) 155e164.
[9] K.S. Hariharan, *J. Power Sources* **227** (2013) 171e176.
[10] L. Cai, R.E. White, *J. Power Sources* **196** (2011) 5985e5989.
[11] R. Mahamud, C. Park, *J. Power Sources* **196** (2011) 5685e5696.
[12] C.W. Park, A.K. Jaura, *SAE Technical Paper*, 2003, 2003-01-2286.
[13] G. Karimi, X. Li, Int. *J. Energy Res*. (2012), http://dx.doi.org/10.1002/er.1956.
[14] K. Yeow, H. Teng, M. Thelliez, E. Tan, *SAE Int. J. Alt. Power* **1** (2012) 65e78.
[15] Y. Li, H.Q. Xie, J. Li, *Appl. Mech. Mater*. **271** (2012) 182e185.
[16] Z. Rao, S. Wang, M. Wu, Z. Lin, F. Li, *Energy Convers. Manage*. **65** (2013) 92e97.
[17] R. Sabbah, R. Kizilel, J.R. Selman, S. Al-Hallaj, *J. Power Sources* **182** (2008)630e638.

IOP Conf. Series: Earth and Environmental Science **312** (2019) 012025 doi:10.1088/1755-1315/312/1/012025

[18] Z. Rao, S. Wang, Renew. *Sustain. Energy Rev.* **15** (2011) 4554e4571.
[19] M.S. Wu, K.H. Liu, Y.Y. Wang, C.C. Wan, *J. Power Sources* **109** (2002)160e166.
[20] A.A. Pesaran, Battery Thermal Management in EVs and HEVs: Issues and Solutions, in: *Advanced Automotive Battery Conference, Nevada, Las Vegas*, **6e8**February 2001.
[21] G.L.F. Kitanoski, C. Kussmann, *SAE Technical Paper*, 2007, 2007-01-3483.

Design and analysis of power convertors for electric vehicle applications

Safare Sudeep Babu[1,3], S Shiva Kumar Chary[1,4], Syed Muhammad Abdullah[1,5], Elangovan D[2,6], Thundil Karuppa Raj R[3,7]

[1]Student, Department of Automotive Engineering, Vellore Institute of Technology Vellore, Tamil Nadu, India
[2]Head of the Department of Power Electronics Engineering, Vellore Institute of Technology Vellore, Tamil Nadu, India
[3]Head of the Department of Automotive Engineering, Vellore Institute of Technology Vellore, Tamil Nadu, India

[3]safaresudeep.babu2018@vitstudent.ac.in
[4]sshiva.kumarchary2018@vitstudent.ac.in
[5]syedmuhammad.abdullah2018@vitstudent.ac.in
[6]elangovan.devaraj@vit.ac.in
[7]thundil.rajagopal@vit.ac.in

Abstract The surge in the cost of fuels can be accounted as a major trigger which leads to the advent of hybrid vehicles. The hybrid electric vehicles are in the reckoning to address the dangers of the exponential increase in contamination of air and global warming associated with greenhouse gases. A vehicle that is a combination of reversible storage devices like batteries and main energy sources (oil or gas) [1]. A HEV emanates reduced emission from its IC engine when compared to a pure petrol car of the same size. In the context of Hybrid Electric Vehicle, the DC-DC power converters are the electronic devices that condition the source voltage of one level to the output voltage of another level of different DC voltage based on the application provided for the electric vehicle for charging and discharging modes. The need for these power converters arises from the different voltage ratings of the hardware equipment being used. These converters acting importance due to the fact that they are efficient, easy to design and cost effective in the Hybrid electric vehicle application. The paper presents the modelling, design and simulation of buck converter the bidirectional converter based on the transmission line of powertrain of electric vehicle application. The aim is to obtain from a stable input voltage to the output voltage which is lower with high frequency and minimum ripple with specified fixed frequency in the buck operation. In bidirectional converter, the voltage stepdown is required while charging the battery while step up is needed while providing power to the motor from the battery. The design of the above-mentioned converters is performed using the standard method present in the literature. The simulation is done on PowerSim 8.0. The waveforms for the voltage and current for inductor, capacitor and the diode is also obtained. It is found that the simulation results of the voltages are in close agreement with the analytical calculations.

Keywords— Power convertors, Electric Powertrain, PowerSim, Buck, Bidirectional, DC-DC convertors

1. Introduction

The surge in the cost of fuels can be accounted as a major trigger which leads to the advent of hybrid vehicles. The hybrid electric vehicles are in the reckoning to address the dangers of the exponential increase in contamination of air and global warming associated with greenhouse gases. A vehicle that is a combination of reversible storage devices like batteries and main energy sources (oil or gas) [1].
A HEV emanates reduced emission from its IC engine when compared to a pure petrol car of the same size. This is because HEVs gasoline engine is smaller than comparably sized pure gasoline burning vehicle [2-3]. The idle emissions are reduced as the IC engine is shut down and restarted only when necessary. The best characteristics of IC engines and electric batteries are utilized as and when

necessary. This results in an enhanced fuel economy, less emissions, optimum power distribution for driving and auxiliary power requirements.

Figure 1. Electric Vehicle Auto-Rickshaw

In the electric power aspect, the design of power electronics converters play a very important role in an efficient conversion of power as per the ratings of the original power source and the ratings of the equipment used. Till recently, the use of power electronics was avoided in the architecture of electric vehicles owing to the huge cost involved. But presently, the power electronics gain immense importance and the reasons are the following. Firstly, the modern architecture which included the integrated switching and fusing functions into a single component with increased reliability. The power convertors can be used for different control methods Secondly, the conversion of power is realized by using adjustable speed drives. Lastly, the conversion of voltage as per the demand of the application can be efficiently realized by using power electronics. The different components used in the power electronics of an electric vehicle are rectifiers, inverters dc-dc converters etc.

The dc-dc power converters are used to condition the level of the voltage and to deliver a stable dc bus voltage. The input DC voltage is converted to a regulated output DC voltage. This process of conversion is highly efficient with an efficiency percentage of almost 90 %.. These converters operate at high frequencies of the order of 10 kHz - 1MHz. The high frequency ensures that size of the hardware components involved in the converter is small which in turn helps in the EV applications. [11]

The various types of types of DC-DC convertors which are not isolated are: Buck converter, Boost converter and Bidirectional converters.

Wangsupphaphol A et al [10] simulated the powertrain for EV using a bidirectional convertor on MATLAB. The result shows the capability to settling supply a significant amount of power for step load change within few milliseconds. The sudden demand for the power from the load can be obtained from the supercapacitor. This process minimizes the battery stress compared to the pure battery supply system. In this work, the buck and bidirectional converter is designed for conditioning the voltage between the power source, battery and motor. Please refer to the below diagram which indicates the power train diagram of "Electric Vehicle.

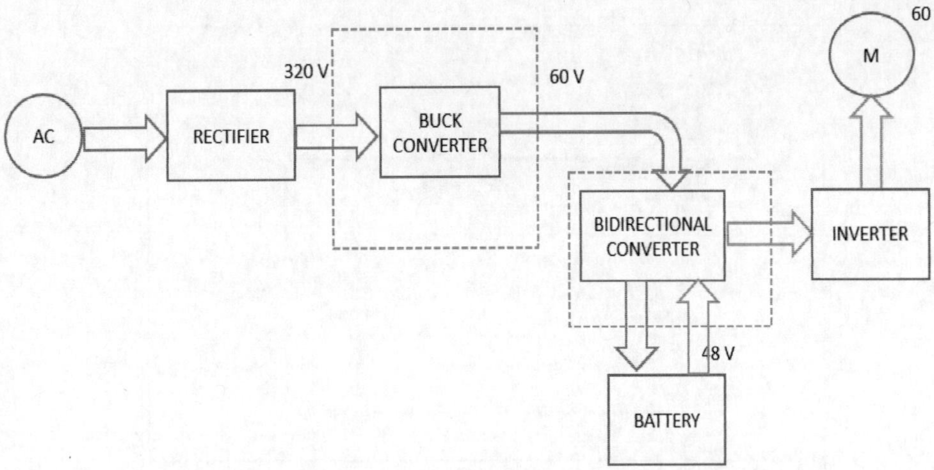

Figure 2. Power train diagram of Electric Vehicle

As evident from the above diagram, the source AC voltage (230V) is rectified to DC using a rectifier. The rms value of the rectified DC voltage is320 V (i.e. √2 * 230). This voltage is reduced to 60 V using a buck converter. Further, the 60 V is reduced to 48 V which is the rating of the battery. This is the procedure of the charging the battery. The 48V battery drives the 60 V using the bidirectional converter which now acts like a boost converter. The analytical design is performed using the formulae available in the literature. The simulation is performed on PowerSim software. The results are validated with the analytical design.

2. Analysis and Design

Buck Converter

The buck converter produces a pure DC voltage which is less than the input voltage in magnitude. To accomplish this action, the converter uses a combination of inductor-capacitor (LC) low pass filter. This low-pass filter is assumed to be ideal. The function of the buck converter is to provide an output voltage less than that of input voltage. The values of output current and voltage in the inductor is obtained for the switch open and switch closed states. In the steady state condition, the gross change for the current in the inductor is zero for a period. The average value of the voltage across the inductor is zero.

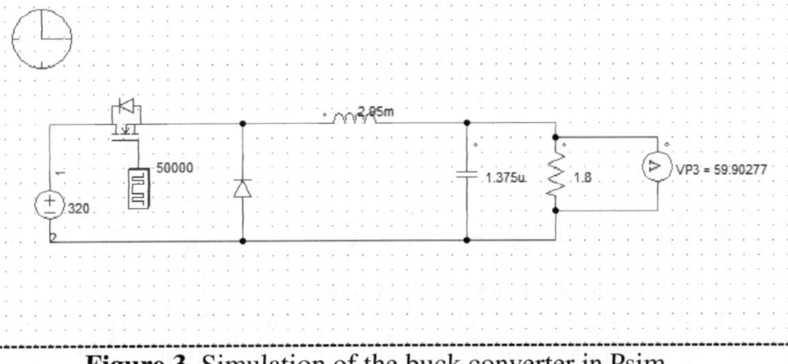

Figure 3. Simulation of the buck converter in Psim

IOP Conf. Series: Earth and Environmental Science **312** (2019) 012026 doi:10.1088/1755-1315/312/1/012026

Operation - switch closed

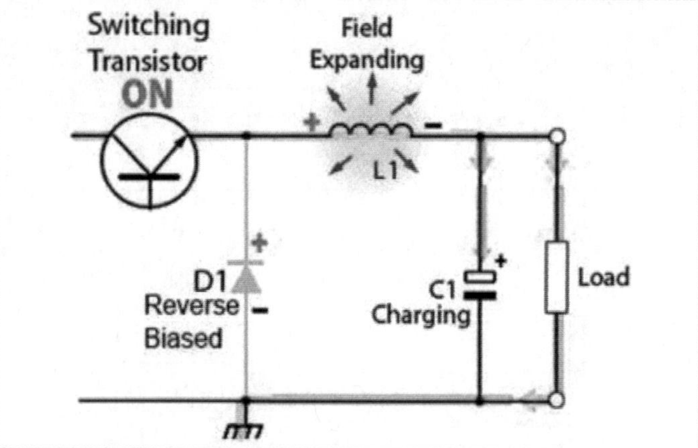

Figure 4. Operation of closed switch buck converter

When the switch is closed, the diode does not allow current to pass through it (reverse biased). The magnitude of the voltage in this condition across the inductor is $V_L = Vs - Vo$. As the rate of change of current is positive, the current increase in a linear mode. The net change in current when the switch is closed is obtained as per the below equation.

Note: D is duty cycle

$$(i_L)_{closed} = DT$$

Operation - switch open

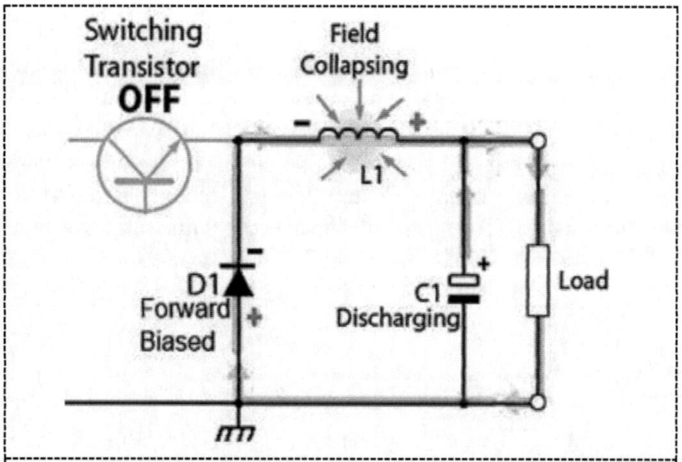

Figure 5. Operation of open switch buck converter

Similarly, in the open state of the switch, the diode allows current to pass through it (forward biased) because the polarity of the inductor changes. The inductor acts like the main source of power to the circuit. The magnitude of the voltage across the inductor in the open state of the switch $V_L = - Vo$. The rate of change of current in the open state of the switch is negative. Hence the current is observed to decrease in a linear pattern. The difference in the current from the original is as below.

ANALYSIS PART

Buck converter Inputs for

Power = 2kW

Input Voltage V_s = 320V

Output Voltage V_o = 60V

Frequency f = 50kHz

The average Inductor Current and output ripple Voltage is no more than 1%

Diode (D) = $\frac{60}{320} = \frac{Vo}{Vs}$

= 0.1875

Power (P) = $V_o^2 * R$

Resistance (R) = V_o^2 / P

= $60^2/2000$

= 1.8Ω

Inductor Current (I_L) = Output Current (I_o)

I = V_o /R

= $60^2/1.8$

= 33.33A

Average Inductor Current ΔI_L = 1% I_L

= 0.33A

Average Voltage = 1% V_o

= 0.6V

Inductor (L) = $\frac{(Vs-Vo)D}{\Delta IL * f}$

= $\frac{(320-60)0.1875}{0.33*50000}$

= 2.95 mH

Capacitor (C) = $\frac{\Delta IL}{8*f*\Delta V}$

= $\frac{0.33}{8*50000*0.6}$

= 1.75 μF

Bidirectional Converter

The bidirectional converter consists of two MOSFET switches i.e. the switch1 operates the voltage source to the battery whereas as switch2 operates the motor through the battery. It is known as bidirectional converter. In this operation when the switch1 is turned ON the buck mode will operate and supplies the voltage from source to battery whereas switch2 is turned ON the boost mode will operate and supplies the voltage from battery to the motor to run the wheels by keeping the power as constant throughout the circuit.

IOP Conf. Series: Earth and Environmental Science **312** (2019) 012026 doi:10.1088/1755-1315/312/1/012026

Simulation of Powertrain

The DC-DC converters for the electric powertrain are simulated by using one buck converter and one bidirectional converter.

The conditions of the analytical buck converter are shown in the below table.

The values of inductance, capacitance and resistance are calculated using the standard method available in the literature.

Table 1. Analysis calculation values of Buck Converter

PARAMETER	VALUE
INPUT VOLTAGE	320 V
OUTPUT VOLTAGE	60 V
DUTY CYCLE	0.1875
FREQUENCY	50 kHz
INDUCTANCE	2.95 mH
CAPACITANCE	1.75 μF
RESISTANCE	1.8 Ω

The resultant of the voltage obtained from the buck converter is applied to the bidirectional converter which will charge the battery in the buck action. Consequently, the battery powers the motor in the boost action of the bidirectional converter and thereby discharges itself.

BATTERY CHARGING MODE

In the battery charging mode, the bidirectional act as buck operation which is opearted in two modes i.e switch1 is ON mode whereas switch2 is OFF mode. The energy from the primary source of power is stored in the inductor in the form of magnetic field. This cycle repeats when switch2 is ON position.Hence the inductor recieves energy from the primary source when the switch is ON.

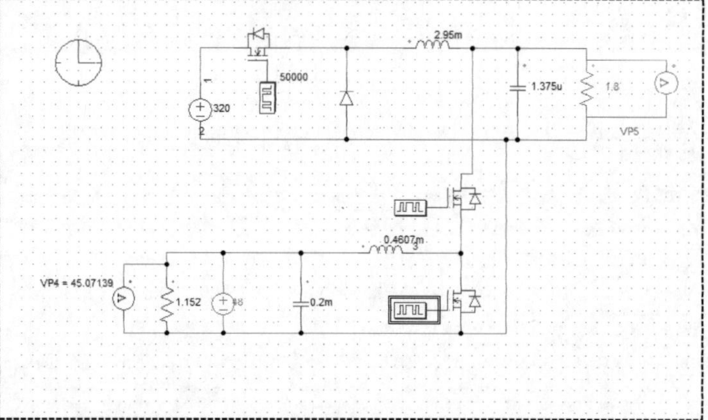

Figure 6. Simulation of Battery charging mode in Psim

The values of inductance, capacitance and resistance are calculated using the standard method available in the literature.

IOP Conf. Series: Earth and Environmental Science **312** (2019) 012026 doi:10.1088/1755-1315/312/1/012026

BATTERY DISCHARGE MODE

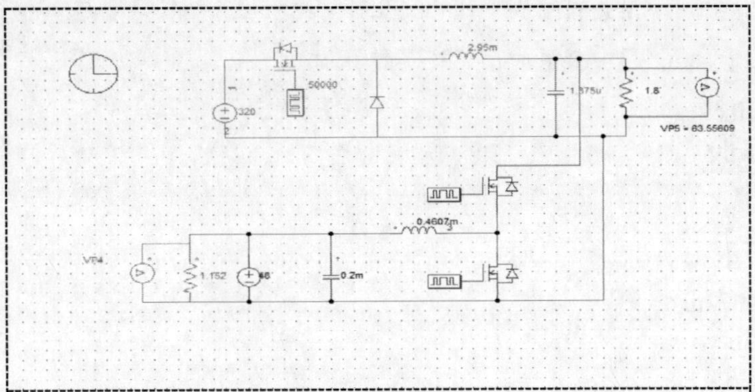

Figure 7. Simulation of Battery discharging mode in Psim

In this mode, the bidirectional act as boost operation by closing the switch2 and by opening switch1. In this operation the bidirectional will boost up the voltage from battery to the motor to run the wheels. The average current through the inductor in this mode is zero.

Analysis Part

The values of inductance, capacitance and resistance are calculated using the standard method available in the literature.

Bidirectional converter Inputs

Power = 2kW

Buck Voltage V_s = 60V

Battery Voltage V_o = 48V

Frequency f = 50 kHz

The average Inductor Current and output ripple Voltage for both buck and boost operation is no more than 1%.

BATTERY CHARGING MODE

Diode (D) = $\frac{Vo}{Vs} = \frac{48}{60}$

= 0.8

Power (P) = $V_o^2 * R$

Resistance (R) = V_o^2 / P

= $48^2/2000$

= 1.152Ω

Inductor Current I_L = Output Current I_o

$I_L = V_o / R$

= 48 / 1.152

= 41.67A

IOP Conf. Series: Earth and Environmental Science **312** (2019) 012026 doi:10.1088/1755-1315/312/1/012026

Average Inductor Current ΔI_L = 1% Inductor Current I_L

$$= 0.4167$$

Average Voltage = 1% Output Voltage V_o

$$= 0.6V$$

$$\text{Inductor (L)} = \frac{(Vs-Vo)D}{\Delta IL*f}$$

$$= \frac{(60-48)0.8}{0.4167*50000}$$

$$= 0.4607 \text{ mH}$$

$$\text{Capacitor (C)} = \frac{\Delta IL}{8*f*\Delta V}$$

$$= \frac{0.4167}{8*50000*0.48}$$

$$= 2.17\ \mu F$$

BATTERY DISCHARGING MODE

Power (P) = $V_o^2 * R$

Resistance (R) = V_o^2 / P

$$= 60^2/2000$$

$$= 1.8\Omega$$

$$\text{Capacitor (C)} = \frac{D}{R*f*(\frac{\Delta Vo}{V})}$$

$$= \frac{0.8}{1.8*50000*0.01}$$

$$= 0.888 mF$$

In Discharging mode only resistance and capacitor values change remain all same as a Charging mode. The conditions of the Analytical part of bidirectional converter in buck and boost action are shown in the below table

Table 2. Analysis calculation values of Bidirectional Converter

PARAMETER	ANALYTICAL (BUCK)	ANALYTICAL (BOOST)
RESISTANCE	1.152 Ω	1.8 Ω
INDUCTOR CURRENT	41.19 A	41.19
INDUCTANCE	0.4607 mH	0.4607 mH
CAPACITANCE	2.17 μF	0.2 mF
OUTPUT VOLTAGE	46.08 V	64.23 V

IOP Conf. Series: Earth and Environmental Science **312** (2019) 012026 doi:10.1088/1755-1315/312/1/012026

3. Results and discussion

The simulation results for the buck converters and bidirectional converters with the respective peak values are shown below.

Buck Converter Waveforms

The following table shows the values of the results obtained in simulation using PSIM.

Table 3. Simulation values of Buck Converter

PARAMETER	SIMULATION
CAPACITOR CURRENT	0.05788 A
DIODE CURRENT	3.38 A
INDUCTOR CURRENT	3.38 A
CAPACITOR VOLTAGE	60.2249 V
RESISTOR CURRENT	3.34 A
OUTPUT VOLTAGE	60.2249V

The following waveforms shows the results obtained in simulation using PSIM.

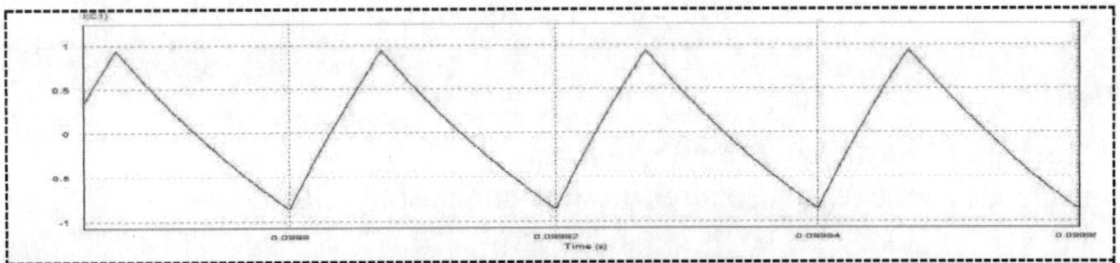

CAPACITOR CURRENT
PEAK VALUE: 0.5788A

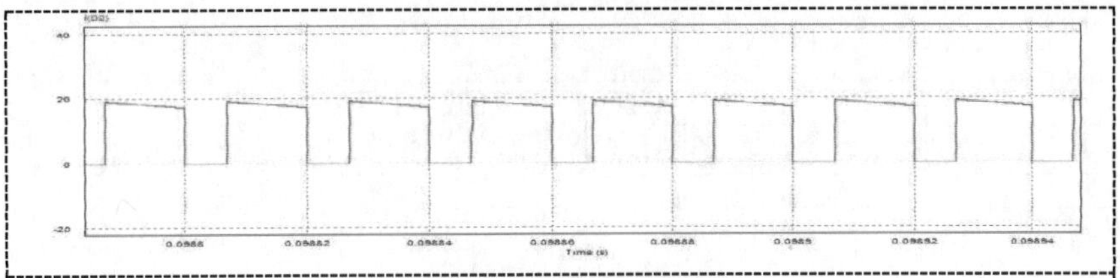

DIODE CURRENT
PEAK VALUE: 3.38A

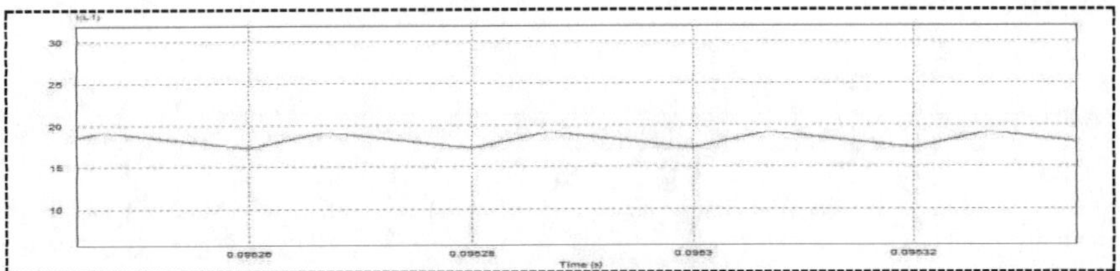

INDUCTOR CURRENT
PEAK VALUE: 3.38A

IOP Conf. Series: Earth and Environmental Science **312** (2019) 012026 doi:10.1088/1755-1315/312/1/012026

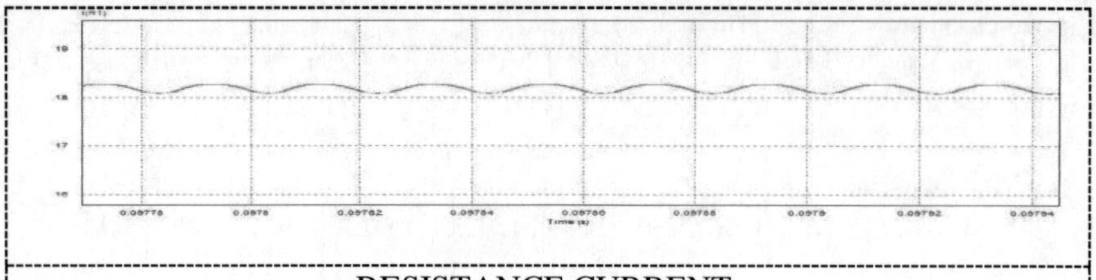

RESISTANCE CURRENT
PEAK VALUE: 3.34A

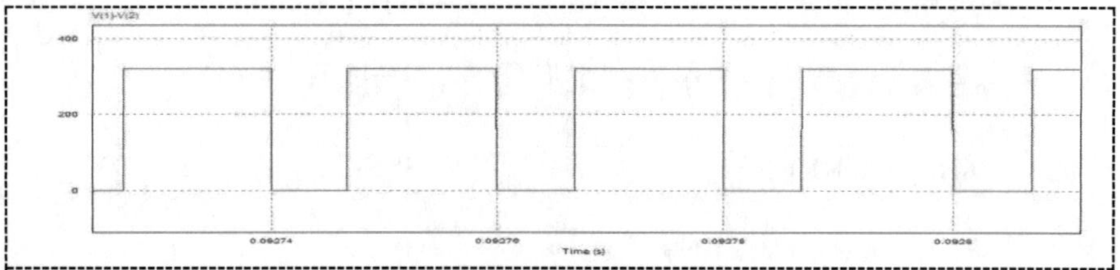

SWITCH CURRENT
PEAK VALUE: 3.38A

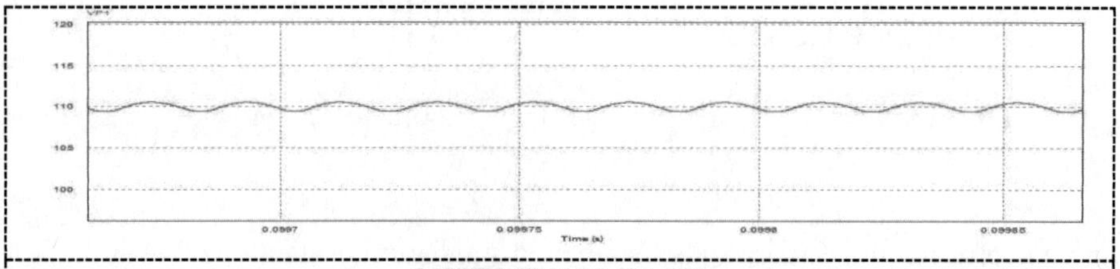

OUTPUT VOLTAGE
PEAK VALUE: 60.224V

Bidirectional Buck and Boost Converter Waveforms

The following table shows the results obtained in simulation using PSIM.

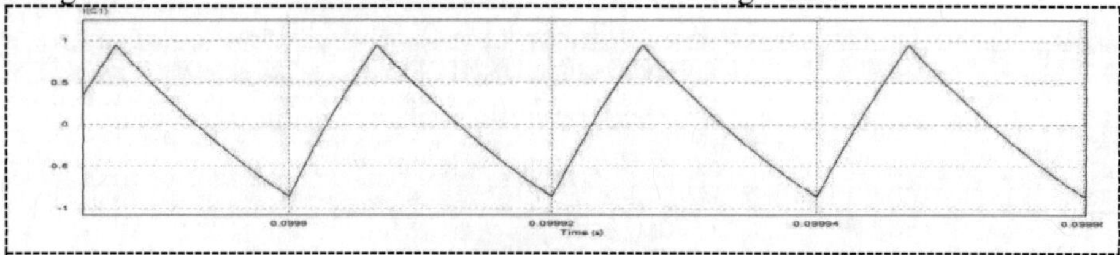

CAPACITOR CURRENT
PEAK VALUE: 0.5788A

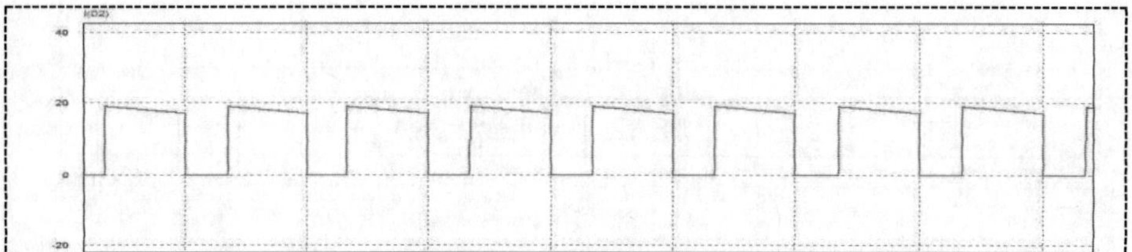

DIODE CURRENT
PEAK VALUE: 3.38A

IOP Conf. Series: Earth and Environmental Science **312** (2019) 012026 doi:10.1088/1755-1315/312/1/012026

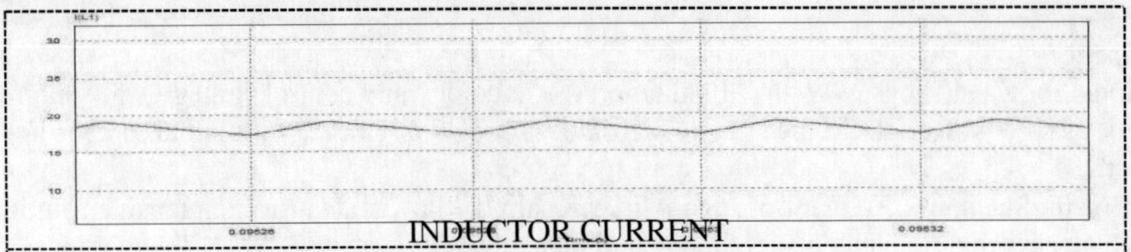

INDUCTOR CURRENT
PEAK VALUE: 3.38A

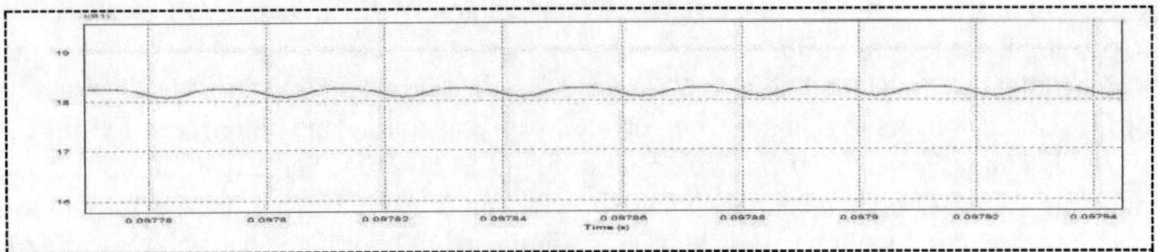

RESISTOR CURRENT
PEAK VALUE: 3.34A

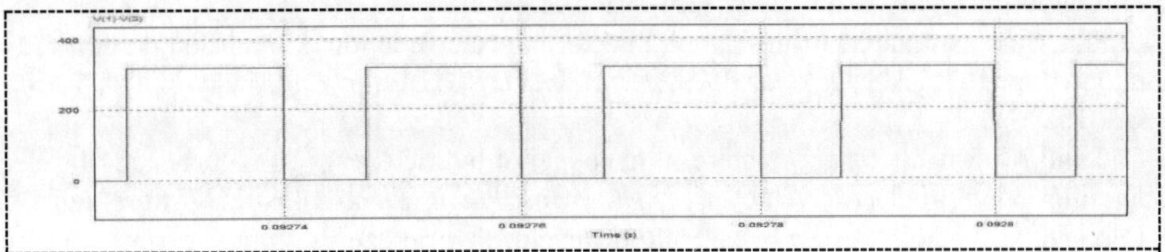

SWITCH CURRENT
PEAK VALUE: 3.38A

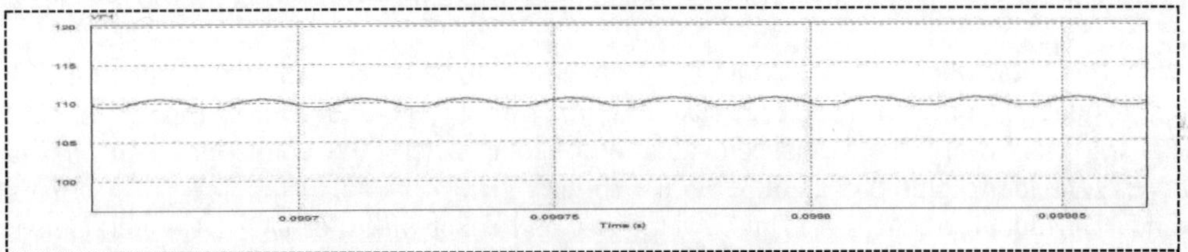

OUTPUT VOLTAGE
PEAK VALUE: 60.224V

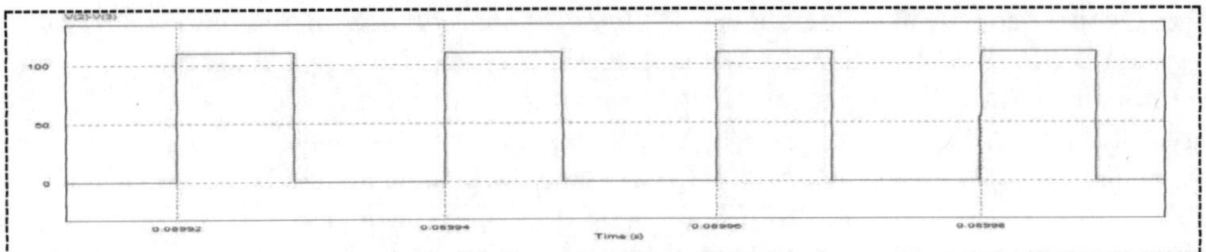

SWITCH2 CURRENT
PEAK VALUE: 3.272A

4. Conclusion

The DC-DC converters for the electrical powertrain applications is successfully designed. The results of simulation are in close agreement with the analytical solutions. This validates the credibility of the software. The designed converters consider the AC power grid as the primary source of power. This investigation can further be carried out by the solar source and the converters can be designed as an extension of the same work.

IOP Conf. Series: Earth and Environmental Science **312** (2019) 012026 doi:10.1088/1755-1315/312/1/012026

5. References

[1]Yantono Song and Bingsen Wang, "Evaluation methodology and control strategies for improving reliability of HEV power electronic system," *IEEE Trans. Veh. Technol.*, **vol.6, no.8**, pp.3661-3676,Oct.2014.

[2] Ali Emadi, Sheldon S. Williamson, and AlirezaKhaligh, "Power electronics intensive solutions for advanced electric, hybrid electric, and fuell cell vehicular power systems," *IEEE Trans. Power Electron..,* **vol. 21, no. 3**, pp. 567-577, may 2006.

[3] Hanna Plesco, JurgenBiela, JormaLuomi, and Johann W. Kolar, "novel concepts for integrating the electric drive and auxiliary DC-DC converter for hybrid vehicles," *IEEE Trans. Power Electron.,***vol. 23, no. 6**, pp. 3025-3034, Nov. 2008.

[4] MamadouBailoCamara, Hamid Gualous, Frederic Gustin, and Alain Berthon, "Design and new control of DC/DC converters to share energy between supercapacitors and batteries in hybrid vehicles," *IEEE Trans. Power Electron.,* **vol. 57, no. 5**, pp. 2721-2735, Sep. 2008.

[5] QuXiandong, Wang Qingnian, and Yu YuanBin, "Power demand analysis and performance estimation for active- combination energy storage system used in hybrid electric vehicles," *IEEE Trans. Veh. Technol.,* **vol. 63, no. 7**, pp. 3128-3136, Sep. 2014.

[6] A. Haque, "Maximum Power Point Tracking (MPPT) Scheme for Solar Photovoltaic System," *Energy Technol. Policy*, **vol. 1, no. 1**, pp. 115–122, 2014.

[7] Zaheerudin, Sukumar and M. Ahteshamul, "Performance evaluation of modified perturb & observe maximum power point tracker for solar PV system," *Int. J. Syst. Assur. Eng. Manag.*, **vol. 7, no. 1**, pp. 229–238, 2015.

[8] Jian Cao and Ali Emadi, "A new battery/ultra capacitor hybrid energy storage system for electric, hybrid, and plig-in hybrid electric vehicles," *IEEE Trans. Power Electron.,* **vol. 27, no.1**, Jan. 2012.

[9] Jian Cao and Ali Emadi, "A new battery/ultra capacitor hybrid energy storage system for electric, hybrid, and plig-in hybrid electric vehicles," *IEEE Trans. Power Electron.,* **vol. 27, no.1**, Jan. 2012.

[10] B. Sahu and G. A. Rincon-Mora, "A low voltage, dynamic, noninverting, synchronous buck boost converter for portable applications," *IEEE Trans. Power Electron.*, **vol. 19, no. 2**, pp. 443-452, Mar. 2004.

[11]Bellur DM, Kazimierczuk MK. DC-DC converters for electric vehicle applications. In2007 Electrical Insulation Conference and Electrical Manufacturing Expo 2007 Oct 22 (pp. 286-293). *IEE*

[12] R.Divya Mageswari ,"Performance evaluation of four switch buck converter" , *An International Journal (ESTIJ), ISSN*: 2250-3498,**Vol.3, No.1**, February 2013.

[13] Nanda R Mude, Prof. Ashish Sahu," Adaptive Control Schemes For DC- DC Buck Converter ", *International Journal of Engineering Research and Applications (IJERA)* ISSN: 2248-9622 ,**Vol. 2,** Issue 3, May-Jun 2012, pp. 463-467

[14] Nittala S K Sastry, Dr. Swapnajit Pattnaik, Mrs. Varsha Singh," Reduction of Ripple insingle phase buck converter by Fuzzy logic control ",*International Journal of Engineering Research and Applications (IJERA)* ISSN: 2248-9622 **Vol. 2**, Issue 3, May-Jun 2012, pp.2202-2204.

[15] B. Bryant and M. K. Kazimierezuk, "Derivation of the buckboost PWM DC-DC converter circuit topology," in Proc. *Int. Symp. Circuits* Syst., May 2002, **vol. 5**, pp. 841-844.

[16] Y. Zhang and P. C. Sen, "A new soft-switching technique for buck, boost, and buck-boost converters," *IEEE Trans. Ind. Appl.*, **vol.39, no. 6**, pp. 1775-1782, Nov./Dec. 2003

[17] Miro Milanovic, Mitja Truntic, Tine Konjedic ," Digital Current Mode Control for Buck-Converter Based on

Average Inductor Current Measurement ", *Transactions on Electrical Engineering*, **Vol. 1** (2012), No. 1

Numerical Validation and Study of Particulate Flow in Cyclone Separator using Commercial Computational Fluid Dynamics Code

Mayuresh Mutha[1], Vineet Katkamwar[1], Bageerathan Thirunavukkarasu[1], Thundil Karuppa Raj R[1*] and Sivamurugan P[2]

[1] School of Mechanical Engineering, VIT University, Vellore, Tamil Nadu, INDIA – 632014

[2] Vel Tech Rangarajan Dr.Sagunthala R & D Institute of Science and Technology, Chennai, Tamilnadu, India.

*Corresponding author email id – thundilr@gmail.com

ABSTRACT: The cyclone separator is widely used for particle separation in industries because of ease in manufacturing and maintenance free working capability. It has also been an automobile component used in air intake system. It's necessary to understand the flow of air and particles inside the cyclone separator for the design of cyclone separator. Its performance depends upon operational and geometrical parameters. This study is focused on geometric parameter in which a numerical simulation has been carried out to understand the effect of air and particle flow on varying the cone height of cyclone separator using ANSYS 15.0. The analysis involves determining pressure drop at different inlet velocity with corresponding change in cone height of the cyclone separator. The results were obtained for different velocity condition with maximum of 8% of error. Moreover, collection efficiency for cyclone separator was also determined by varying particle size from 0.01 to 10 micron. Decreasing cyclone height led to increase in tangential velocity thus increasing the cyclone efficiency.

Keywords: CFD, Cyclone separator, Particle separation, collection efficiency, pressure drop

1. Introduction

Cyclone separators are widely used for particulate emission control in industries because of their simple design and low maintenance working capability. Cyclone separator is a stationary device which is used to separate the dust or liquid particles from the air or gas. Being an Inertial type of separator, it works on the principle of centrifugal and gravity action. In cyclone separator, the flow enters in the separator tangentially due to which the centrifugal

IOP Conf. Series: Earth and Environmental Science **312** (2019) 012027 doi:10.1088/1755-1315/312/1/012027

action is generated and because of which the particles or liquid gets trapped at the cyclone surface. Due to the spiral action or moment, the particles moves down and scatter in the hopper located underneath and this outlet is called as particle outlet. The gas with lower density pressurized in cyclone collector is escaped from the outlet which is situated at the top position of the cyclone collector and is called pressurized outlet. Cyclone collector are mostly used in ship unloading installation, power station, spray drier, detergent production units etc.

There are various techniques available for particle separation but cyclone separator is the only one with minimum operating cost and with least maintenance requirement hence preferred the most wherever possible. Selection of the right technique being the most critical part, depends on the size of particles needed to be separated. Figure 2 below shows different types of techniques available and the size range they cater to. Cyclone collectors are used for particle size 3 microns and above. They are considered a better option to filters that require particle separation above 3 micron especially in case of high collection requirement.

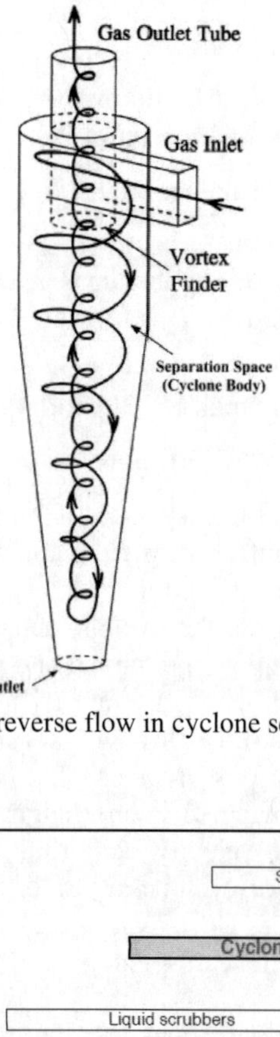

Figure 1. Typical reverse flow in cyclone separator.

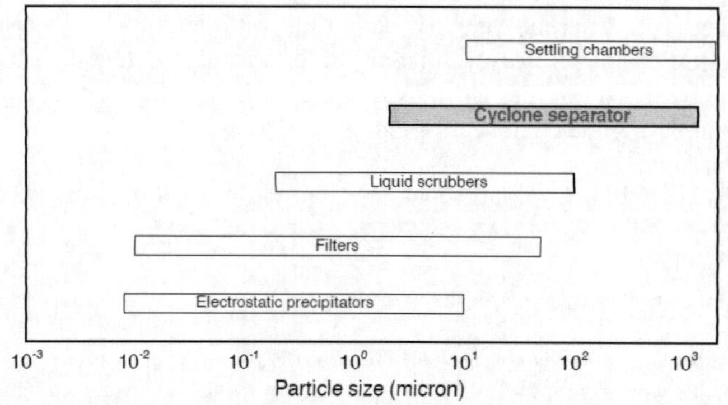

Figure 2. Various particle separation techniques for different particle size

IOP Conf. Series: Earth and Environmental Science **312** (2019) 012027 doi:10.1088/1755-1315/312/1/012027

The inlet for cyclone collector is rectangular in cross section with dimension a x b as shown in Figure 6. The gas swirl inside the cyclone collector because of the conical characteristic of cyclone. It receives the axial movement and escapes from the top of cyclone. Once manufactured, it is very difficult to change construction of such structures and hence the effect of geometrical parameters needs to be studied.

2. Cyclone Performance

Parameters affecting cyclone performance include cyclone dimensions, gas properties, particle properties and shape of vortex finder. Cyclone dimensions being major part of study found in literature includes effect of cyclone diameter, inlet height, inlet width, vortex finder diameter, vortex finder length, cylinder height, cyclone total height and cone tip diameter. Gas properties like velocity, density, viscosity, temperature and pressure along with particle properties like density, shape, mass loading, diameter and distribution also affect the cyclone performance. Xiang et al. [1] has conducted the investigation on the performance of the cyclone seperator with varying geometry profiles. They studied the effect of varying bottom diameter and cone size on collection efficiency of the cyclone separator. The authors have validated with the Barth's, Leith's and Licht's theory. They found that none of the theories have better prediction over effect of the cone size. Chuah T.G et al. [2] has conducted CFD studies on the cyclone separator with varying cone dimensions. The modelling of high swirl cyclone separator is critical. The authors have an algebraic turbulent model to overcome the problem.

The authors have validated the results with the experimental data available in literature with acceptable error deviations. This model has the benefit of enlarging or reducing of the cone profile of cyclone separator. The author have shown that CFD can be used to predict the flow regime of the cyclone separator. Hideto yoshida et al. [3] studied the effect of the height of the cone on the collection performance of the cyclone separator. The authors have conducted experimental and numerical simulations and validated. The auhor have studied the effect of velocity, resistance of incoming particles by varying the apex cone angle of the cyclone separator. They found the change in the velocity in the upper part of the separator with the optimum cone height configuration. Ting Zhang et al. [4] conducted CFD studies on the cyclone separators with square, hexagonal and circular cross section cyclone separators. The authors have designed optimised cyclone separators with three parameters. They found that the hexagonal outer body has better efficiency and lower wear rate. Yaxin Su et al. [5] has numerically simulated the gas-solid turbulent flow in cyclone separator with CFD method. The author has modelled the fluid with Reynolds Stress Model (RSM) and traced the particle with Lagrangian model. Exhaust gases have been used as the gas-solid medium with a circular and two square cross section model. The performance characteristics of the cyclone separator were compared and pressure drop of the three models were evaluated. They found that square cross section has better efficiency than circular cross section.

Yaxin Su et al. [6] used the three dimensional particle dynamic analyser to trace the particle in the square cyclone separator with downward gas exit. The authors have conducted cases with varying inlet velocities and different particle concentrations. They used glass beads as the particle with diameter from 30 microns to 40 microns. The author has evaluated parameters like flow vector, turbulent intensity, mean velocity and kinetic energy of both the gas and solid particles. The author has found that the elevation of temperature of the cyclone separator has better uniform flow when compared to the case at room temperature. The laminar motion of the particles has increased the turbulent kinetic energy of the particle-particle collisions and particle wall collisions. The authors have found that the corners of the cyclone separator has contributed towards the increase in the kinetic energy of the particle. Yuki Wakizono et al [7] has conducted CFD simulations on cyclone separators with the ring configuration to the upper outlet region. The pressure drop is less due to the ring configuration used. The author has parameterized the cut off ratio, wall thickness, pressure drop and collection efficiency. The author has found that the ring has contributed

IOP Conf. Series: Earth and Environmental Science **312** (2019) 012027 doi:10.1088/1755-1315/312/1/012027

over the collection efficiency at the bottom of the cyclone separator. The numerical simulations were validated along with experimentation with allowable error percentage. Yifang Zhu et al [8] has conducted experimental studies on high flow rate conditions in small cyclone models. The authors have conducted studies on seven different cyclone models with varying flow rate conditions and particle sizes. The authors have varied the height of cylinder and length to the diameter of the exit tube to study the cyclone collection efficiency. The authors have found that the increase in cylinder height is efficient to some extent. The author concludes that cylinder height and exit tube length affects the collector efficiency of the cyclone separator. Eflita Yohana et al. [9] has numerically modelled three dimensional new cyclone separator model and studied the particle separation with varying inlet velocity configurations. The author has used the Reynolds averaged Navier Strokes equation with Reynolds Stress Model to solve finite volume method based on the SIMPLE algorithm model. The author concludes that the newly modelled cyclone separator has better efficiency than the conventional cyclone separators. Ali Sakin et al. [10] has investigated cyclone separator in air intake system of automobile and studied the influence on the engine performance. The author has preferred the cyclone separator in place of air filter, as the filter acts as porous medium and pressure drop arises. The author has conducted performance studies on the engine with both filter and cyclone separator intake systems. Numerical simulation were validated with experimentation and found to be satisfactory.

3. CFD Simulation:

In this paper, CFD simulation is done to validate the results of experimental setup by Xiang et. al. and the analysis done by Chuah et. al. [2]. O grid blocking done in ICEM CFD as shown in figure 3 below and Hexa meshing is done as in the figure 4. FLUENT was selected for analysis.

Boundary conditions include inlet air being at 8 m/s or 30 lpm, 10.66 m/s or 40 lpm, 13.33 m/s or 50 lpm and 16 m/s or 60 lpm. Pressure drop being the key parameter, is validated and found to be within 8 percent, error being maximum at 16 m/s velocity or 60 lpm discharge. Reynolds Stress Model (RSM) with discrete phase modelling was used for particle simulation.Figure 7 represents the pressure contour of the case of cyclone separator with height of 36 mm and alpha angle 5 degree. Figure 8 shows the tangential velocity of the case with height 36 mm and 5 degree alpha angle.

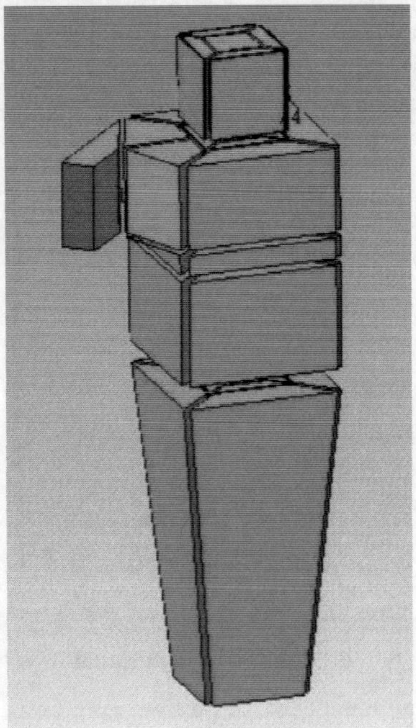

Figure 3. Blocking of cyclone separator with O- Grid

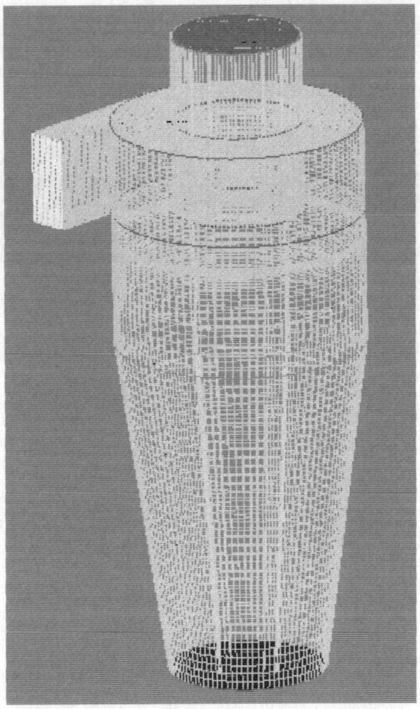

Figure 4. Hexa meshing in ICEM CFD

IOP Conf. Series: Earth and Environmental Science **312** (2019) 012027 doi:10.1088/1755-1315/312/1/012027

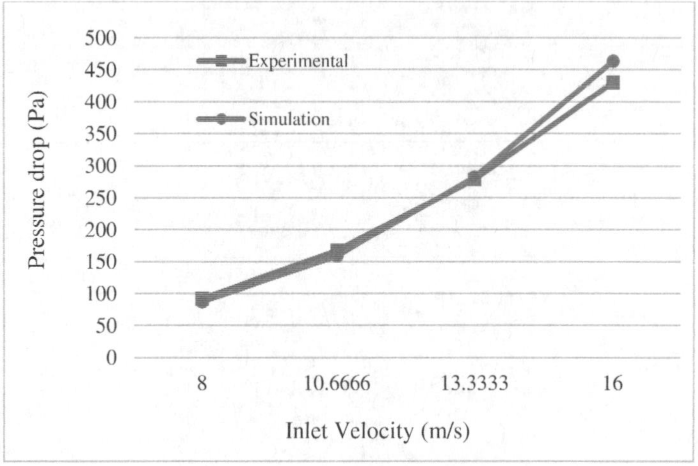

Figure 5. Pressure versus velocity for experimental and simulation values

Table 1. Pressure drop validation of Experimental Cyclone by Xiang et.al.

Inlet Velocity (m/s)	Pressure		Percentage error (%)
	Experimental (Pa)	Simulation (Pa)	
8	92.18	87.5	5.08
10.66	166.71	158.20	5.17
13.33	278.51	281.80	1.18
16	430.51	464	7.78

IOP Conf. Series: Earth and Environmental Science **312** (2019) 012027 doi:10.1088/1755-1315/312/1/012027

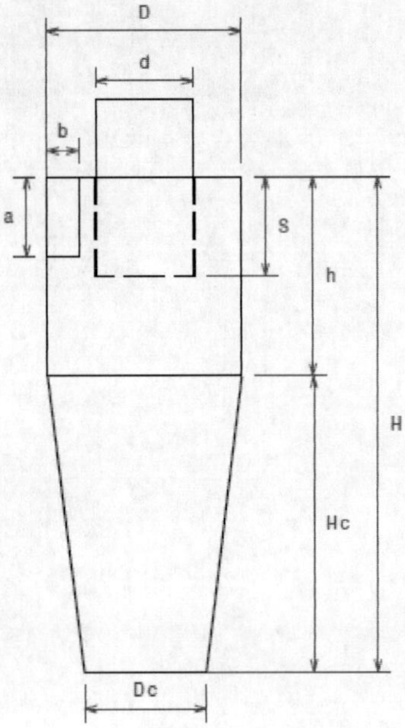

Front view

Figure 6. Geometrical layout of Cyclone

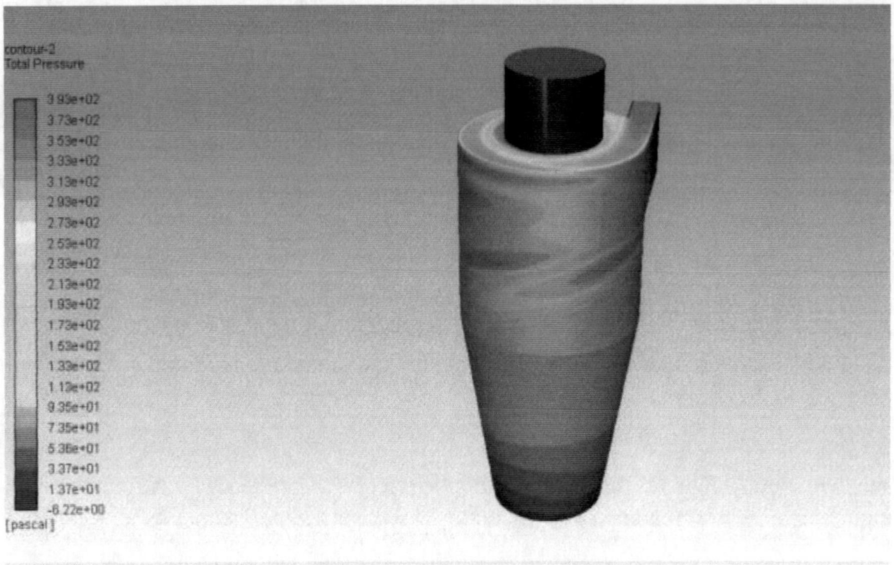

Figure 7. Pressure contour of case with H = 36 and α = 5

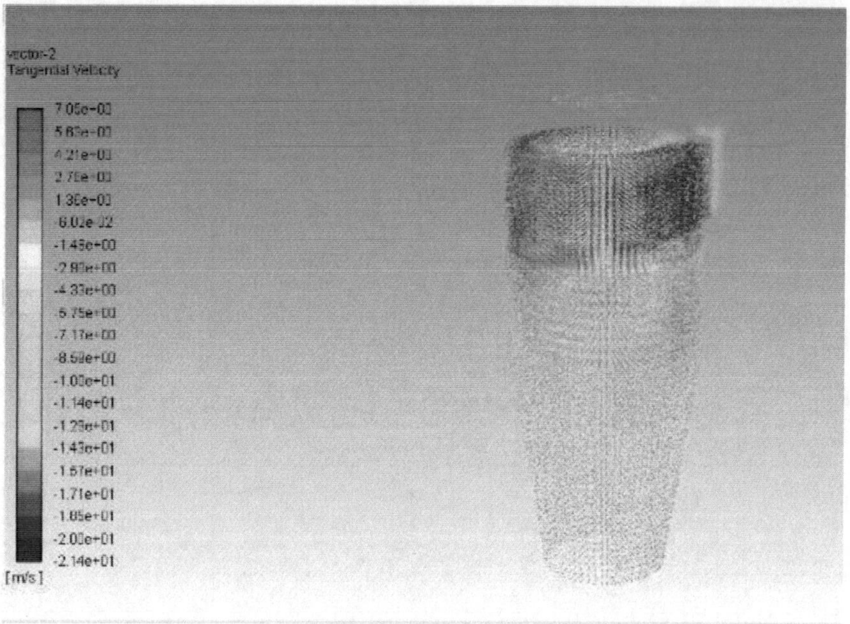

Figure 8. Tangential velocity vector with H = 36 and α = 5

Table 2. Geometrical dimensions of cyclone studied

Dimension		Value (mm)
Body Diameter (D)		31
Gas Outlet Diameter (d)		15.5
Inlet Height (a)		12.5
Inlet Width (b)		5
Gas Outlet Duct Length (S)		15.5
Cylinder Height (h)		31
Cyclone Height (H)		77
Cone Height (Hc)	Cyclone 1	46
	Cyclone 2	36
Cone Outlet Diameter (Dc)		19.4

The cone height of cyclone was also varied and the effect was studied. It was found that with decrease in cone height, the tangential velocity of air increases and the efficiency increases.

4. Results and Discussion:

4.1 Effect of inlet velocity, alpha angle and height of the cyclone separator on collection efficiency

Numerical simulations were conducted on the cyclone separator with varying inlet velocity and particles with 0.01 microns to 10 microns. The particles allowed at inlet and the particles collected at outlet is traced. The collection efficiency is calculated by the number of particle traced. When the inlet velocity of the particles is increased, the collection efficiency is also increased. When the velocity at the inlet is increased, the total pressure drops down which makes the gas to escape out easily through the upper outlet. The velocity of the particle at inlet is significant in determining the colletor efficiency. The trend of velocity is shown in figure 10.

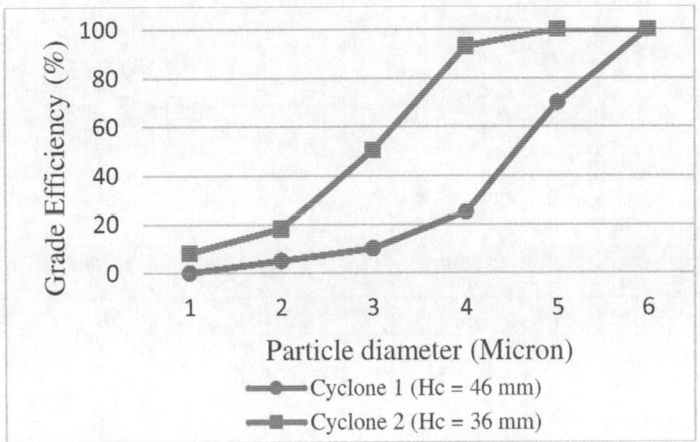

Figure 9. Effect of Cone height variation on Grade or Collection efficiency for different particle size

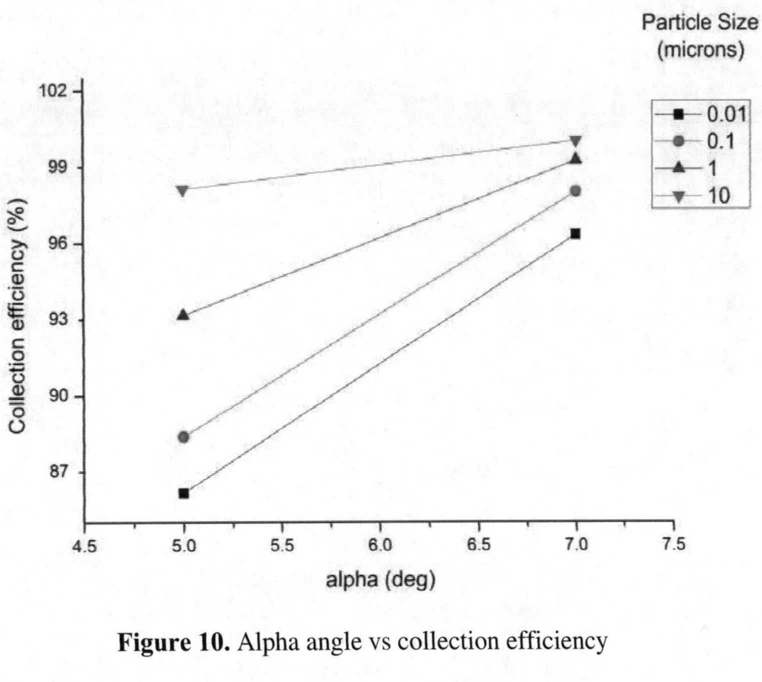

Figure 10. Alpha angle vs collection efficiency

IOP Conf. Series: Earth and Environmental Science **312** (2019) 012027 doi:10.1088/1755-1315/312/1/012027

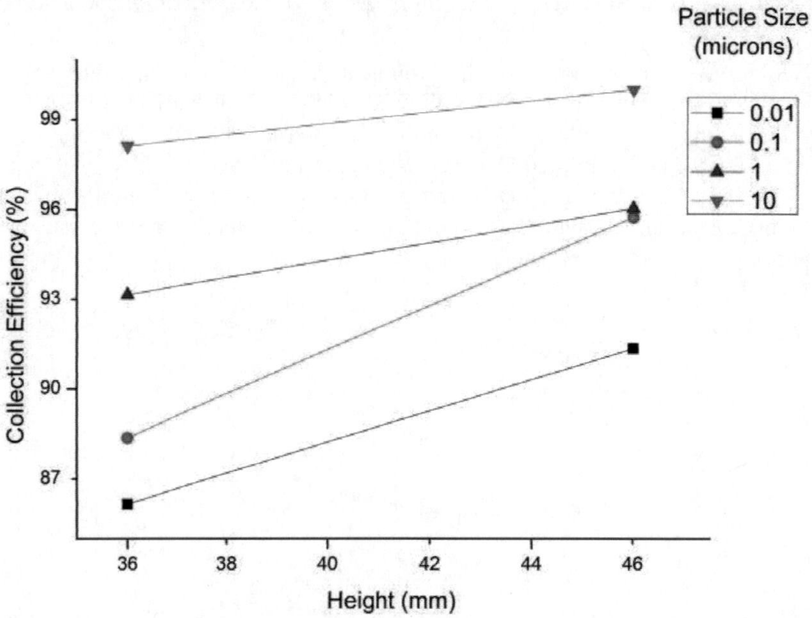

Figure 11. Height vs collection efficiency

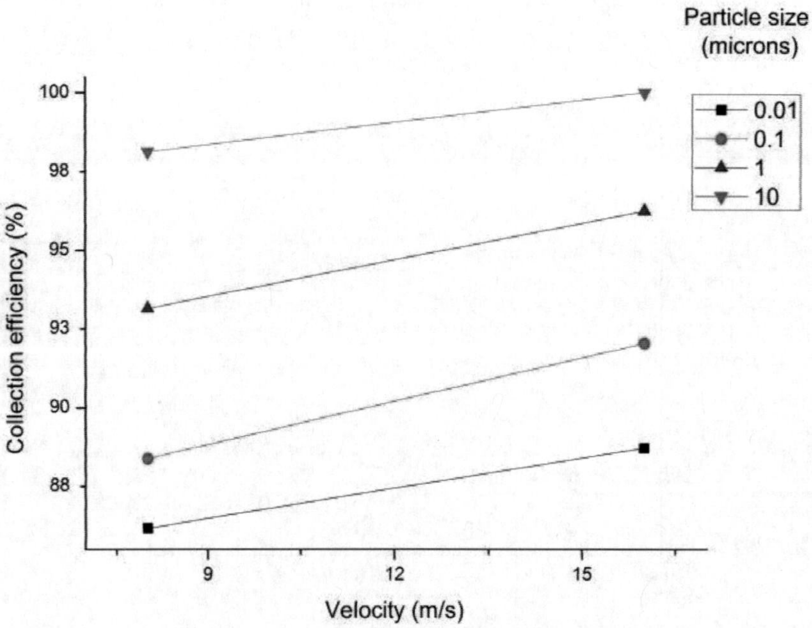

Figure 12. Velocity vs collection efficiency

Simulations were conducted with varying alpha angle values of 5 degree and 9 degree angles with height of 36 mm and 46 mm. When the alpha angles have been increased, there is significant increase in the collector efficiency along with increase in height. The height of the cyclone separator needs effective alpha angle where the collector efficiency is maximum. The trend is shown in figures 8 and figure 9. The particle-particle interaction and particle-wall interaction in cases with less alpha angle along less height is efficient because the proportional alpha angle and height allows the particle-gas mixture to move in a streamlined flow where the particle gets collected in the bottom outlet and gas escapes through top outlet.

5. Conclusion:

Cyclone separator is observed to separate particles above 1 micrometre but it is found to be more effective in separation of particles above 5 micrometre. Cone dimensions have great impact on cyclone performance. It was found that the efficiency increases with decrease in cone bottom diameter. Higher the inlet velocities, better is the grade efficiency and thus the cyclone can separate particles with lower particle size too. Decreasing the cyclone height increases tangential velocity and leads to increase in the efficiency. Thus cyclone separator design is very critical and the cone dimensions need to be decided based on particle size to be separated and inlet velocity of the air.

6. Nomenclature:

$\dot{m}$	-	Mass flow rate, g/sec
P	-	Pressure, Pa
T	-	Temperature, K
U	-	Superficial velocity, m/s
ρ	-	Density of air, kg/m^3
ΔP	-	Pressure drop, Pa
L	-	Substrate length, mm

7. References:

[1] Xiang, R., Park, S.H. and Lee, K.W., 2001. Effects of cone dimension on cyclone performance. Journal of Aerosol Science, 32(4), pp.549-561.

[2] Chuah, T.G., Gimbun, J. and Choong, T.S., 2006. A CFD study of the effect of cone dimensions on sampling aerocyclones performance and hydrodynamics. Powder technology, 162(2), pp.126-132.

[3] Yoshida, H., Kwan-Sik, Y., Fukui, K., Akiyama, S. and Taniguchi, S., 2003. Effect of apex cone height on particle classification performance of a cyclone separator. Advanced Powder Technology, 14(3), pp.263-278.

[4] Zhang, T., Liu, C., Guo, K., Liu, H. and Wang, Z., 2015. Analysis of flow field in optimal cyclone separators with hexagonal structure using mathematical models and computational fluid dynamics simulation. Industrial & Engineering Chemistry Research, 55(1), pp.351-365.

[5] Su, Y., Zhao, B. and Zheng, A., 2011. Simulation of turbulent flow in square cyclone separator with different gas exhaust. Industrial & Engineering Chemistry Research, 50(21), pp.12162-12169.

[6] Su, Y. and Mao, Y., 2006. Experimental study on the gas–solid suspension flow in a square cyclone separator. Chemical Engineering Journal, 121(1), pp.51-58.

[7] Wakizono, Y., Maeda, T., Fukui, K. and Yoshida, H., 2015. Effect of ring shape attached on upper outlet pipe on fine particle classification of gas-cyclone. Separation and Purification

IOP Conf. Series: Earth and Environmental Science **312** (2019) 012027 doi:10.1088/1755-1315/312/1/012027

Technology, 141, pp.84-93.
[8] Zhu, Y. and Lee, K.W., 1999. Experimental study on small cyclones operating at high flowrates. Journal of Aerosol Science, 30(10), pp.1303-1315.
[9] Yohana, E., Diana, A.E., Maulana, A.F., Tauviqirrahman, M. and Choi, K.H., 2019, April. Inlet velocity influence on particle separation in three-dimensional new cyclone separator using CFD modelling. In AIP Conference Proceedings (Vol. 2097, No. 1, p. 030067). AIP Publishing.
[10] Sakin, A., Karagoz, I., Ergul, M., Demirtas, U. and Savas, F.H., 2018. An investigation into the use of a cyclone separator in the intake air system and its influence on the engine performance. Proceedings of the Institution of Mechanical Engineers, Part D: Journal of Automobile Engineering, 232(5), pp.667-678.

IOP Conf. Series: Earth and Environmental Science **312** (2019) 012028 doi:10.1088/1755-1315/312/1/012028

PSO Based Dispatch and Control of Hybrid Power System with PV-Diesel Generator- A Prototype

Bharathi Kannan K[1], Elavarasan E[2], Aavala Yashoda[1], Sovana Kumari[1], Ram JethMalani C H[1].

[1]Department of Electronics and Electrical Engineering
[2]Department of Mechanical Engineering,
National Institute of Technology Puducherry, Karaikal – 609 609,
U.T. of Puducherry, INDIA.

Email: bharathikkannan@gmail.com, aemaero16@gmail.com

Abstract. A micro-grid is a small-scale power grid that can operate independently or in conjunction with the area's main electrical grid. Micro-grid paves way to power remote locations and effectively integrate various sources of distributed generation, including renewable. In this paper, an economical way of dispatching PV power and power generated by generators in islanded micro-grid is proposed. Here, part of the micro-grid demand is supplied by PV power station. The demand which cannot be supplied by the PV panel, are dispatched between DGs by solving unit commitment and economic dispatch problems. Unit commitment is carried out using priority list method and economic dispatch particle Swarm optimization. A hardware prototype for the proposed micro-grid dispatch methodology has been developed. Here, commitment and de-commitment of the sources and demands is carried out by generating control signals denoting the ON/OFF status to the sources for different intervals.

Keywords: Hybrid Power Generation System; Micro-grid; Photovoltaic Panel; Unit Commitment; Particle Swarm Optimization; Audrino based ON/OFF control.

1. Introduction

Electricity has become an essential commodity nowadays. However, many rural and remote places are not accessible by the power distribution companies yet. There are two ways of providing electricity to rural communities, either by extending the existing grid or by developing an isolated off-grid power generation entity. Many of the developing countries have accepted these off-grid power generation solutions as the cheapest and most sustainable solutions particularly for rural areas. A MICRO GRID is a small set of loads and distributed energy sources which are inter and intra connected within a small area and that entity can be singly controlled with respect to the main grid. Micro-grids usually consist of distributed generation sources such as diesel generators, solar panels and wind turbines [1]. In this context, this paper investigates a dispatch and control mechanism used in a diesel generator and photovoltaic system based micro grid. The schematic diagram of the proposed micro grid is given in figure 1.

Unit commitment (UC) is a process of turning ON and OFF of the units which are available to supply the forecasted load for particular hour, meanwhile capacity requirement should also be considered [2]. After the decision of UC, economic load dispatch (ELD) is carried out performed the units which are committed. ELD is performed to determine the power output of each and every committed units. It is also to calculate the optimal fuel cost for the thermal units while gratifying the constraints associated with them.

IOP Conf. Series: Earth and Environmental Science **312** (2019) 012028 doi:10.1088/1755-1315/312/1/012028

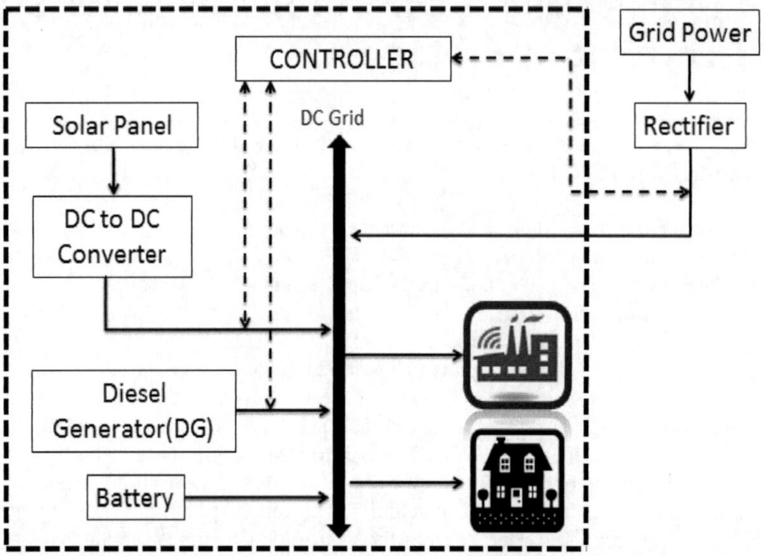

Figure 1. Schematic diagram of the proposed micro-grid

2. Literature review

Priority list is one of the methods by which UC is normally performed and that can also be applied for larger power systems. Priority list technique is the simplest method among all methods to understand. And also it is very fast to compute. In this method the ON and OFF state of the units are considered IC value of each unit is considered and incremental cost (IC) of each unit is computed [2][3].

Karamjeet Kaur and Amarjet Kaur have discussed solving of unit commitment using improved priority list method by which the most economic unit will be committed first followed by the other units in the same hierarchy which is better than other existing priority list methods. Their proposed solution includes the minimum up/down time constraints which are usually neglected in the existing methods. Economic dispatch of the committed units is then done using lambda iteration method. The fast calculation of ED and the simplicity of the proposed priority method leads to competent and methodological approach compared to other conventional method [3].

Md. SajidAlam and Durga Hari Kiran B have performed UC which includes renewable sources by priority list [4] based particle swarm optimization (PL-PSO) [5] as well as using priority list based genetic algorithm optimization (PL-GA) [6] wherein a unit is committed depending on incremental cost and economic load dispatch by particle swarm optimization and comparison of both the methods is done. Here, both GA and PSO converge to similar optimal solutions. However, in case of PSO the computation time is decreased and good convergence will also be observed.

3. Problem Formulation

The objective of a unit commitment problem (UCP) is to get an optimized schedule to commit the available generating units. This gives optimized minimal operation cost (1) to supply the forecasted system load while considering the power balance constraint (2), unit constraints (3), ramp-rate constraints (4), and minimum ON/OFF constraints (5).

$$\text{Minimize OC} = \sum_{t=1}^{T} \sum_{i=1}^{ng} f_i(P_{i,t})\, I_{i,t} + SU_i^{\,t} I_i^{\,t}(1 - I_{i,t-1}) + SD_i^{\,t}(1 - I_{i,t}) \tag{1.a}$$

where,

$$f_i(P_{i,t}) = a_i + b_i P_{i,t} + c_i P_{i,t}^{\,2} + |e_i \sin\left(f_i(P_{min,i} - P_{i,t})\right)| \tag{1.b}$$

Subject to:

$$\sum_{i=1}^{ng} P_{i,t} = P_{D,t} \tag{2.a}$$

$$P_{min,i} \le P_i \le P_{max,i} \text{ for i} = 1,2\dots\text{ng} \tag{3}$$

$$P_{min,i,t} = max\{P_{min,i}, P_{i,t-1} - Ud_i\} \tag{4.a}$$

$$P_{max,i,t} = min\{P_{max,i}, P_{i,t-1} + Up_i\} \tag{4.b}$$

$$X_{i,t}^{ON} \ge T_{i,t}^{ON} \rightarrow I_{i,t-1} - I_{i,t} = 1 \tag{5.a}$$

$$X_{i,t}^{OFF} \ge T_{i,t}^{OFF} \rightarrow I_{i,t} - I_{i,t-1} = 1 \tag{5.b}$$

In this work generating units are considered to be diesel generators, for which the start-up costs and shut down costs are zero.

4. Solution methodology

4.1. Priority list approach for unit commitment:

Priority list method determines the priority rank of each generating units based on their Average Full Load Cost (AFLC).

$$AFLCi = \frac{Ai * Pimax^2 + Bi * Pimax + Ci}{Pimax} \tag{6}$$

The procedure for obtaining unit commitment status by priority list [4] method is given below:

- Determine the priority list. Here, the unit with the least value of AFLC tops the priority list, followed by the other units in ascending order of AFLC.
- Set t=1.
- Set the rank of generator k=1 and the maximum amount of power supplied by the combination of generators P_{ms}= P_{max}k.
- If $P_{supplied}$>P_{net} demand, then go to step 5. Else, increment k, $P_{supplied}$=$P_{supplied}$+P_{max}k and go to step 3.
- Solve economic dispatch considering the committed units.

IOP Conf. Series: Earth and Environmental Science **312** (2019) 012028 doi:10.1088/1755-1315/312/1/012028

- If$t = T$, then stop execution. Otherwise, increment$t$, update the operating limits using equations (5), (6) and (7), and go to step 3.

4.2. Economic dispatch by PSO

PSO is a population-based stochastic optimization technique modeled on swarm intelligence. Here, each particle flies in the search space in certain velocity. The particle's flight is influenced by cognitive and social information attained during its exploration. The PSO can generate highly excellent solutions taking small time for calculation and stable convergence characteristic than many other stochastic optimization methods. The PSO model consists of a swarm of particles moving in a D-dimensional real-valued space of possible problem solutions. Every particle has a position and a flight velocity. During each time step$k$, the velocity is updated and the particle is made to move to a new position. This new position is calculated as the sum of the previous position and the new velocity. The step procedure in implementing PSO to solve ED is:

- Read system data, demand profile, tolerance in power balance mismatch (ε), number of particles (N), initial inertia weight (WI), final inertia weight (WF), social scaling factor (c_1), cognitive scaling factor (c_2) and maximum steps (K).

- Initialize iteration count (k) as 1 and velocity (V) of each particle as 0.

- N Numbers of initial particles are randomly generated. The position of each of the particles is represented in $'ng'$ dimensional space as a vector, $X_p = \left[P_I{}^1, P_I{}^2 P_p{}^i P_I{}^{ng}\right]$. Each element of $X_p$represents the real power generated by corresponding generator. Then, the swarm can be expressed as in equation (7). Here, the particles are generated in a way such generator operating limit constraints are satisfied. The position of the particle (p) in dimension (i) is calculated as in equation (8).

$$X = \begin{bmatrix} P_1{}^1 & P_2{}^1 & \dots & P_p{}^1 & \dots & P_N{}^1 \\ \cdot & \cdot & \dots & \cdot & \dots & \cdot \\ P_1{}^i & P_2{}^i & \dots & P_p{}^i & \dots & P_N{}^i \\ \cdot & \cdot & \dots & \cdot & \dots & \cdot \\ P_1{}^{ng} & P_2{}^{ng} & \dots & P_p{}^{ng} & \dots & P_N{}^{ng} \end{bmatrix} \tag{7}$$

$$P_p{}^i = u * \left(P_{max,i,t} - P_{min,i,t}\right) + P_{min,i,t} \tag{8}$$

- Calculate the fitness of each particle using equation (9).

$$fit(X_p) = \frac{1}{\left(\sum_{i=1}^{ng} F_i(P_{i,t}) + 1\right)} \tag{9}$$

- Update the fitness of the local and the global best particle using the equations (10) and (11), respectively.

$$Pbfit_p{}^k = min\{Pbfit_p{}^{k-1}, fit_p{}^k\} \tag{10}$$

$$Gbfit = min\{Pbfit_1{}^k, \dots Pbfit_p{}^k, \dots Pbfit_{Np}{}^k\} \tag{11}$$

- Updating the local best position of particles $({Pb_p}^k)$ and global best particle (Gb).
- Update the velocity for each of the particles using equation (12).

$$V_p^{k+1} = WV_p^{k} + C_1u_1({Pb_p}^{k} - X_p^{k}) + C_2u_2(Gb - X_p^{k}) \tag{12}$$

- Update position of each particle using equation (13).

$$X_p^{k+1} = X_p^{k} + V_p^{k+1} \tag{13}$$

- If the iteration count k = K (maximum number of iterations), then increment $t = t + 1$ and go to step 2. Else, go to step 3.

5. Prototype development

A prototype of Micro-grid is constructed in laboratory environment and is tested with all the requirements in the real time Micro-grid. The Micro-grid considered in our project utilizes the available solar energy for meeting the load demand and the remaining part of the load i.e., the net demand (gross demand-Solar power generated) is met by the group of three Diesel Generating units in which minimization of cost [7] is the main objective to be achieved. In the prototype developed the solar power generating unit is replaced by a PV panel [8] and the three diesel generators are made analogous to three transformers. In real time scenario, the grid is an AC grid so the Micro-grid developed has to be converted to AC grid at the end terminals by using inverter circuits. But in the case of prototype the DC Micro-grid is developed and DC loads are considered. DC load demand and solar power generating output for six intervals is taken and is simulated using Matlab codes for UC chart. The schematic diagram, the circuit diagram and the ratings of the prototype are given in Figure 2 and Table 1, respectively.

Arduino is an open source computer hardware and software. This board has a variety of microprocessors and controllers which contribute to design the boards. The boards are furnished with sets of digital as well as analog input/output (I/O) pins. These pins can be interfaced with different expanding boards and any circuit. The features of the board are that the Universal Serial Bus (USB) can also be interfaced by serial communication.

Table 1. Rating of the prototype

S.No	Component	Rating
1	Solar Panel	10W , 12V
2	Voltage Regulator	L7812 IC
3	Boost Converter	12V TO 24V
4	DG1(Transformer) –24W	230/12-0-12V , 1A
5	DG2(Transformer) – 12W	230/12-0-12 , 500mA
6	DG3(Transformer) – 5.4W	230/9-0-9, 300mA
7	Load (Resistive)	10W , 100 Ω
8	Rectifier & Filter Circuits	12V, 1A
9	Fuse	0.75A, 0.5A,1A

IOP Conf. Series: Earth and Environmental Science 312 (2019) 012028 doi:10.1088/1755-1315/312/1/012028

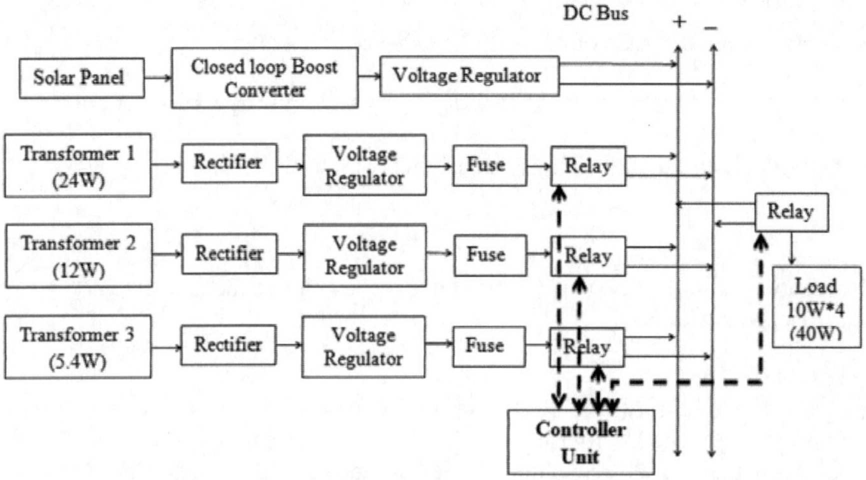

Figure 2. Block diagram of the prototype

6. Results and discussion

Scheduling and control of an islanded micro-grid comprising of the renewable solar power plant and convention diesel generating units are carried out. A hardware prototype to represent the same and the results are discussed in the following sections. Here, one solar photovoltaic power generation plant along with three diesel generators is considered. The load demand data and solar power insolation data used for the validation are given in Table 2. The capacity of solar PV system is taken as 10 KVA. The cost coefficients of diesel generating units are given in Table 3.

The solar power developed by the solar PV generating unit for the given schedule time is estimated using a model discussed [8]. Here, the Matlab SIMULINK model is developed and the values of solar power generated for each insolation value is obtained. Based on the estimated values of real power generated by the solar power generating unit, the net demand, to be supplied by the diesel power generating units, are calculated.

Table 2. Gross Demand and Solar Insolation Data

Interval No.	Time (IST) (hh:mm)	Gross demand (P_D)(kW)	Solar insolation (W/m^2)
1	08:00	1200	500
2	10:00	1950	1400
3	12:00	2370	2000
4	14:00	1550	1800
5	16:00	1429	1000
6	18:00	970	350

Table 3. Cost coefficients of Units

Sources	a	b	c	P_{min}	P_{max}	e	f
Unit 1	0.0037	2	10	0	200	40	0.08
Unit 2	0.0175	1.75	12	0	180	30	0.09
Unit 3	0.0625	1	9	0	150	30	0.1
Unit 4	0.00834	3.25	10	0	300	20	0.075

Once the net demand is obtained, the combination of generation units committed at each of the schedule interval is obtained using the priority list method. Values of AFLCs and the priority rank of each unit are given in Table 4. The commitment status of each unit for each schedule interval is given in Table 5.

Table 4. Prioritized units based on AFLC

Unit	AFLC	Priority rank
1	2.7900	1
2	4.9667	2
3	5.7853	4
4	10.4350	3

Table 5. Unit Commitment chart

Interval	Net Demand	Unit 1	Unit 2	Unit3	Unit4
1	552	1	1	0	1
2	285.6	1	1	0	0
3	104	1	0	0	0
4	-112	0	0	0	0
5	594	1	1	0	1
6	829.4	1	1	1	1

Once the commitment schedule is obtained, units economic dispatch for each schedule interval is carried out using PSO. The economic dispatch output obtained using PSO is given in Table 6 and the economic dispatch output obtained using lambda iteration method is given in Table 7.

It has to be noted that the cost obtained using PSO is lesser than the cost obtained using lambda iteration method.

All the components of micro-grid are connected together in parallel i.e., one PV panel of 10W and three units of transformers (5.4W, 12W, 24W) [9] are connected in parallel. In our work six intervals in a day are considered and the load demand for each interval of time and three units of transformers are controlled according to the unit commitment chart developed and is implemented by using Arduino generated signals.

IOP Conf. Series: Earth and Environmental Science **312** (2019) 012028 doi:10.1088/1755-1315/312/1/012028

Table 6. ED schedule using particle swarm optimization

Interval	Net Demand	Unit 1	Unit 2	Unit3	Unit4	Cost(Rs)
1	552	194.7986	143.6857	0	213.5706	2258.3
2	285.6	196.3796	89.1849	0	0	871.71
3	104	104.0415	0	0	0	289.13
4	-112	0	0	0	0	41
5	594	173.1899	171.5669	0	249.3328	2642.5
6	829.4	200	180	150	300	4752.9
Total Cost			Rs. 10855.54			

Table 7. ED schedule using lambda iteration

Interval	Net Demand	Unit 1	Unit 2	Unit3	Unit4	Cost(Rs)
1	552	200	142.6	0	209.4	2240.7
2	285.6	200	85.6	0	0	867.02
3	104	1240	0	0	0	284.22
4	-112	0	0	0	0	41
5	594	200	156.2	0	237.8	2533.8
6	829.4	200	180	150	300	4752.9

The gross load demand and PV panel output power in each interval along with the net demand (in prototype simulated data is scaled down) is shown in the Table 8.

The unit commitment chart developed for the net demand data in Matlab for six intervals of time is scaled down and is implemented in prototype for all intervals. The Unit commitment chart implemented for controlling the transformers is tabulated in Table 9.

In the considered micro-grid scheduling and controlling of diesel generators has to be done for the demand which is not met by the solar power generating units. Scheduling is done by solving Unit Commitment Problem using priority listing method based on which DG's are ranked based on the AFLC.

Table 8. Net demand in Prototype

Interval Number	Gross demand (W)	PV panel output power (W)	Net demand (W)
1	10	3	7
2	20	5	15
3	40	6	34
4	30	5	25
5	20	4	16
6	10	2	8

Table 9. Unit Commitment Status of the prototype

Interval number	Net demand (W)	Unit 1 (24W)	Unit 2 (12W)	Unit 3 (5.4W)
1	7	0	0	1
2	15	0	1	1
3	34	1	1	1
4	25	1	0	0
5	16	0	1	1
6	8	0	0	1

Economic Dispatch problem is solved using two different methods namely Lambda iteration method and Particle Swarm optimization method [5] [10]. Both UCP and Economic dispatch problem are solved in Matlab platform and the scheduling results are discussed in the above sections. Controlling of micro-grid includes the controlling of three available DG's (transformers in prototype) according to the Unit commitment chart developed in the Matlab and Unit commitment is implemented in the prototype by using control signals which are generated by the Arduino. Thus scheduling of DG's is carried out in Matlab and control of the micro-grid is carried out in the prototype developed.

7. Conclusion

It can be observed from the results that PSO [5] & [10]method provides a better converged solution without any initial conditions as compared to lambda iterative method .The optimized cost in PSO is slightly greater than lambda iterative method since it takes into consideration the valve point coefficients for better convergence. It is shown that unit commitment schedule can be translated into control signal by turning ON/OFF different LEDs analogous to generating units. Therefore, it is seen that Arduino boards are viable to control a generator as per the UCP schedule.

References

[1] Anilkumar TT Paper (open well PSO hybrid) Anilkumar, T. T., and Sishaj P. Simon. "Experimental İnvestigation on a Prototype Solar-Wind Hybrid System with a Pico Hydro Turbine", *International Journal of Emerging Electric Power Systems* 18.5 (2017).

[2] Divya Ananthan, Prof. S. Nishanthinivalli, "Unit Commitment Solution Using Particle Swarm Optimisation", *IOSR Journal of Engineering* (IOSRJEN), ISSN (e): 2250-3021, ISSN (p): 2278-8719 Vol. **04**, Issue 03 (March. 2014), ||V1|| PP 01-09.

[3] Neha Thakur, L. S. Titare, "Optimal Unit Commitment Based on Economic Dispatch Using Improved Particle Swarm Optimization Technique", *International Journal of Recent Research in Electrical and Electronics Engineering* (IJRREEE) Vol. **3**, Issue 1, pp: (50-56), Month: January - March 2016

[4] Karamjeet Kaur, Amarjeet Kaur, "Unit Commitment by an improved Priority List Method", *International Journal of Innovations & Advancement in Computer Science IJIACS* ISSN 2347 – 8616 Volume **6**, Issue 8 August 2017.

[5] Md. Sajid Alam, Durga Hari Kiran B, MatamSailajaKumari, "Priority list and particle swarm optimization based Unit commitment of thermal units including renewable uncertainties", Power System Technology (POWERCON), 2016 IEEE International Conference on 28 Sept.-1 Oct. 2016, DOI: 10.1109/POWERCON.2016.7753989.

[6] K.S. Swarup, S. Yamashiro, "A generic Algorithm approach to generator unit commitment", *International Journal of Electrical Power and energy Systems*, Vol. **25**, No. 9, pp. 679-687, November 2003.

[7] Anilkumar, T. T., Sishaj P. Simon, and Narayana Prasad Padhy. "Residential electricity cost minimization model through open well-pico turbine pumped storage system", *Applied energy* 195 (2017): 23-35.

[8] Subarto Kumar Ghosh, Mohammad Hasanuzzaman Shawon, Ashifur Rahman and Rifat Abdullah, "Modeling of PV Array and Analysis of Different Parameters", *International Journal of Advancements in Research & Technology*, Volume **2**, Issue 5, May-2013 358 ISSN 2278-7763.

[9] F. Katiraci and M.R. Iravani, "Power Management strategies for a micro-grid with multiple distributed generation units", *IEEE Trans. On Power Systems*, vol. **21**, no.4, pp. 1821-1831, Nov. 2006.

[10] C.H. Ram Jethmalani, Sishaj P Simon, Kinattingal Sundareswaran, P. Srinivasa Rao Nayak, Narayana Prasad Padhy, "Auxillary Hybrid PSO-BPNN-Based Transmission System Loss Estimation in Generation Schduling", *IEEE transactions on industrial informaics*, Vol. **13**, No. 4, August 2017.

IOP Conf. Series: Earth and Environmental Science **312** (2019) 012029 doi:10.1088/1755-1315/312/1/012029

Heliotropism of solar PV using the 4-bar mechanism analysis

Anuj Bimal Desai and Sendhil Kumar Natarajan*

Department of Mechanical Engineering,
National Institute of Technology – Puducherry, Karaikal 609609
U.T of Puducherry, INDIA.

Email:drsendhil1980iitmuk@gmail.com

Abstract. Solar energy is the most abundant source of renewable energy upon which most conversions of energies could be possible. Solar tracking can be one method through which the energy reception could be increased. Heliotropism is a phenomenon that is observed in various botanical species like leaves, flowers, stems, etc, where there is reaction observed by various parts of the plant corresponding to a solar stimulus. The classic example of this phenomenon is the sunflower effect wherein the sunflower tends to orient itself towards the sun all throughout the day and receives maximum energy at its surface. Similarly, such an effect can be produced using the conventional four bar mechanism which would toggle around the spectrum of the sunlight received to at various positions at various durations of the day. This format of receiving energy would not only increase the quanta of packets being received on an average throughout the day but would also not waste the energy falling on the earth's surface due to the misalignment of Photovoltaic (PV) cells usually installed in just one direction.

Keywords: Solar; Heliotropism, 4-bar mechanism, PV cells, Sunflower effect.

1. Introduction

As the world's population inflation is at its never seen before increasing rate, so does the simultaneous need for the availability of the sustainable energy sources increases. There are various sources of energy both renewable and non-renewable. Non- renewable energy comes from a variety of sources such as natural gas, fossil fuels, coals etc. The depletion of such valuable energy sources is now at a high alarming rate and is degrading both climate as well as the vulnerability of mankind [1]. Despite it being a very valuable and rich source of energy there is continuously increasing demands for the potentially replenishable and renewable energy. The solar energy is one such abundant energy source that may not fade out in a faraway future and remain the best form of the ultimate source of energy. The Solar energy is an ultimate, continuous and a never-ending source of renewable energy and can be captured and converted into usable electricity by various methods. One such method is the usage of the PV cells where in the solar energy reflects onto the surface of the PV and that incident energy is converted into usable electricity. The solar energy reaching the surface converts as high as 30% of the energy into electricity and rest is reflected and absorbed by various water bodies [2]. The energy here however remains convertible to only that amount of time when the energy is directly falling over the solar PV. The rest energy is not taken into account. To solve this issue the modern-day concept of solar tracking and heliotropism is being employed. In this adopted method the aim is that the solar PV module or panel orients itself towards the direction of illumination so as to achieve the maximum energy incident to the surface of the PV cell. The heliotropism is majorly inspired from the sunflower effect wherein the

IOP Conf. Series: Earth and Environmental Science **312** (2019) 012029 doi:10.1088/1755-1315/312/1/012029

sunflower has its orientation towards the sun and keep receiving direct normal energy. This effect of response to solar stimulus is due to the plant hormone known as auxin and various other photoreceptors and is explained in detail in [3]. This type of a method is taken into account and various methods have been employed to achieve the heliotropism phenomenon in the conventional solar PV cells. One such method to achieve heliotropism in the solar PV cells is the usage of Carbon Nano Tubes (CNT) and the Liquid Crystalline Elastomer (LCE) to achieve the maximum power input. CNT and LTE have very good optical and thermal absorption respectively and can be used as self-tracking systems for the maximum power by the proposed mechanism in [4]. Another method of achieving the solar tracking system is by the microchip controller PIC 18F452 [5]. The system in here uses two different methods of solar tracking out of which the first method is the search mechanism (PILOT) and the other is the intelligent panels. Herein the method to achieve maximum energy output lies in the orientation of the PILOT. The intelligent panels only align themselves with the pilot if there is a maximum energy condition and this could happen only when the sun is facing towards the panel. There are various other methods of achieving this technique and another such method is by using the IBL 2403 drive unit [6], which connects to a dc motor and through the help of light intensity sensors the solar tracking could be achieved. These were some of the various methods by which the solar tracking could be achieved. In the current paper a different technique is adopted to incorporate heliotropism. Unlike the other stated few techniques, this does not include any micro-controlling devices and is based on the conventional 4-bar mechanism. It works according to the Grashof law and the crank rocker mechanism is obtained. The solar panel's motion is controlled by the rocker motion and it oscillates in the range where the maximum solar energy spectrum is obtained. This form of technique is different than most cases, as the energy obtained on the surface of the panel in them is due to the steady tracking whereas the mechanism followed here covers a spectrum of energy that falls on the surface of the panel. Here there is another feature, which distinguishes it from the general 4 bar mechanism. The toggling of the solar panel can be divided into three major segments wherein it would be oscillating in the bands of energy at various durations of the time of the day. All in all, it is another, not very efficient but an innovative way of tracking the solar energy and can be implied in regions abundant of maximum solar powers. The further sections discuss regarding the construction of the mechanism, operations of the mechanism, materials used in the mechanism and also the scope of future development and research can be seen.

2. Methodology

2.1. Working Requisites

The proposed mechanism should satisfy the following technical aspects under its working condition:

- The toggle obtained of the solar panel must be able to cover the entire spectrum of the energy falling over the PV panel
- The energy obtained through this method should be greater than the normal conventional installed solar PV cells.

2.2. Structural Setup

As stated above, that the mechanism obtained is the crank rocker mechanism, the structuring of the setup could be done in the following way:

- The entire mechanism is divided into 3-stage process where the rocker mechanism will be obtained at three different orientation depending upon the location of the sun.
- The major segment of time for the operation of the mechanism is considered to be between 8 A.M. to 5 P.M. where the Sun will be delivering maximum energy onto the Earth's surface.
- According to calculations the rocker motion can be segmented into three slots of 45° each and that is divided into time slots of 3 hours each.

2.3. *Constructional Aspects*
For the construction of the mechanism, the following aspects need to be considered:

- The proposed mechanism is the 4-bar mechanism, wherein there is a crank, coupler, rocker and a fixed link.
- Since the link lengths need to be changed at various time slots, the crank, coupler and rocker are segmented.
- This segmentation can provide us with a suitable length configuration that can be applied to achieve the desired oscillation of the rocker between a particular location at a particular time.
- There are majorly 3 configurations that can be derived out of the setup as shown below.

2.4. *Structural arrangements*
The mechanism obtained should be majorly divided into 3 segments and the transition between each setup can be done manually. The 4-bar mechanism is arranged as follows:

- The fixed link is placed in the direction parallel to the rays of the sun.
- The crank depending upon the duration of the time can be attached to the fixed link at one end.
- Based on the design, the coupler and the rocker can be fixed.
- Finally, the rocker is connected at the end of the fixed link.
- An electric faculty such as a DC drive system can be applied to the crank, such that, a constant RPM of the crank is obtained.
- In correspondence to the same, the rocker mechanism toggles in an orientation based on its connection.

The detailed mechanism is shown in the next section.

3. Mechanism
The mechanism of the entire set up as discussed earlier can be broadly divided into three segments:

- 8 A.M. – 11 A.M
- 11 A.M. – 2 P.M
- 2 P.M – 5P.M

The position of the sun in these intervals are broadly in three different positions as conventionally observed. Hence, to cover up all these segments, the extreme positions of the solar panel in the proposed mechanism are shown.

3.1. *First Phase*

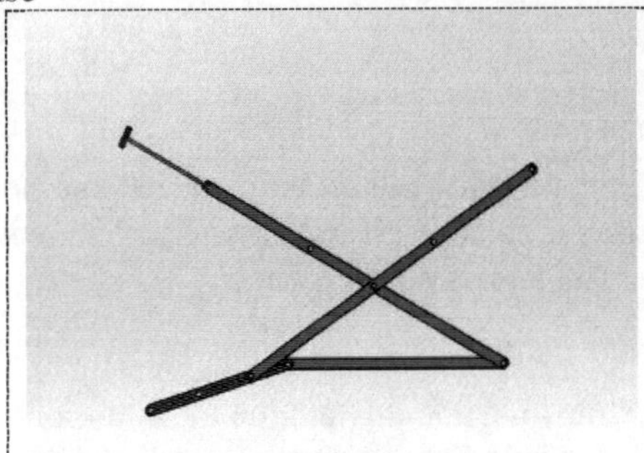

Figure 1. Left Extreme Position

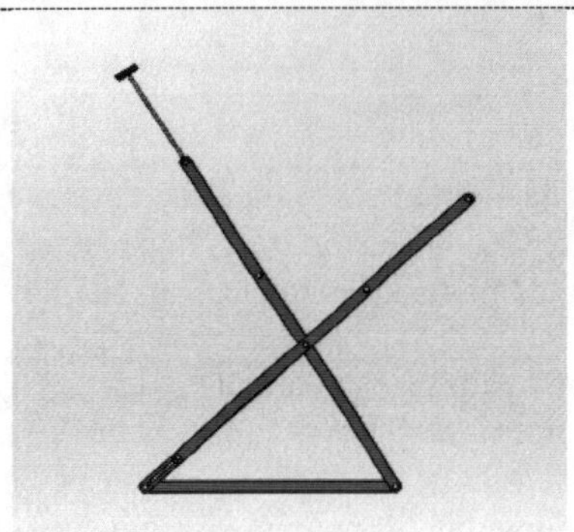

Figure 2. Right Extreme Position.

These are the extreme positions in the automation and covers the maximum energy band falling on the surface of the earth. The energy obtained on the earth is assumed to be constant and around 1000 – 1300 kW. This configuration will only imply until the end of first hour, after which the configuration suitable to the next phase will have to be mechanised manually.

IOP Conf. Series: Earth and Environmental Science **312** (2019) 012029 doi:10.1088/1755-1315/312/1/012029

3.2. Second phase

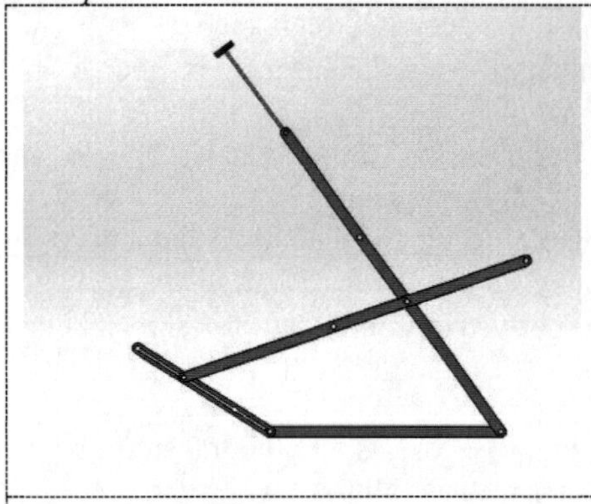

Figure 3. Left Extreme.

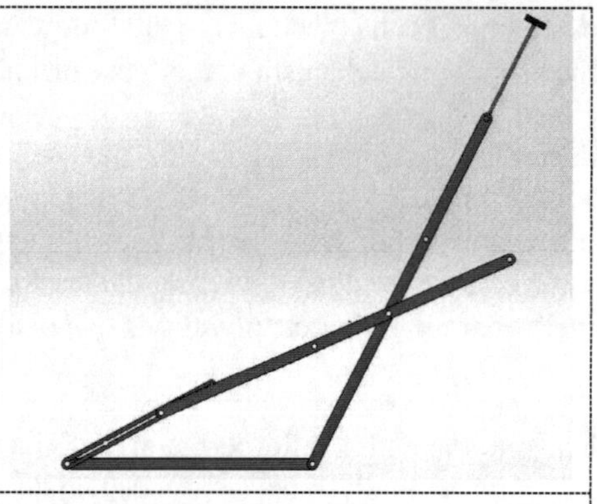

Figure 4. Right Extreme.

The above images depict the extreme positions that can be employed during the second phase. The suns position is almost over head during this duration and the energy band obtained as per general observation are almost normal and high in both temperature and energy content. The toggling of the rocker almost covers the entire spectrum of band during the second phase interval. At the end of the second phase, the process will be needed to repeat of interchanging the links so as to obtain the 3rd inversion that follows next.

3.3. Third Phase

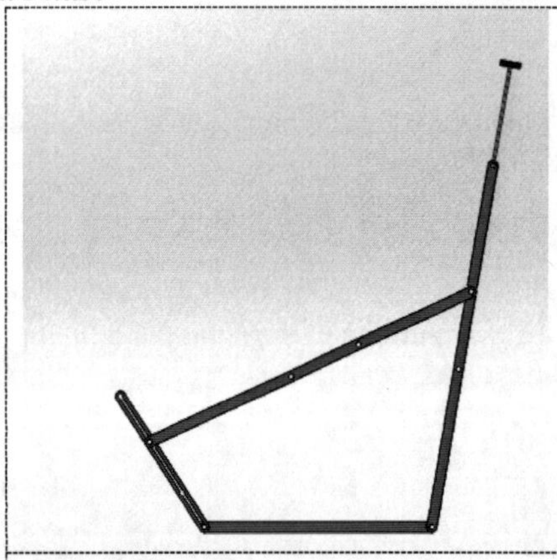

Figure 5. Left Extreme Position.

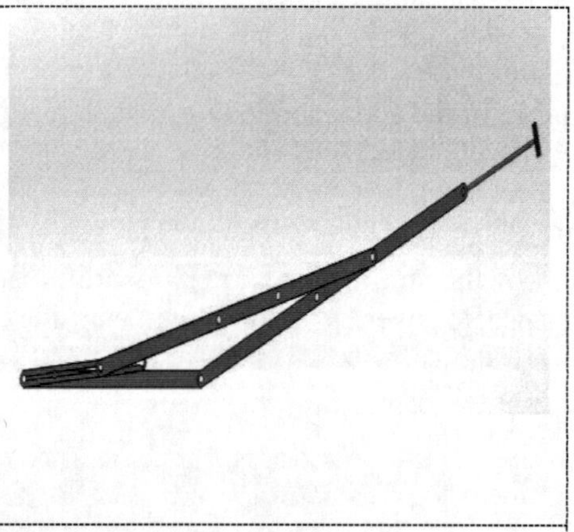

Figure 6. Right extreme position.

This the last phase that has to be employed after the second phase, and the extreme positions of the rocker are shown above. In this the evening spectrum is considered and the energy obtained in this phase is relatively lower. The mechanism can be further stopped as not much energy can be expected.

4. Scope of further research

The control of the crank in this case is by an electrical DC faculty and the transition of various phases is manual as of now. However various techniques can be employed to automate it. Some of them are as stated below:

- The usage of microprocessors in automating the crank and other transitions of the mechanism can reduce the mechanical operating of the system.
- Redesigning the segmentation and making it continuous and uniform in reference to the suns position using microcontrollers can evict he three stage system and the whole set up will be corresponding uniformly with the suns mean positioning.

- Addition of artificial intelligence to the core operation of the system to auto-calibrate it to the desired location can reduce all chances of energy losses.

In terms of design, there could be other permutations of linkage mechanism that could facilitate better toggling of the solar PV.

5. Conclusions

The paper proposes a way to provide a better surface area to the solar PV by the crank rocker mechanism of the 4-bar linkage system. It ensures that the entire band of useful energy is covered during the oscillatory movement of the rocker and no energy is much wasted. This action also helps in increasing the usability of the PV cell and also raises the amount of electrical energy so produced. The main components of the set up are the segmented bar mechanism that can be used in various configurations to attain the desired output angle of the rocker. The fundamentals behind the entire mechanism lies in the basic Grashoff's law that provides this type of a suitable mechanism that is employed to gain the usability of the solar PV. This is one of the many ways to achieve the source concept of solar heliotropism.

References

[1] Höök M and Tang X 2013 Depletion of fossil fuels and anthropogenic climate change-A review *Energy Policy* **52** 797–809

[2] Shaikh M R S 2017 A Review Paper on Electricity Generation from Solar Energy *Int. J. Res. Appl. Sci. Eng. Technol.* **V** 1884–9

[3] Holland J J, Roberts D and Liscum E 2009 Understanding phototropism: From Darwin to today *J. Exp. Bot.* **60** 1969–78

[4] Sharma D K and Purohit G 2014 Nature Inspired Solar PV Panel Tracking using Sunflower based Heliotropism for Higher Efficiency of Solar PV Power System *6th World Conf. Photovolt. Energy Convers.*

[5] Ghassoul M 2008 Design of an Automatic Solar Tracking System to Maximize Energy Extraction *Int. J. Emerg. Technol. Adv. Eng.* **3** 453–60

[6] Tudorache T and Kreindler L 2010 Design of a Solar Tracker System for PV Power Plants *Acta Polytech. Hungarica* **7** 23–39

IOP Conf. Series: Earth and Environmental Science **312** (2019) 012030 doi:10.1088/1755-1315/312/1/012030

Techno-economic analysis of solar trigeneration system

B Anand[1]* and S Murugavelh[2]*
[1, 2] CO_2 Research and Green Technologies Centre
Vellore Institute of Technology, Vellore, Tamil Nadu, India
Email: [1]anandgbea@gmail.com, [2]murugavelh.s@vit.ac.in

Abstract. In this work, a techno-economic analysis of concentrated photovoltaic/thermal collector (CPVT) driven desalination and cooling plant is presented. The system comprises of a CPVT collector, a humidification-dehumidification (HDH) desalination unit and vapour compression refrigeration (VCR) unit. The CPVT collector having 10 m^2 aperture area and a triple junction PV panel placed at the receiver supplies required hot water to the HDH desalination system. Chilled water generated from 0.5 hp compressor operated VCR unit is supplied to the dehumidifier of HDH desalination. The process of cooling and dehumidification using chilled water result in increased desalination yield and resulting air is suitable for air conditioning purpose. The work is aimed to evaluate the annual performance of the trigeneration plant and to estimate the unit cost of fresh water and payback period. The annual performance of the plant is evaluated for the typical metrological year data for the location of Vellore, India. Results show that the annual desalination, electricity and cooling output are 4.2 m^3/ yr and 3.46 MWh/yr and 0.788 MWh/yr respectively. The daily average desalination and cooling output are 11.12 L/day and 2.47 kWh/day respectively. The annual electricity consumption of the solar cogeneration plant is 6570 kWh/yr. The estimated unit cost of fresh water is ₹10/L. By considering the interest rate of 9 % and total life time of the plant as 20 years, the payback period is estimated to be 6.5 years.

1. Introduction

The solar cell efficiency drops with increase in temperature of solar cell. The solar cell efficiency also depends on solar radiation. The commercial solar cell efficiency varies from 10 % to 20 % and remaining energy is wasted as heat loss [1]. A well proven method to improve the conversion efficiency is recovering the waste through process integration [2]. A Photovoltaic/Thermal panel (PVT) simultaneously generates electricity and thermal energy. This is achieved by circulating a cooling medium such as air, water behind the PV panel [3]. The cogenerated heat from the PVT panel is also used in domestic application such as space and water heating [4]. A concentrated PVT system is combination of solar concentrating collector and PVT panel. The PV panel in CPVT collector is triple junction PV panel and it can withstand temperature up to 240 °C [5]. The CPVT coupled with water desalination performance better than other solar desalination technologies [6]. A solar heating and cooling system based on CPVT operated vapor absorption chiller resulted in high primary energy savings [7]. A novel heat pump integrated with PVT system, result in average COP of 4.8 while generating hot water at 70°C [8]. A high concentrating CPVT collector integrated with space cooling system having combined efficiency of 87.5 % [9]. From the literature studies, the CPVT technology perform better than individually operated system. Further, very few studies are carried out to integrated cooling and desalination. This result in increased desalination and resulting air suitable for air conditioning. But, these technologies require hot water for desalination and electricity to run VCR (Vapor Compression Refrigeration) unit. In such cases PVT collector will require more land are and the hot water from PVT panel are at low temperature. In this case, CPVT collector with high conversion efficiency, and high hot water temperature will make the plant self-sustained. In this work, a detailed mathematical modelling and CPVT operated desalination and cooling plant is carried out. Further, a detailed economic analysis of the plant is also carried out to evaluate the unit cost of fresh water and payback period.

2. Plant description:

The schematic process flow diagram of CPVT operated desalination and cooling plant is shown in figure 1. The plant consist of a CPVT collector, a HDH desalination unit, a vapour compression refrigeration unit, a blower and pump. The CPVT collector is made of 10 m^2 aperture area, a triple junction solar cell placed at the receiver. The HDH unit consists of a humidifier and dehumidifier. The humidifier receives required hot water from CPVT collector and the dehumidifier receives chilled water from the VCR unit. Initially the temperature of the solar cell is high as the concentrated solar radiation is focused on it. At high temperature the electrical conversion efficiency of the solar cell is very low. In order to reduce the cell temperature a cooling medium is supplied to the heat transfer pipes attached back of the PV panel. This result in increased electrical conversion efficiency and hot water suitable for thermal desalination process. The CPVT system is modelled based on the hot water and electricity requirement of the desalination and cooling plant. The temperature of the solar cell and hot saline water to humidifier can be controlled by controlling the saline water flow rate. A blower supplies air at ambient condition from bottom of the humidifier. After heat transfer in CPVT collector, the hot saline water enters the humidifier where simultaneous heat and mass transfer occurs between air and hot water. This result in heating and humidification of air. The hot humid air leaving the humidifier is then entering the dehumidifier. In dehumidifier, the moisture content in the hot humid air is cooled and dehumidified using chilled water. This result in increased desalination yield and air coming out of the dehumidifier is at low temperature suitable for air conditioning.

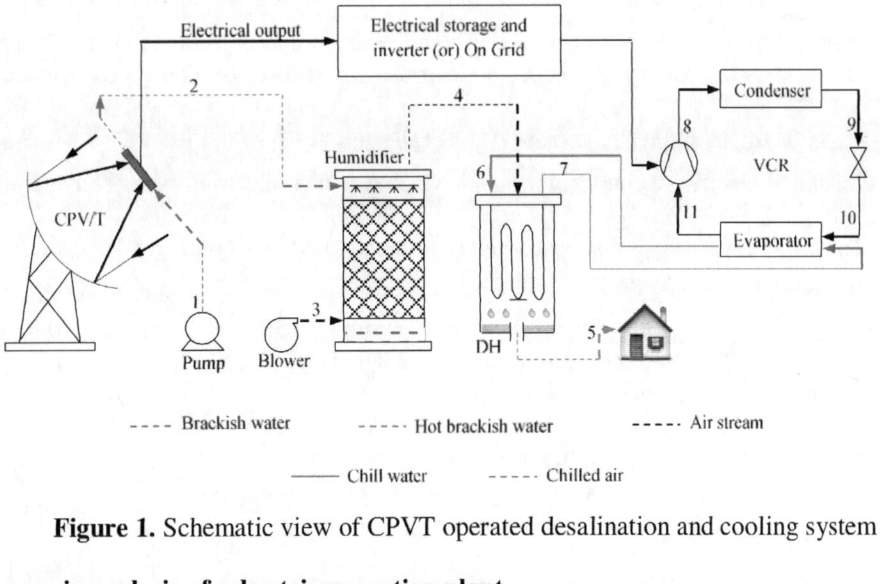

Figure 1. Schematic view of CPVT operated desalination and cooling system

3. Economic analysis of solar tri-generation plant

In this section, a detailed economic model is developed to estimate the unit cost of fresh water payback period of CPVT operated desalination plant. The total direct capital cost the plant is given in table 1 and the details of financial parameters considered in this work is shown in table 2. The unit cost of the fresh water (C_{FW}) is the ratio of annual total cost of the plant (ATC_P) to the annual fresh water yield (FW_y) and it is expresses as

$$C_{FW} = \frac{ATC_P}{FW_y} \tag{1}$$

The annual total cost of plant includes annual capital cost (AC_P), operation & maintenance cost (OMC_P) and annual salvage value of the plant (ASV)

$$ATC_P = AC_P + OMC_P - ASV \tag{2}$$

The annual capital cost is product of total capital cost (M) and CRF

$$AC_P = M(CRF) \tag{3}$$

Where CRF is the Capital Recovery Factor and it is calculated from eq. (4)

$$CRF = \frac{i(1+i)^n}{(1+i)^n - 1} \tag{4}$$

Since the plant also produces cooling output, the equivalent electrical energy (AER_C) required to produce the same annual cooling output (AQ_C) from conventional air conditioner is calculated from eq. (5) by considering COP_{ac} as 2.5

$$AER_c = \frac{AQ_c}{COP_{ac}} \tag{5}$$

The annual salvage value of the plant (ASV_P) is product of future salvage value (SV_P) and sinking fund factor (SFF_P)

$$ASV_P = SV_P(SFF_P) \tag{6}$$

Considering future salvage value of the plant as 20 % of total capital cost (M), the sinking fund factor is estimated from eq. (7)

$$SFF = \frac{i}{(1+i)^n - 1} \tag{7}$$

Further, the operation and maintenance cost of the plant includes energy cost, labor cost, periodic cleaning of salt deposition etc. Considering the cost of electricity as 6.5 /kWh the O&M cost is calculated as

$$OMC_P = C_e(AEC_P - (AEG_P + AER_c)) + 0.15(AC_P) \tag{8}$$

where, AEC_P is the annual electricity consumption and AEG_P is annual electricity generation from the plant.

Table 1. Components of solar tri-generation plant and its capital cost

Components	Description	No's	Cost
CPVT collector	CPVT collector with aperture area of 10 m^2 and triple junction solar panel placed at the receiver.	1	1,10,474[a]
Humidifier	• Made of SS 304 material of 1.5 m height and 0.30m in diameter. • Corrugated plastic sheet is used as packing material	1	23,000
Water cooled Dehumidifier	Shell and tube heat exchangers with tube diameter of 0.012 m and 0.5 m length.	1	10,000
Water chiller	0.5 hp compressor operated chiller unit	1	25,000
Air blower	Centrifugal blower with 500 m^3/h discharge capacity	1	3,000
Pump	Discharge: 33.6 LPM, Head:6-26 m	2	6,000

IOP Conf. Series: Earth and Environmental Science **312** (2019) 012030 doi:10.1088/1755-1315/312/1/012030

Piping cost	All the pipelines are constructed using CPVC pipes	-	3,000
Others	Rotameters, valves, gasket, bolt and nuts	-	6, 250
	Total direct capital cost		1, 86, 724
Unit cost of fresh water			₹ 10/L
Payback period			6.5 years

[a]Total cost including 30 % subsidy offered from MNRE, India.

Table 2. Financial parameters of solar tri-generation plant

Parameters	Description
Direct capital cost	1,86,724
Service life of the plant	20 years
Interest rate	13 %
subsidy	30 %
Transportation and insurance	5 % direct capital cost
Cost of contingency	5 % of direct capital cost
Salvage value	20 % of direct capital cost
Total capital cost of the plant	2,05,396

4. Results and discussion

The monthly average global horizontal irradiation and mean ambient temperature in Vellore, India is shown in figure 2. The monthly average radiation in the month of January is around 5.5 kWh/m^{2}day and it increases gradually and reaches highest value of 7 kWh/m^{2}day in April month. Then the radiation and ambient temperature drops towards the December month.

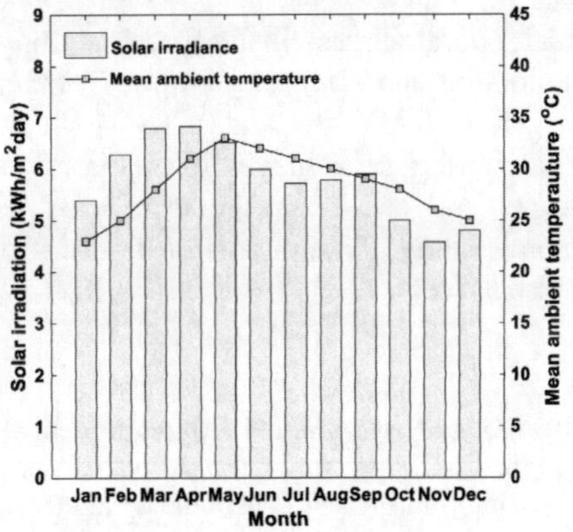

Figure 2. Monthly averaged GHI and ambient temperature in Vellore, India

The annual desalination yield, electricity production and cooling output are shown in figure 3. The total desalination yield and electricity production are high during the months of higher solar radiation. This is because higher solar radiation results in higher hot water temperature into the humidifier. Meanwhile more electricity is generated as high amount of radiation is available for the electrical energy conversion. The cooling output is also high at higher solar radiation. At higher radiation the hot humid air leaving the humidifier is at high temperature. This increases load on dehumidifier and resulting air is also at high temperature. Meanwhile, the temperature difference between ambient and chilled air is high this result in high cooling output. In winter, the ambient temperature itself at low level, so the changes in temperature of air is not much high compared with summer time. The highest average desalination yield is achieved in the months of April and May.

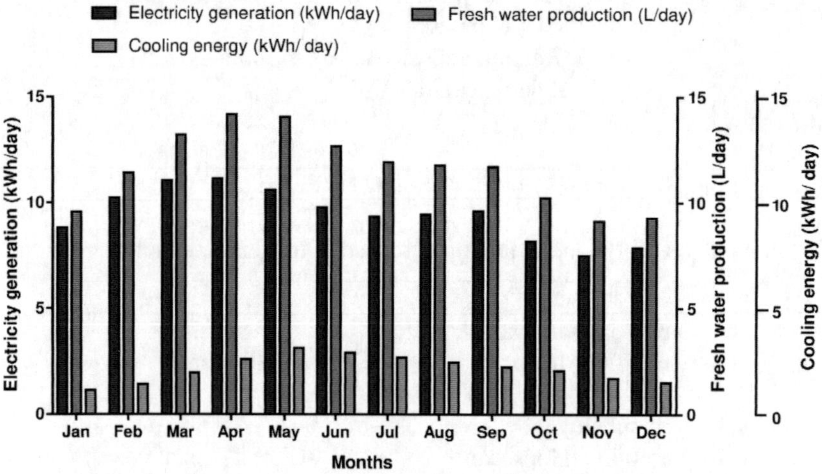

Figure 3. Monthly averaged desalination output, electricity production and cooling output from the CPVT operated system

The detailed economic model is presented in section. 3, based on the model developed and total capital cost the unit cost of fresh water and payback period are calculated. Assuming the interest rate as 9 % and energy cost as ₹ 6.5/kWh the unit cost of electricity and payback period are calculated as ₹ 10/L and 6.5 years.

5. Conclusion

The concentrated photovoltaic/thermal operated desalination and cooling plant is modelled and simulated for the annual desalination, cooling and electricity output. The annual desalination, cooling and electrical energy output are 4.2 m^3/yr, 3.46 MWh/yr and 0.788 MWh/yr respectively. The highest desalination yield, electrical output and cooling are achieved in the month of April/May. The highest monthly average desalination, electricity and cooling outputs are 14 L/day, 11.5 kWh/day and 4 kWh/day respectively. From the economic study, the unit cost of desalination and payback period are calculated to be ₹ 10/L and 6.5 years respectively.

References

[1] Gul M, Kotak Y and Muneer T 2016 *Energy Exploration & Exploitation* **34** 485–526.
[2] Wu D. W and Wang R Z (2006) *Progress in Energy and Combustion Science* **32** 459–495.
[3] Charalambous P G, Maidment G G, Kalogirou S A and Yiakoumetti K 2007 *Applied Thermal Engineering* **27** 275–286.
[4] Chow T T, Ji J and He W 2007 *Journal of Solar Energy Engineering* **129** 205.
[5] Mittelman G, Kribus A and Dayan A 2007 *Energy Conversion and Management* **48** 2481–2490.
[6] Mittelman G, Kribus A, Mouchtar O and Dayan A 2009 *Solar Energy* **83** 1322–1334.
[7] Buonomano A, Calise F and Palombo A 2013 *Applied Energy* **103** 588–606.
[8] Xu G, Zhang X and Deng S 2011 *Applied Thermal Engineering* **31** 3689–3695.
[9] Garcia-Heller V, Paredes S, Ong C L, Ruch P and Michel B 2014 *Renewable and Sustainable Energy Reviews* **34** 8–19.

IOP Conf. Series: Earth and Environmental Science **312** (2019) 012031 doi:10.1088/1755-1315/312/1/012031

Bio oil production from Agro waste residue: Thermochemical conversion to improve oil quality

K Midhun Prasad[1] K Murugavelh[2*]
[1]School of Mechanical Engineering
[2]CO_2 Research and Green Technology Centre
Vellore Institute of Technology, Vellore – 632014

E-mail: midhunprasad.k@vit.ac.in, Murugavelh.s@vit.ac.in

Abstract. Energy is the symbol of economical developments and human settlements. This study is carried out to find the effect of oil yield by optimization of temperature and residence time for the thermochemical conversion of solid waste into the liquid fuel by pyrolysis reaction in an auger reactor made of stainless steel with the heating rate of 10 °C/min with the maximum temperature range of 1000 °C. The agro waste of Tomato plant are left in land after harvesting which can be converted into usable carbon neutral liquid fuel by further processing. The experiment is carried in a different temperature ranges from 450, 500, 550, 600, 650 °C. Tomato plant is collected, dried in open solar drying until the sample weight remains constant to remove the bound moisture content. Then crushed into small pieces, and subjected to thermogravimetric analysis, proximate and ultimate analysis. Further the bio oil is characterised with GC-MS and FTIR to find the compounds present. The moto of this study is to find the convert the waste into high energy content liquid fuel.

1. Introduction

World today focused in the search of alternative fuel to overcome the usage of fossil fuels. The culture nowadays is infrastructure with using of huge amount of fossil fuels without knowing the drawbacks [1]. Increase in the threat on environmental pollution and other issues becomes the serious effects nowadays. The major cause of environmental pollution results in the exhaust from automobiles because of using of fossil fuels which associates with CO_2, NO_x and SO_x. Future is dependent on alternative fuel which can helps in production of clean energy. One of the solutions for the production of alternative fuel is based on the biomass and agricultural waste residue [2]. Biomass can be used as an alternative source for the energy fuel generation in the present scenario. Pyrolytic conversion of waste left out after crop harvesting can be utilised. Tomato is harvested in other regions all around the world, the plant was then left of the land for the natural decomposition or burned in the land. Burning also create the problem with environmental issues, so collecting those and using is user friendly way can be approached [3]. Pyrolytic conversion of waste biomass to liquid fuels has a significant in the climate change problems which associated with reducing the CO_2 emission also consider as a better replacement for fossil fuels.

India is a country which major population depend on agricultural activities focusing on the waste generated can help in the larger quantity production of valuable by products, which can be of solid, liquid and gases [4]. The obtained liquid is pyrolytic oil which can directly use in boilers and turbines on upgradation this can be use as a high-quality hydrocarbon fuel [5]. Whereas biomass also emits little amount of sulphur and nitrogen, which is less when compared to the fossil fuel emissions.

Among the other thermochemical conversion process pyrolysis is one of the better energy production process. Thermal degradation of waste solids and biomass in the absence of oxygen in a closed reactor is called pyrolysis. In olden days pyrolysis is used for the charcoal production [6]. More attention has been gained to the pyrolysis process due to it efficiency in converting solid to liquid fuels, also this technology helps in generation of high fuel to feed ratio. There are basically three types of energy extraction process combustion, gasification and pyrolysis [7]. In pyrolysis the effect of temperature starts around 350 ˚C to 550 ˚C and keep on increasing to the maximum temperature of 900 ˚C in an absence of oxygen [8]. During the high temperature the biomass starts cracking long molecular chain of carbon, hydrogen and oxygen breakdown into short chain compounds in gaseous and char form [9].

In this study optimization of temperature for the efficient liquid yield was carried out using tomato plant waste in a auger reactor at the retention time of 2 min with varying temperature ranges from 450 ˚C, 500 ˚C, 550 ˚C, 600 ˚C and 650 ˚C, the experiment was carried out at the heating rate of 20 ˚C min^{-1}. The obtained liquid is collected and analysed with GCMS and FTIR.

2. Experimental Setup

Pyrolysis process is carried out using a closed auger reactor made of stainless steel with the maximum feedstock holding capacity of 5 kg specifically designed for batch process, with a variable heating rate

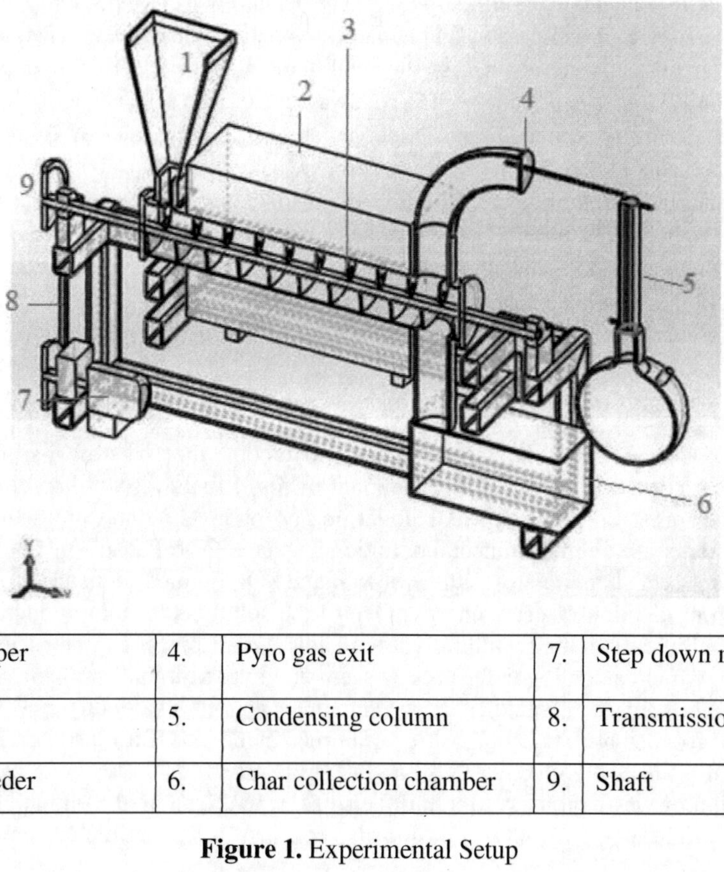

1.	Feed hopper	4.	Pyro gas exit	7.	Step down motor
2.	Furnace	5.	Condensing column	8.	Transmission chain
3.	Screw feeder	6.	Char collection chamber	9.	Shaft

Figure 1. Experimental Setup

IOP Conf. Series: Earth and Environmental Science **312** (2019) 012031 doi:10.1088/1755-1315/312/1/012031

and temperature range up to 1000 ˚C. The entire setup is compactly packed the feed is feeded at the hopper and char is collected the chamber below shown in figure 1. The shaft speed is adjusted by step down gear box. At the end of the reactor the char is collected down and the gas produced at the top. The gas obtained is then flow through a water-cooled condenser column where the temperature of the water is maintained at 8 ˚C for the proper condensation.

3. Materials and method

The experiment is carried out using tomato plant waste, the biomass is collected from local farmers from Vellore. The sample is then washed to remove sand and dirt present and dried in open solar drying until the weight drop occurs. Then the sample is milled to the size less than 2 mm for the analysis.
The elemental analysis of tomato plant (leaf, steam and root) was carried out using PerkinElmer 2400 model asper ASTM standards. Proximate analysis was equipped to find the moisture content, ash content, fixed carbon and volatile matter by ASTM norms using hot air oven, muffle furnace and weighing balance. Cellulose, hemicellulose and lignin content present in the sample was analysed by Van Soest method. Thermogravimetric analysis of feedstock is analysed using Mettler Toledo in a nitrogen atmosphere at the flow rate of 20 ml min^{-1}. The TGA experiment was carried out from the range of atmospheric temperature to 800 ˚C with a constant heating rate of 20 ˚C min^{-1}. The oil obtained is then analysed with GCMS and FTIR to find the functional group and compounds present in the sample, calorific value of the sample is found by bomb calorimeter.

The operation parameter of the experiment was optimized the temperature range from 450 ˚C to 650 ˚C and fixed other parameters for the other set of experiments, retention time of 2 min with the fine particle size (2 mm) and constant heating rate of 10 ˚C min^{-1}.

4. Results and Discussion

Table 1. Biomass characterisation

Proximate Analysis		Ultimate Analysis		Composition analysis	
Moisture (wt%)	10.2	C (wt%)	38	Hemicellulose (wt%)	28.8
Ash (wt%)	8.08	H (wt%)	5.2	Cellulose (wt%)	39.1
Fixed Carbon (wt%)	51.06	N (wt%)	2.7	Lignin (wt%)	12.1
Volatile Content (wt%)	30.12	S (wt%)	1.1	-	-
Calorific Value	21 MJ/kg	O (wt%)	41.1	-	-

Pyrolysis of tomato plant waste is carried out at the different temperature range trials from 450 ˚C, 500 ˚C, 550 ˚C, 600 ˚C and 650 ˚C to find the better liquid yield, at the initial stage the reactor is set to the temperature and the sample is feeded into the reactor in an inert atmosphere absence of oxygen, the retention time of the reaction was optimized to 2 min. The gas produced was condensed and the liquid and char obtained was collected then measured, uncondensable gas calculated by difference Table1 describes the proximate which shows the presence of moisture content, ash content, fixed carbon, volatile matter and ultimate analysis of the sample shows the presence of carbon, hydrogen, nitrogen, sulphur, oxygen presence in the feedstock sample. The biomass is also analysed with hemicellulose, cellulose and lignin content present shown in table 1.

4.1 TGA

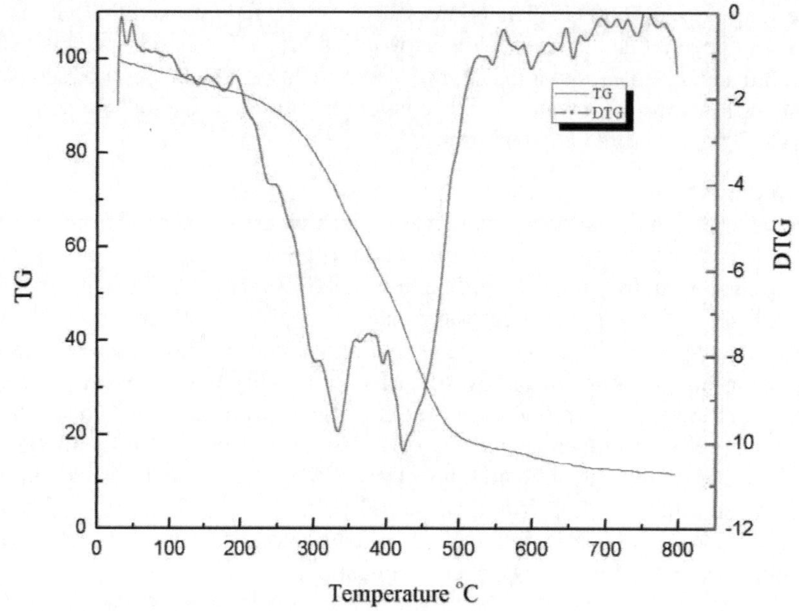

Figure 2. TGA

Thermogravimetric analysis of tomato plant waste was carried out to find the behaviour of thermal decomposition. Figure 1 shows the degradation curve of the feedstock at 20 °C min^{-1}, at the initial stage 10 mg of sample is taken in a crucible and the weight loss at the varying temperature is noted. A slight drop around 100 °C denotes the presence of bound moisture. Thermal degradation at the second zone starts at 200 °C to 500 °C. And the final stage in the degradation occurred at the temperature range of 500 °C to 650 °C. After reaching at the high temperature the curve become constant. The three zones in degradation denotes the decomposition of cellulose, hemicellulose and lignin.

4.2 Influence of temperature

Change in temperature has a direct impact on the by products yield form the tomato plant pyrolysis. From the analysis it is noted that increase in the temperature leads to the decrease in the liquid and char yield this impact in the increase in the gases yield. The decrease in liquid yield leads to the thermal cracking of biomass at the higher temperature, also increase in this temperature results in higher gas production. Table 1 and figure 3 shows the yield percent of liquid, solid and char. It is observed that pyrolysis at 600 °C is the efficient range to obtain higher liquid yield and for better gases yield 650 °C is the optimum range where it produces up to 45.4 wt% of gases compounds.

Table 2. Effect of Temperature

Yield	450 °C	500 °C	550 °C	600 °C	650 °C
Solid	56.5	54.1	39.8	30.0	29.8
Liquid	26.44	24.0	21.5	33.6	24.8
Gases	17.06	21.9	38.7	36.4	45.4

IOP Conf. Series: Earth and Environmental Science **312** (2019) 012031 doi:10.1088/1755-1315/312/1/012031

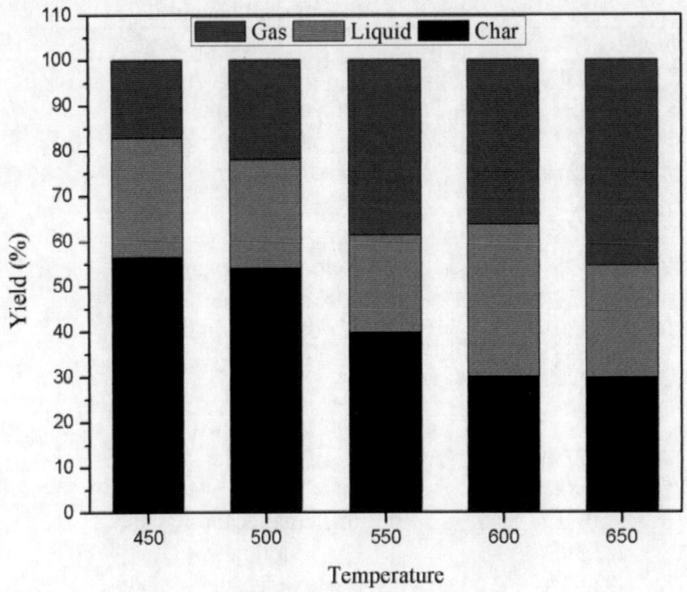

Figure 3. Yield Percentage

4.3 GCMS analysis of bio oil

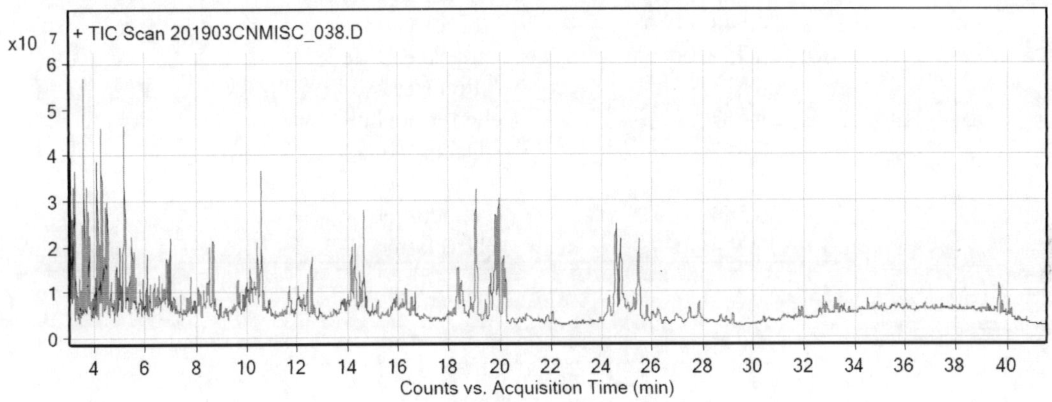

Figure 4. GCMS of bio oil

GCMS of the oil sample was analysed to find the compound group present in the sample, this helps in analysing the compound group present in the sample. It is equiped with GCMS- QP 2010 SHIMADZU analyser, operates at the temperature range from 70 ˚C to 300 ˚C in an helium atmosphere. From the figure 4 it is observed that most of the peak found at the range of 1 to 6 and minor peak found at the entire range, the compounds present in the sample was given in table 2 below. The compound present in the oil sample contains mostly flamable compounds like indane, indene, benzene, p – xylene, Silane etc, most of the compounds found are highly flammable in nature.

Table 3. GCMS compound table

IOP Conf. Series: Earth and Environmental Science **312** (2019) 012031 doi:10.1088/1755-1315/312/1/012031

S.no	RT	Compound Name
1	3.124	Benzene
2	3.238	p-Xylene
3	3.432	Cyclopentene
4	3.703	Octane
5	3.927	Indene
6	4.023	Indane
7	4.128	Phenol
8	4.735	Cyclopentasiloxane
9	5.233	Naphthalene
10	5.489	Cyclododecane
11	6.617	1H-Indene
12	6.923	Silane
13	7.396	Naphthalene
14	8.663	Tetradecane
15	9.482	Bicyclo
16	10.41	n-Pentadecanol
17	12.393	Fluorene
18	12.934	Hexadecane
19	14.65	Heptadecane
20	16.679	Hexadecane
21	19.837	Oleanitrile
22	21.849	Cyclohexadecane
23	24.548	Eicosane
24	25.454	Octadecanenitrile
25	26.213	Methyl stearate
26	28.667	Behenic alcohol
27	28.892	Docosane
28	31.756	Tricosane
29	33.554	Tetracosane
30	35.784	Pyrrolidine

4.4 FTIR

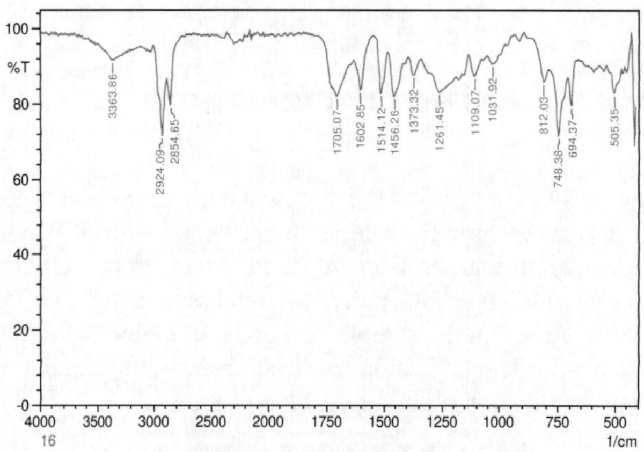

Figure 5. FTIR of bio oil

IOP Conf. Series: Earth and Environmental Science **312** (2019) 012031 doi:10.1088/1755-1315/312/1/012031

FTIR of the bio oil sample is analysed to found the organic, inorganic and polymers present in the sample. FTIR spectra was carried out from 500 – 4000 cm^{-1} equiped with Perkin Elmer analyser. From the table it is observed that peak corresponds to the vibrations. The peak at 505 denotes the presence of aromatic compounds, peak at 1031 denotes the presence of epoxi group, peak at 1373 coresponds to the presence of alkenes group and the peak at the range 1700 represents the presence of carboxylic acids, finally the peak at 3363 denotes the presence of N-H bending denotes the presence of amines and amides.

Table 4. Functional group using FTIR

Wave number (cm^{-1})	Vibration	Functional group
400 - 600	C-H Bending	Aromatic
980 -1200	C-O Stretching	Epoxide
1250 - 1300	C-O Stretching	Carboxylic Acid
1350 - 1500	C=C Bonding	Aromatic ring
1500 - 1550	C=C Bending	Alkenes
1600 – 1620	C=O Bonding	Aldehydes, Ketones, Carboxylic acids
1700 - 1725	C-OH Stretching	Carboxylic Acid
2854	C-H Bending	Alkanes
2924	C-H Stretching	Alkanes
3363	N-H Bending	Amines, amides

5. Conclusion

This study was carriedout to find the optimization of temperture for the better oil yield using tomato plant waste left after harvesting. The experiment was conducted with different temperature ranging from 450 °C to 650 °C, it is noted that 600 °C is the maximum optimum temperature for better oil yield. The calorific value of the oil was found to high compared to the biomass. GCMS and FTIR reported that the oil has more number of flamable compounds like benzene, p-xylene, cyclopentene, octane, indene, indane and Naphthalene. Presence of this compounds helps in increasing the calorific value of the oil sample

Reference

[1] Li Z, Li N, Yi W, Yongjun P F and Bai X 2017 *Fuel Pro.Technology* **161** 182 -92
[2] Tinwala F, Mohanty P, Parmar S, Patel A and Pant K K 2015 *Bio. Technology* **188** 258-264
[3] Aboagye D N, Banadda N K and Kabenge I 2016 *Renw. and Sus.Energy Reviews* **70** 814-21
[4] Balat M, Kırtay E and Balat H 2009 *Ene. Conv. and Management* **50** 3147-57
[5] Loannidou and O Zabaniotou A 2007 *Renw. and Sus.Energy Reviews* **11** 1966-05

IOP Conf. Series: Earth and Environmental Science **312** (2019) 012031 doi:10.1088/1755-1315/312/1/012031

[6] Bonelli P R, Buonomo E L and Cukierman A L 2007 *Jou. Ene. Sou., Part A: Reco. Utilia. And Envir. Effects* **29** 731 - 40
[7] Kothari K, Shah A and Murugavelh S 2015 *Int. jou. Of Chem. Tech Research* **8** 349-57.
[8] Bridgwater A V *Jour. of Anal. and App. Pyrolysis* 1999 **51** 3-22.
[9] Bridgwater A V, Peacocke G V C 2000 *Renw. and Sus.Energy Reviews* **4** 1-73.

Acknowledgments
The project was funded by Science and Engineering Research Board (SERB) file no ECR/2016/001304.

A Numerical Study of fluid flow and Pressure Drop analysis of Catalytic Converter using Commercial CFD code

Aniket Patil[1], Mayuresh Takale[1], Bageerathan Thirunavukkarasu[1], Thundil Karuppa Raj R[1*]

[1] School of Mechanical Engineering, VIT University, Vellore, Tamil Nadu, INDIA - 632014

*Corresponding author email id – thundilr@gmail.com

ABSTRACT: In recent years catalytic converter is a necessary device to achieve low emissions in all the vehicles. This study of the after treatment device is to understand the fluid flow and to optimize the design. It is important to know the pressure drop taking place inside the catalytic converter in order to design the catalytic converter. In this paper, a numerical study has been conducted to study the effect of fluid flow inside the catalytic converter and effect of change in geometry using commercial CFD tool. The working fluid considered in this study is air. A 3D simulation is carried out having straight inlet pipe and with change in inlet pipe angle of the catalytic converter. The monolith substrate region is modeled as a porous medium. The governing equations namely conservation of mass, momentum are used for solving the analysis. The analysis involved determining pressure variation across the geometry of the converter system for different mass flow rates and change inlet pipe angle. The numerical results were used determine the pressure drop across the substrate and the flow characteristics inside the catalytic converter. The numerical study shows good conformance with the experimental results with maximum deviation of error 2.54% for straight inlet pipe configuration. The effect of change in inlet pipe angle, effect of inlet cone angle and effect of splitting the substrate into sections are also presented.

Keywords: Catalytic converter, Pressure Drop, CFD analysis, substrate, ANSYS CFX, monolith.

1. Introduction

In recent past decade automotive emission standards have become stricter and there are different methods and efforts have been presented to find the cause of emissions and the development of new technologies for controlling and regulating the emissions. An Automotive catalytic converter usually made of a round shaped or oval shaped substrate encased in a metallic shell. The monolith substrate is having tubular channels along the length and connected to the exhaust system through inlet and outlet cones. Different methodologies and efforts are investigated in the recent years which will use the maximum volume of the catalytic converter. The design of any after treatment device is largely depends upon the pressure drop occurring in the device. The reason behind that is when your pressure drop is more then back pressure is created on engine exhaust gas and engine performance gets affected. So it is important to know the pressure drop occurring in catalytic converter. The uniform flow distribution assures the maximum utilisation of the catalyst volume inside the catalytic converter. Therefore, most modern catalytic converters have configuration geometry having inlet and outlet pipe, inlet and outlet cone, long substrate to ensure smooth uniform flow between these different cross sections. J. S. Howitt and T. C. Sekella [1] have experimentally conducted durability studies on monolithic catalytic converters having cylindrical geometry with symmetric cone profiles. They performed the test on the engines with dynamometer but not on the on-road actual vehicle test. They found that the ability of the converter to reach the active temperature and capability of the converter to convert high percentage of the pollutants has more influence on the effectiveness of the converter. They suggested that the small angle deflectors can result in balanced flow profile where the effectiveness is increased. A. P. Martin et al., [2] investigated the performance of catalytic converter under different conditions of flow distributions. CFD techniques were utilised to model the maldistribution of aging and light off conversion performing steady tests. They have conducted emission testing on a vehicle throughout ECE and EUDC test cycles. They found insignificant effects on later stages of ECE cycle. Kwangsup Hwang at al., [3] has conducted dynamic flow studies on the catalytic converter with flow visualization techniques. They studied the flow patterns of the normal flow conditions using Laser Doppler Velocimetry and other visualization techniques. They found that the geometry of the catalytic converter influences the effective surface area of the converter. Also the space between the catalyst brick has influenced the result. Seokhwan Lee et al., [4] conducted the studies on the catalytic converters by varying the engine loading conditions from no-load to full-load conditions. They found that the temperature beyond 1050 degree Celsius can spoil the characteristics of the catalyst in the converter. The increase in the misfire has added up temperature to the exhaust gas which ruins the catalyst in catalytic converter. Yong-Seok Cho et al., [5] has developed an alternative method to control the emissions from the catalytic converter by Cranking Exhaust Gas Ignition method. This method uses the unburnt mixture to flow through the catalytic converter for 10 seconds and the mixture is burnt with the help of glow plugs. The thermal energy released during the operation has warmed up the catalyst and efficiency is increased. They also found that the warm-up process can be faster by using a leaner air-fuel mixture. Exhaust emissions were controlled when this method was used. Gregory et al., [6] conducted the emission studies on the tuned exhaust of a four-cylinder engine. Two coupled catalyst is used in this study and comparative studied were carried out with single catalyst system. Under steady state conditions, both the catalyst performs the same. During higher flow rate conditions, the peak flow rate remains the same. The authors have found that the catalyst sizing influences the efficiency of the system. Onorati et al., [7] has conducted 1D modeling of flow reactions in pre-catalyst and main catalyst using a numerical model GASDYN. The chemical reactions were taken into account along with the thermal as well as mass transfer. The influence of insulating mat and cerium oxides has also been considered. The catalyst contains two close coupled pre-catalyst along with a main catalyst where the simulated results were in good agreement with the experimental results. Soo-Jin Jeong and Woo-Seung Kim [8] has conducted 3D numerical study on the warm-up catalyst with the emission performance. Compressible flow with unsteady conditions was considered to analyse the impact of the warm-up catalyst over the main catalyst and emissions of the tail pipe. The authors have found that the early warming of the catalyst has no effect on the exhaust gas. The main catalyst has less efficiency during the early warm up period, which is increased with gradual hot exhaust gas flow. Jinke Gong at al., [9] considered a three way catalyst to perform the study on the monolith geometric parameters on influencing the emission conversion characteristics. The authors have used the commercial CFD code

FLUENT to solve the governing equations by considering the chemical reaction mechanisms. The decrease in the ratio of the elliptical geometry model has efficient catalytic conversion results. They found that the increment in length of the converter has better conversion efficiency. The catalyst with higher surface area and lesser wall thickness has better conversion rates and efficient. Bassem H. Ramadan et al., [10] has conducted a study on the character of the internal flow in the catalytic converter. The authors have used the CFD technique to perform the 3D calculations of the flow through the catalytic converter. The authors have conducted experimentation on the catalytic converter to study the flow analysis, velocity profile at various locations. The author has found that the geometry profile has influence over the effectiveness of conversion in device. The authors have considered the back pressure, mass flow, aspect ratios, and inlet and outlet cone angles and offset conditions. The profile over the angled pipe configuration has non-uniform flow conditions; increase in mass flow conditions had influenced the uniform flow conditions. The author has found that the increase in the cone angle has increased the pressure drop. Xiaogang Zhang and Paul Tennison [11] investigated a numerical three dimensional fluid flow and pressure drop characteristics for the catalytic converter geometry of two substrate inside. They carried out this study for different positions of two substrate considered and air gap width between them. The study shows the effects of relative positions of flow uniformity inside the catalytic converter with two substrates. Julia Windmann at el., [12] investigated and predicted tail pipe emissions based on numerical simulation of catalytic converter. They carried out three dimensional transient numerical studies showing the influence of velocity in front of inlet face of three way catalytic converter. They investigated and discussed the differences of chemical behaviour and thermal behaviour because of the shape and velocity distribution. Sophie Salasc at el., [13] carried out a study that aims on highlighting the impact of inlet manifold design on flow distribution focussing on different manifold designs configurations. The results obtained shows good trends with the design of closed coupled catalytic converter and manifold for improved emission performance. Thundil Karuppa Raj R. and Ramsai R [14] have conducted numerical analysis of flow of fluid in the catalytic converter using commercial CFD code. They assumed air as the working fluid. The authors have considered a rotationally symmetric geometry model with a porous medium to analyze the effect of inlet pipe angle. They have validated the numerical results with the experimental results from literature. The authors have varied the mass flow rates and inlet pipe angles to study the back pressure across the converter. These results were used to effectively design the catalytic converter to perform at maximum efficiency. Soo-Jin Jeong and Woo-Seung Kim [15] study shows the light of performance of monolith catalytic converter with reduced pressure loss to both balanced space velocity and usage of geometric surface area of the channels. The study also shows thermal durability based on thermal shock parameter, the cell combined shows weaker thermal resistance than conventional monolith.

2. Motivation and Objective:

Catalytic converter is an after treatment device which is widely used in automotive industries to achieve lower levels of emissions. With the use of the device pressure drop takes place across the porous zone which is modelled with catalyst materials for chemical reaction. The porous zone thus offers frictional effects and there is a loss of velocity and pressure drop takes place across the length of substrate. This causes back pressure on the exhaust gas which is entering into the catalytic converter. So in this case the engine has to more work in order to remove out the exhaust from the tail pipe. So the pressure drop analysis becomes essential. The good optimal design of a catalyst inside catalytic converter needs a good trade-off between the pressure drop and flow uniformity in the monolith substrate. This numerical analysis is not considered with respect to chemical behaviour inside the catalytic converter but it concerns with pressure drop analysis across the substrate. This numerical study involves computational simulations to perform three dimensional calculations inside the catalytic converter using CFD.

3. Numerical Methodology:

3.1 Geometry & modeling:

The design configuration considered for this numerical study is shown in figure 1. The substrate is considered is having circular cross section. A typical catalytic converter consists of catalyst substrate, mat-insulation material, and metallic outer cell. The porous zone which is a monolith substrate consists of 350 channels per square inch. Here the pressure drop study is carried out by changing the inlet pipe cone angle. Figure 1 (b) and figure 1 (c) shows the truncated inlet geometrical configuration at 45°and 60° respectively. Working fluid considered is air and it is at 873K. Following Table 1 shows the dimensions of the of zero degree inlet geometry.

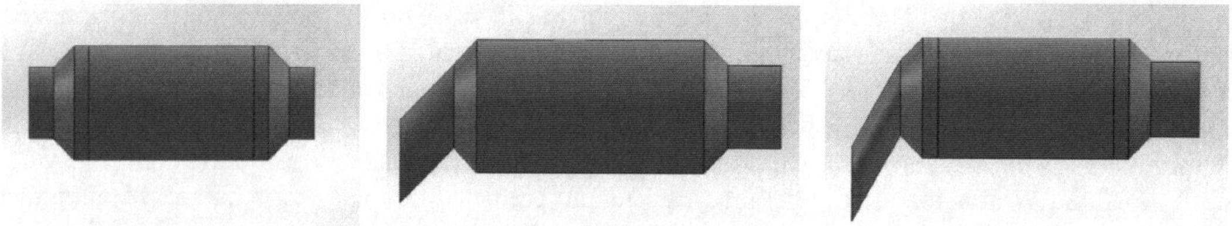

(a) 0° inlet pipe (b) 45° inlet pipe (c) 60° inlet pipe

Figure 1. Three dimensional model configuration of with 45° cone angle.

Table 1. Dimensions of the geometry considered 0° inlet pipe

Section	Dimensions
Inlet	1.875 inches
Cone Angle	45°
Substrate Diameter	3 inches
Substrate Length	4.5 inches
Pipe Length	1.24 inches

3.2 Discretization:

All the 3-D geometries are modelled in solid works 14 software and meshed in ICEM CFD 18.1 pre-processing software tool. The entire domain is discretized by using hexagonal mesh elements. The reason behind is hexagonal mesh gives accurate results and it involves less computational efforts. The geometry is divided into main four parts namely inlet and outlet pipe, inlet and outlet cone and porous zone or monolith substrate for the sake of this numerical study. The minimum determinant quality of 30% is considered to be better for the meshed geometry and minimum angle. The maximum angle for ICEM CFD meshing is 44.19° maintained with a maximum determinant quality of 82.8 %. Number of nodes are around 1262144 and number of elements are 1232065.

3.3 Mathematical modeling – solving technique:

The Shear stress transport model is used in this study. The 3-D analysis is based on steady conditions and considering conservation of mass, momentum and energy using ANSYS code. ANSYS CFX 18.1 is used as solver in this work. All the governing equations are shown in (1), (2), (3) and (4) respectively. The simulations are done based on steady flow conditions. The convergence criterion for the solution is set under specific value decided by user and those criteria are normalized to absolute residual pressure drops below 10^{-4}. The fluid considered is isothermal condition and so heat transfer effects are neglected. The flow in the inlet section and outlet section is turbulent type. The monolith substrate which is a porous medium and the flow inside porous medium is considered as laminar in nature. The hydraulic diameter of the channel is about 1.167 mm for 350 cpsi and the corresponding Reynolds number results in a laminar flow inside the substrate. So the flow inside the porous zone can be considered as laminar flow. The modelling of monolith substrate is done by two resistances called linear and quadratic resistances. From figure 2 [14] linear and quadratic resistances are calculated to model the substrate inside catalytic converter using Darcy's relation from the experimental data.

Table 2. Experimental pressure drop across substrate

Mass flow rate(g/sec)	Superficial Velocity(u)(m/sec)	Pressure Drop(Pa)
10	5.4	690
20	10.77	1400
40	21.38	2900
80	42.1	6160
110	57.16	8800
150	76.59	12550
190	95.27	16500
220	108.8	19560
250	121.9	22700

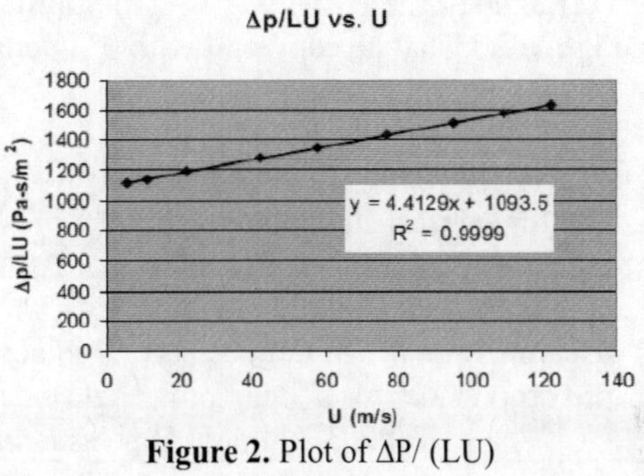

Figure 2. Plot of ΔP/ (LU)

Conservation of mass:

$$\nabla.(\rho V) = 0 \tag{1}$$

Conservation of x-momentum:

$$\nabla.(\rho\ u V) = -\frac{\partial p}{\partial x} + \frac{\partial Txx}{\partial x} + \frac{\partial Txy}{\partial y} + \frac{\partial Txz}{\partial z} + \rho.g \tag{2}$$

Conservation of y-momentum:

$$\nabla.(\rho\ v V) = -\frac{\partial p}{\partial y} + \frac{\partial Txy}{\partial x} + \frac{\partial Tyy}{\partial y} + \frac{\partial Tyz}{\partial z} + \rho.g \tag{3}$$

Conservation of z-momentum:

$$\nabla.(\rho\ w V) = -\frac{\partial p}{\partial z} + \frac{\partial Txz}{\partial x} + \frac{\partial Tyz}{\partial y} + \frac{\partial Tzz}{\partial z} + \rho.g \tag{4}$$

The experimental values of pressure drop from zero degree inlet pipe angles are shown in table 2 [14]. These values are considered for conformance to the validation of this numerical study. Assuming the walls to be having smooth surface. Radiation effects and buoyancy effects are neglected during the analysis.

3.4 Boundary conditions and simulation:

The catalytic converter simulation is done in ANSYS CFX software 18.1. In CFX-Pre, the various domains of the catalytic converter such as monolith substrate, fluid, substrate wall are defined. The mass flow rates are varied for each inlet pipe angle geometry and the numerical analysis is carried out. Mass flow rate is applied at the inlet section of catalytic converter. The simulations are done based on steady flow conditions. Outlet section is kept at zero relative pressure condition. The linear and quadratic resistances required are calculated using Figure 2. Figures 3 (a), (b) and (c) shows substrate section and modelling configuration and splitting configurations considered for analysis. In post-processing phase, the results of the numerical study has been carried out which include the velocity contours, pressure contours, 3D flow characterises.

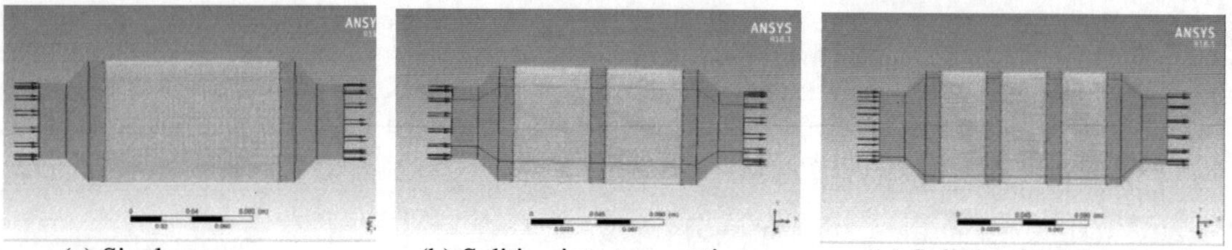

(a) Single zone (b) Spliting into two sections (c) Splitting into three sections

Figure 3. Catalytic converter with 0° inlet pipe configurations showing modeling of porous zones

4. Validation, Results and Discussion:

4.1 Effect of change inlet pipe angle and inlet cone angle:

This CFD numerical study shows that the pressure drop occurs to be 699.66 Pa. If the mass flow rate at inlet increases then pressure drop also increases. The computational model results good conformance with the experimental value shown in table 3 with maximum deviation around 2.54 %. For 0° inlet section and 150 g/sec mass flow rate the pressure drop across the substrate found to be 12303.94 Pa. Table 3 shows pressure drop values for 0° inlet pipe angle calculated numerically varying mass flow rates.

Table 3. Numerical pressure drop across substrate at 0° inlet pipe angle.

Inlet pipe Angle	Mass flow rate (g/sec)	Pressure at substrate inlet (Pa)	Pressure at substrate outlet (Pa)	Pressure drop across substrate (Pa)
0°	10	720.999	21.3324	699.66
	20	1504.08	80.8137	1423.98
	80	7293.52	1284.95	6008.59
	150	17340.2	5036.26	12303.94

Tables 4 and 5 show the pressure drop values of 45° and 60° inlet pipe angle configuration respectively for the corresponding mass flow rates. It can be seen that there is no any significant change in pressure drops.

Table 4. Numerical pressure drop across substrate at 45° inlet pipe angle.

Inlet pipe Angle	Mass flow rate (g/sec)	Pressure at substrate inlet (Pa)	Pressure at substrate outlet (Pa)	Pressure drop across substrate (Pa)
45°	10	685.275	20.7929	664.4821
	20	1356.86	80.4807	1276.37
	80	6870.77	1225.13	5645.64
	150	15976.1	4128.28	11847.82

Table 5. Numerical pressure drop across substrate at 60° inlet pipe angle.

Inlet pipe Angle	Mass flow rate (g/sec)	Pressure at substrate inlet (Pa)	Pressure at substrate outlet (Pa)	Pressure drop across substrate (Pa)
60°	10	664.534	22.0482	642.48
	20	1292.56	87.8615	1204.7
	80	6747.6	944.223	5803.37
	150	15737.6	3127.49	12610.11

Table 6 shows the numerical analysis of 0° inlet pipe angle with 60° inlet cone configuration. It is found that there is no significant difference in the pressure drop as compared to experimental values of 45° inlet cone angle.

Table 6. Numerical pressure drop across substrate at 0° degree inlet pipe angle & 60° inlet cone angle.

Inlet Cone Angle	Mass flow rate (g/sec)	Pressure at substrate inlet (Pa)	Pressure at substrate outlet (Pa)	Pressure drop across substrate (Pa)
60°	10	726.593	25.9275	700.66
	20	1529.88	100.375	1429.505
	80	7866.39	1597.71	6268.68
	150	18739.4	6076.32	12662.68

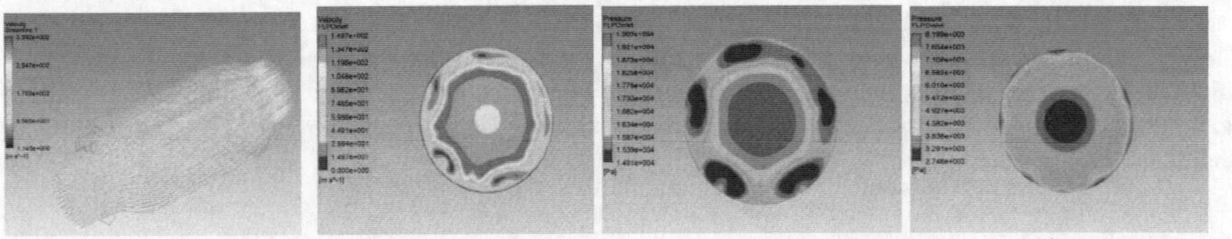

(a) (b) (c) (d)

Figure 4. (a) Fluid flow and streamline characteristics of 0° inlet pipe angle geometry at 150 g/sec, (b) Velocity distribution at substrate inlet of 0° inlet pipe angle geometry at 150 g/sec, (c) Pressure distribution at substrate inlet of 0° inlet pipe angle at 150 g/sec, Pressure distribution at substrate outlet of 0° inlet pipe angle at 150 g/sec

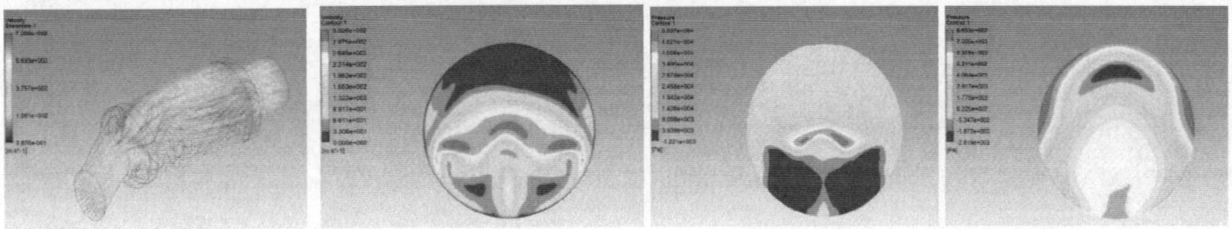

(e) (f) (g) (h)

Figure 5. (e) Fluid flow and streamline characteristics of 45° inlet pipe angle geometry at 150 g/sec, (f) Velocity distribution at substrate inlet of 45° inlet pipe angle geometry at 150 g/sec, (g) Pressure distribution at substrate inlet of 45° inlet pipe angle at 150 g/sec, (h) Pressure distribution at substrate outlet of 45° inlet pipe angle at 150 g/sec

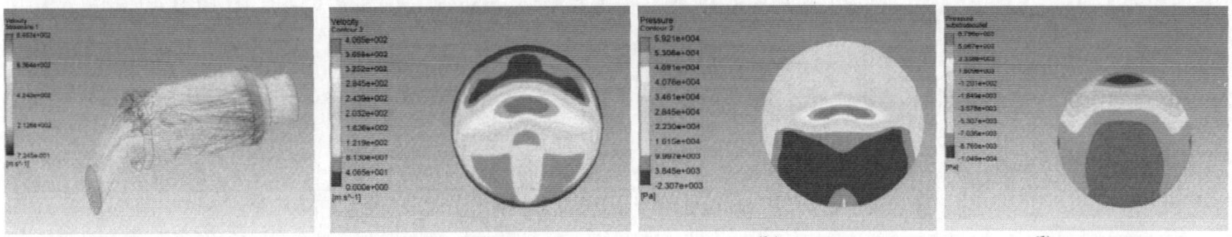

(i) (j) (k) (l)

Figure 6. (i) Fluid flow and streamline characteristics of 60° inlet pipe angle geometry at 150 g/sec, (j) Velocity distribution at substrate inlet of 60° inlet pipe angle geometry at 150 g/sec, (k) Pressure distribution at substrate inlet of 60° inlet pipe angle at 150 g/sec, (l) Pressure distribution at substrate outlet of 60° inlet pipe angle at 150 g/sec

For 45° and 60° inlet pipe angle configuration velocity distribution for the mass flow rate of 150 g/sec is shown in figure 5 (f) and figure 6 (j) respectively. The flow inside the truncated inlet configuration is non-uniform and that can be seen from figure 5 (e) and figure 6 (i) which show the streamline characteristics. It is found that for low mass flow rate the flow found to be uniform. But for increased mass flow rate the non-uniformity of the flow increases. The velocity distribution at inlet of substrate shows maximum velocity occurs at middle portion and at the periphery velocity is zero due to no slip condition. Towards peripheral surface velocity drops down to zero. As mass flow rate increases pressure drop across the substarte increases. The experimental and numerical pressure drop values are good conformance with each other and figure 7 (a) shows the variation of pressure across the substrate with mass flow rate. There is no any significant change in pressure drop across the substrate is observed with the change in inlet pipe angle. Figure 7 (b) shows the variation of pressure drop values with change in inlet pipe angle. It can be seen that there is not much difference in pressure drop due to change in inlet pipe and inlet cone angle.

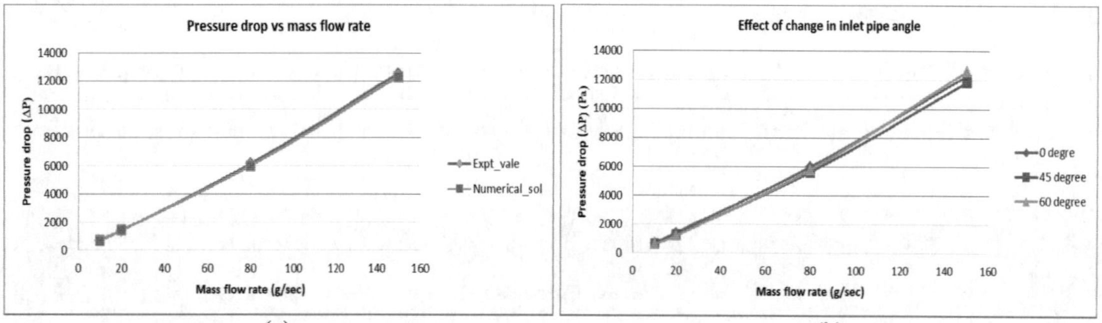

(a) (b)

Figure 7. (a) Variation of pressure drop with mass flow rate (b) Effect of change in inlet pipe angle to pressure drop

4.2 Effect of splitting the substrate with air gap:

Further analysis is carried out by dividing the substrate into two and three sections maintaining all the same dimensions as shown in figure 8. It can be seen that maintaining air gap between substrate sections shows decrease in pressure drop. Table 7 shows the pressure drop across the substrate when the porous zone is splitted into two substrates section as shown in figure 3 (b). It can be observed that as splitting the single substrate into two sections pressure drop across the total section decreases. A 10 mm air gap width between the each substrate section is maintained and length of each substrate is kept 52.15 mm of length. It can be seen that for a mass flow rate of 10 g/sec pressure drop reduced by 60.0977 Pa. Table 7 shows the corresponding decreased in pressure drop values for different mass flow rates.

Table 7. Numerical pressure drop across substrate at 0° inlet pipe angle with 45° inlet cone.

Inlet pipe Angle	Mass flow rate (g/sec)	Pressure at substrate inlet (Pa)	Pressure at substrate outlet (Pa)	Pressure drop across substrate (Pa)
0°	10	660.508	20.9457	639.5623
	20	1383.32	80.2657	1303.0543
	80	6833.83	1274.29	5559.54
	150	15618.3	4950.34	10667.96

Table 8 shows the pressure drop across the substrate when the porous zone is splitted into three substrate zones as shown in figure 3 (b). It can be observed that as splitting the single substrate into three sections pressure drop across the total section decreases. A 10 mm air gap width between the each substrate section is maintained and length of each substrate is kept 31.4333 mm of length. It can be seen that for a mass flow rate of 10 g/sec pressure drop reduced by 123.4866 Pa. Table 8 shows the corresponding decreased in pressure drop values for different mass flow rates.

Table 8. Numerical pressure drop across substrate at 0° inlet pipe angle with 45° inlet cone.

Inlet pipe Angle	Mass flow rate (g/sec)	Pressure at substrate inlet (Pa)	Pressure at substrate outlet (Pa)	Pressure drop across substrate (Pa)
0°	10	600.185	21.0116	579.1734
	20	1261.01	80.4355	1180.574
	80	6293.8	1274.61	5019.19
	150	14539.7	4950.79	9588.91

Figure 8 shows the decrease in pressure drop across the total substrate when it is a single substrate, two substrate and three substrates configuration.

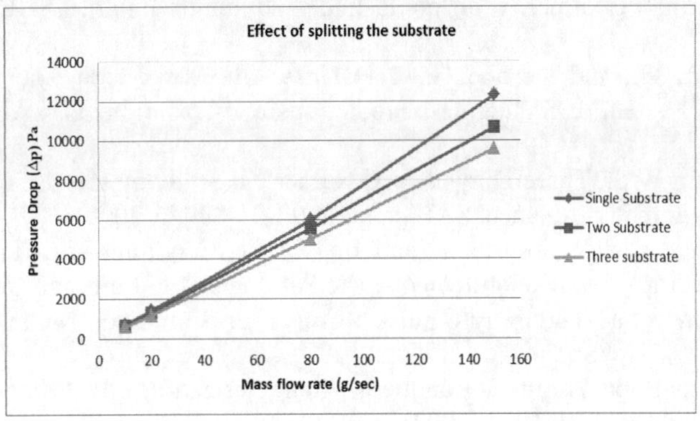

Figure 8. Effect of splitting the substrate into sections

5. Conclusion:

The numerical simulation of pressure drop and flow analysis was completed successful in ANSYS CFX. The simulation results agree with experimental values. The results show that pressure drop geometry has significant effect on flow distribution inside the catalytic converter monolith substrate. In addition to this the flow in the catalytic converter with inlet angles appears to be less uniform. The effect of change in inlet pipe angle and inlet cone angle is studied. As the mass flow rate increases pressure drop across the substrate increases. The flow tends to be more uniform if the angles are closer to inlet cone. As the mass flow rate increases flow non uniformity also increases. It is also found that changing the inlet pipe angle and cone angle does not affect the pressure drop much. So it can be taken as positive side for changing the shape of exhaust manifold as per the space availability from engine to catalytic converter. The effect of splitting the porous zone is presented which shows decrease of the pressure drop across the substrate.

6. Nomenclature:

$\dot{m}$	-	Mass flow rate, g/sec
P	-	Pressure, Pa
T	-	Temperature, K
U	-	Superficial velocity, m/s
ρ	-	Density of air, kg/m^3
ΔP	-	Pressure drop, Pa
L	-	Substrate length, mm

IOP Conf. Series: Earth and Environmental Science **312** (2019) 012032 doi:10.1088/1755-1315/312/1/012032

7. References:

[1] Howitt J.S. and Sekella T.C., Flow Effects in Monolithic automotive Catalytic Converters, SAE paper 740244, (1974)

[2] Martin AP, Will NS, Bordet A, Cornet P, Gondoin C, Mouton X. Effect of flow distribution on emissions performance of catalytic converters. SAE transactions. 1998 Jan 1:384-90.

[3] Hwang K., Lee K., Mueller J., Stuecken T., Schock H. and Lee J.C., Dynamic Flow Study in a Catalytic Converter Using LDV and High Speed Flow Visualization, SAE paper 950786, (1995)

[4] Lee S., Bae C., Lee Y. and Han T., Effects of Engine Operating Conditions on Catalytic Converter Temperature in an SI Engine, SAE paper 2002-10-1677, (2002)

[5] Cho Y.S., Lee Y.S., Kim D.S., Jung S.Y. and Ohm I.Y., An Alternative Method for Fast Light-Off of Catalysts – Cranking Exhaust Gas Ignition, SAE paper 2002-01-1678, (2002)

[6] Gregory D., Read M., Campbell B., Inman G., Nice G., Hims R., Rabinowitz H., Tauster S., and Collin T., Emissions Implications of a Twin Close Coupled Catalyst System Designed for Improved Engine Performance on an In-line 4 Cylinder Engine, SAE paper 2001-01-1092, (2002)

[7] Onorati A., Ferrari G., and Derrico G., 1 D Unsteady Flows with Chemical Reactions in the Exhaust Duct- System of SI Engines: Predictions and Experiments, SAE paper 2001-01-0939, (2001)

[8] Jeong S.J., and Kim W.S., Three-Dimensional Numerical Study on the Use of Warm-up Catalyst to Improve Light- Off Performance, SAE paper 2000-01-0207, (2000)

[9] Jinke Gong, Longyu Cai, Weiling Peng and Jingwu Liu, Yunqing Liu, Hao Cai and Jiaqiang E, Analysis to the Impact of Monolith Geometric Parameters on Emission Conversion performance Based on an Improved Threeway Catalytic Converter Simulation Model, SAE paper 2006-32-0089 (2006)

[10] Bassem H. Ramadan and Philip C. Lundberg, Characterization of a Catalytic Converter Internal Flow, SAE paper, 2007-01-4024 (2007)

[11] Zhang X, Tennison P. Numerical Study of Flow Uniformity and Pressure Loss Through a Catalytic Converter with Two Substrates. SAE Technical Paper; 2008 Apr 14.

[12] Windmann J, Braun J, Zacke P, Tischer S, Deutschmann O, Warnatz J. Impact of the inlet flow distribution on the light-off behavior of a 3-way catalytic converter. SAE transactions. 2003 Jan 1:713-23.

[13] Salasc S, Barrieu E, Leroy V. Impact of manifold design on flow distribution of a close-coupled catalytic converter. SAE Technical Paper; 2005 Apr 11.

[14] Thundil Karupparaj R, Ramsai R. Numerical study of fluid flow and effect of inlet pipe angle In catalytic converter using CFD. Research Journal of Recent Sciences ISSN. 2006:2277-502.

[15] Jeong SJ, Kim WS. A study on the optimal monolith combination for improving flow uniformity and warm-up performance of an auto-catalyst. Chemical Engineering and Processing: Process Intensification. 2003 Nov 1;42(11):879-95.

A Review on Computational Fluid Dynamics Analysis on Greenhouse Dryer

Sendhil Kumar Natarajan* and Elavarasan E

Department of Mechanical Engineering, National Institute of Technology, Puducherry, Karaikal – 609 609, U.T. of Puducherry, INDIA.
*Corresponding Author: Assistant Professor & Head

Email: sendhil80@nitpy.ac.in, drsendhil1980iitmuk@gmail.com

Abstract. Over the decades Solar energy has been utilizing for the conservation of various agricultural products at different zones in the world. Existing literature reveals that open dehydrating is the mostly followed for the desired application and its efficiency limited to certain extent. In order to overcome the limitations, solar dehydrating is proposed and implemented by various researchers. Greenhouse dryer found to be most appropriate technique to preserve different kinds of fruits, vegetables and grain products. In recent years, a great deal of research has been focused to enhance the performance of the selected method. Traditionally it is confined to empirical form and later it is extended to CFD analysis to obtain effective results through simulations. The present article presents the extensive survey over empirical and CFD analysis of Green house dryer application on dehydrating in both active and passive mode of convections.

Keywords: Solar greenhouse dryer; In-direct type greenhouse dryer; Computational Fluid Dynamics; Natural Convection; Forced Convection; Open Sun Dehydrating.

1. Introduction
Greenhouse dryer in agricultural applications widely used to harvest the crops from the preserved condition and protect from the affect made by bug and maladies. Also it serves as a yield dryer to various applications due it robust structure and lesser operating and low maintenance cost compared to other existing dryers. Dehydrating is a process of reducing the moisture content in the food products either by forced or natural convection. Solar dehydrating is unique over the other due to direct exposure of sun light supplies adequate amount of thermal radiation at probable surface area helps essential buoyancy effect to carry out desired effect. The rate of moisture content reduction signifies the effect of the process which involves enormous calculations to draw the relation over the influencing variables at various conditions.

Multiple sensors are used to fix at different locations of the dryer to test air flow. CFD offers greater work space to conduct various simulations for estimating the air velocity in the drying chamber which is very expensive and difficult in practical experimentations. Developing the relations the between obtained results of numerical and experimental leads to develop new theories and further insights in the field. The further sections intend to explain various research approaches in numerical and experimental by different researchers.

2. Review
CFD is most versatile software for the design and performance investigation of green house dryers. It overcomes the traditional complexities of green house dryers with minimal cost and resulting with greater outputs even at which the practical tests are almost impractical [1-3]. Based on the results obtained through CFD, it extremely supports to modify the design parameters to obtain desirable temperature and air flow

based atmosphere [4]. It anticipates mass transfer conduct in food items since the estimated greenhouse dryer microclimate with respect to its design descriptions and utilized apparatus [5-10].

The utilization of the 3D CFD model permits getting a real picture of the whole volume of the greenhouse dryer, including at overhang level. As to outcomes acquired in the past works, numerous investigations have concentrated on the impact of the air flow difference from the temperature appropriation inside the greenhouse [11-13]. Manoj and Manivannan built up a MATLAB-based modelling and simulation framework to predict the wind stream properties, equilibrium wetness substance of the sun based dryer techniques as shown in Fig. 1. Greenhouse dryer is considered for investigation. Wetness expulsion process was investigated for 7 days and 7% of weight decrease was noticed. The performed on MATLAB program and the 3-D model has been created on procedure. The Crank-Nicholson expression has been connected to thermal and FD technique has been utilized to produce for dehydrating cocoa bean. These outcomes demonstrate a period of dehydrating decreases the wetness contains for 50 to7 % and 60 to 8 % of wetness removed of dehydrated bean[14]. Mabrouk et al. [15] observed a empirical model for temperature and mass exchange of granular items in a fixed-bed greenhouse tunnel dryer.

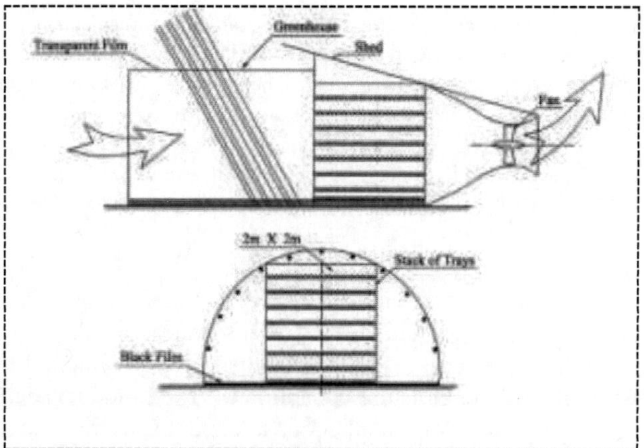

Figure 1. 2D of Green House Solar Dryer [14].

Aghbashlo et al. [16] inspected a greenhouse dryer for chamomile blooms, and they utilized TRNSYS programming. They composed a program in Compaq visual FORTRAN programming language. RMSD was under 9.3% and found that TRNSYS is a decent apparatus for simulation of sun powered dryers. Chen et al. [17] suggest by REA for clarifying distinctive testing of dehydrating techniques. These techniques followed the essential guidelines of chemical reaction engineering to show the physical changes characters of dehydrating [18]. It is demonstrated that the REA techniques is a basic, exact and strong method for displaying the convective dehydrating of eaten and non-eaten materials under state steady natural conditions [19]. Technique likewise utilized for modelling different intermittent dehydrating [20]. Mohammadi et al. [21] have examined about the significance of greenhouse solar dryer. The investigators examined about the controlled strategy, for example, versatile, input and intelligent control for their introduction. The investigators have modeled & numerically valuated within condition of semi solar greenhouse dryer. It has been reported that the dynamic model with beginning qualities used to predict within air and soil temperature. As per the outcomes, the connection between the consequences of dynamic model with experimental value according to RMSE, MAPE and EF demonstrated that, the dynamic technique can gauge within air and soil temperature with 5.3°C, 10.2 %, 0.78 % and obtained 3.45°C, 7.7 %, 0.86 %, separately also similar comparative strategy can be utilized to item fuel utilization in greenhouse.

Simulation output results from CFD should be compared with real dehydrating investigations and if there is an understanding among anticipated and empirical results, at that point the CFD code might be apply for other dehydrating conditions to upgrade the dehydrating execution of the dryer. These have been demonstrated by investigation of numerous analysts [22-28] then demonstrate that the execution of new dryer plans can be anticipated by CFD Simulation output. These examinations additionally demonstrated that the CFD reproduction result is practically identical with real conditions. Taki et al [29] evaluated three dissimilar variables – inside air, soil and plant temperature utilizing ANN & SVM. The investigators have considers influence within temperature as shown in Fig. 2. Among 13 unique calculations along these lines the investigators has determined the best model for gauge of real value in greenhouse which can similar exchange actuality and predict power lost.

Figure 2. Type of selected greenhouse at Shahreza city [29].

Arunsandeep et al [30] simulated heat and mass exchange for sun based dehydrating of circular items, for example, green peas. FD technique with implicit method has been utilized for exposition. The arrangement of FD conditions was resolved by TMA and PC codes in MATLAB were created to explain them. Outcome has been compared for dehydrating rate with the empirical value from the survey and both are noticed to be in better understanding.

Chauhan et al. [31] examined the regions of programming use for sun powered dryer operation. Elements of fluid can be possible by ANSYS and FLUENT. MATLAB and FORTRAN can be useful in numerical demonstrating. For measurable investigation SPSS can be utilized. Malekjani and Jafari et al. [5] studied about the components basic the dehydrating operations in dehydrating out of food and agricultural items. The writers have created propelled modelling of simulation procedures for new dryers. It has been called attention to that nourishment quality is most significant parameters amid dehydrating. It has been reasoned that CFD can be utilized to forecast fluid flow, temperature and mass transfer in dryers. The approval issue of CFD outcomes is additionally exists in the light of complexity of gaining empirical

IOP Conf. Series: Earth and Environmental Science **312** (2019) 012033 doi:10.1088/1755-1315/312/1/012033

information in some dryer structures and conditions. Coupling of CFD with recent methods, for example, response designing methodology discreate component technique as profitable. Kumar et al [32] have utilized CFD to deal with indenting temperature conveyance with keep up in greenhouse dehydrating. CREO 5.0 has been utilized for geometric modelling as shown in Fig. 3. The examination has been done in familiar for forced and natural circulation. Numerically acquired outcomes have been compared shaped and works and with be in correlation.

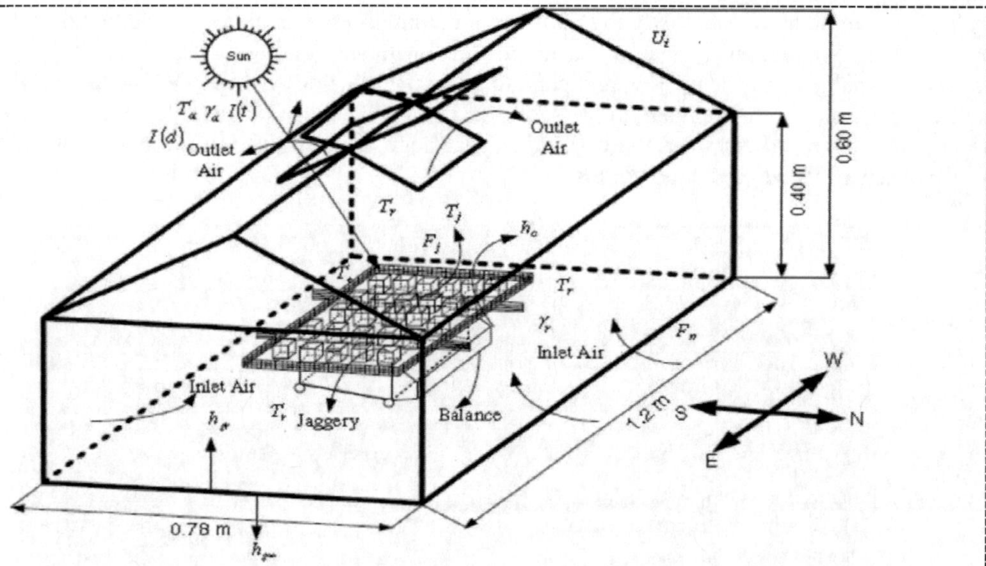

Figure 3. Greenhouse jaggery dehydrating under natural convection mode [32].

Prakash et al. [33] observed on various systems of sunlight based dehydrating where proficiency of dehydrating, expectation of temperature, wetness in yield, rate of dehydrating, fluctuated hues and different in harvests and their quality were examined. CFD software is used for investigation by ANFIS programming. Numerically modelling suggests better results.

By utilizing CFD method, the miniature can be simulated to contemplate the attributes of temperature, air velocity conveyances and heat transfer in the dryer [34-35]. Miniature methods such as fuzzy, artificial neural system, etc can be effective for anticipating the data of reaction factors [36]. Numerical miniatures and simulation procedures are helpful for inspecting the impact of input factors on the yield responses [37]. Harjunowibowo et al. has observed that recent developments in greenhouses for vitality sparing purposes. It has been recommended that fitting greenhouse with LEDS can save power by 75% when compared with expense of artificial lights. It is reported to control framework that introducing automatic framework in greenhouse dryer would create a steady and cost-effective microclimate, subsequently ensuring the quality and amount of the yield [38]. Gupta at el [39] have done CFD investigation of greenhouse dryer as shown Fig. 4. Past exploratory information has been utilized as boundary condition for recreation. Temperature circulation and wind flow example have been investigated about and reported that numerically acquired outcome elucidate local parameter.

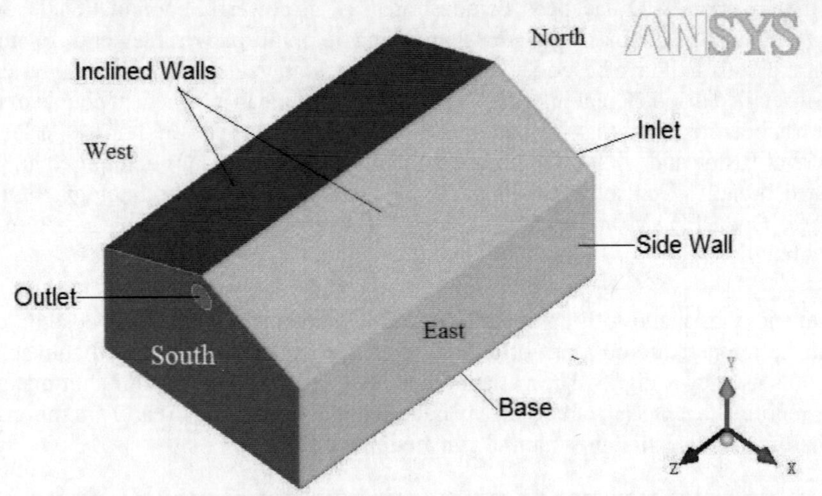

Figure 4. Geometric Modeling For Direct Type Solar Greenhouse Dryer [39].

The apparatus model was assessed by exploitation and victimization commercially which as accessible by CFD code in order to handle the mass and heat transfer inside the apparatus. Natural convection was the sole technique of temperature transportation plan of getting a simple style. A biomass burner was incorporated into the star apparatus intended to oblige the further effect of low buoyancy found in natural convective driers[40]. Numerical simulation of the wind flow inside a half and half sunlight based electrical drier, utilizing a business CFD bundle was showed. With recommended temperature and velocities, the miniature predicts the conduct of the wind flow inside the system [41]. Chauhan et al [42] has concentrated on the expanding ecologically concerns and concentrated on utilization of sun powered vitality as a substitute source. The investigators have recommended utilization of sun powered dryer coordinated with heat energy storage framework. Utilization of intelligent mirror, improvement in the geometry of level plate collector to build heat exchange zone and utilization of selective coatings on safeguard plate to enhance accumulation of sunlight based radiation can improve the thermal performance of the dryer. Mujumdar and Wu [43] have stressed that the CFD approach is the cost-effective arrangements that can push advancement and innovativeness in dehydrating structure and improvement. Since the dehydrating procedure, particularly for huge scale dryer, estimation of air speeds and temperature is tedious, since a few sensors should have been install at different areas and directions of wind flow, along these lines CFD software might be utilized to predict the air speeds and temperature in the dehydrating chamber. Jyoti et al. [44] described about the utilization of parabolic trough in sunlight based collector. It is reported that the effectiveness of this type of collector rely upon the heat power of sun. Survey on the parabolic trough sunlight based collector has been introduced. It is reported that the investigators have utilized different advancement method, and approved the same by utilization of devices likes ANSYS, Fluent and so forth.

Designing appropriate geometric arrangements of the dehydrating chamber, the forecast of the wind stream and temperature inside the dryer streamlines the structure and improving the dehydrating procedure before the real dryer was assembled. Numerical strategies have been utilized to numerically show the conduct of the wind stream inside a few sorts of dryers [15, 45 & 46]. This procedure is tedious and includes a complex scientific definition. CFD is a broadly acknowledged procedure and instrument for numerical recreations of such frameworks. CFD had been demonstrated as an powerful computational device for anticipating the flow behaviour and mass transfers happening in multi-part frameworks in numerous enterprises [47-48]. Kant et al [49] have described about the power storage materials sunlight based dryer. The investigators have detailed that different investigations have been done in most recent couple of decades for the equivalent. It has been reported that for better heat execution of sun powered dryers, a PCM with high latent temperature of fusion and with a vast ground zone for thermal exchange is required and is as of now being investigated broadly. Comsol Multi-Physics, CFX advanced solver technology, FORTRAN programming language, Open FOAM and Fluent – Ansys, to improve the dryer configuration similar utilization of business bundle.

Purusothaman and Valarmathi [50] have performed CFD investigation of greenhouse dryer when communicated to surface temperature on a hot bright day with the expectation of forced convection and natural convection forms as shown Fig. 5. The variation has been made in thickness of shut material and mass stream rate. It is reported that the sun radiation expanded shapely from 10am to early afternoon. Higher temperature is seen in forced convection then natural convection by 41%.

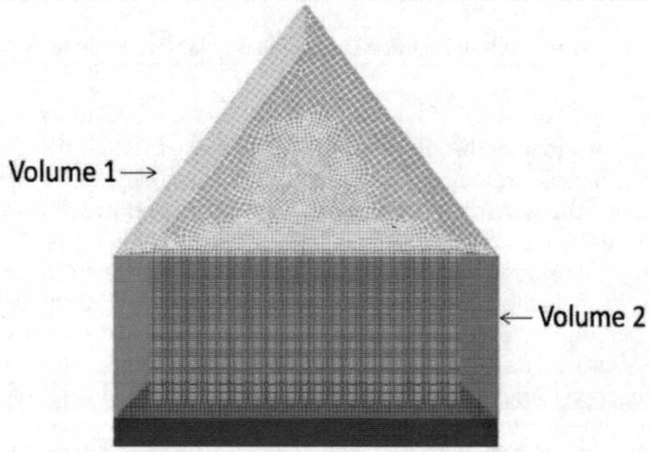

Figure 5. Volume mesh of dryer [50].

Ingle et al [51] has been numerically simulated sun based collector for grape dehydrating. The investigators have utilized CFD to see better heat transfer capacity. Unstructured framework has been created utilizing ICEM CFD. For investigation Fluent bundle has been utilized. They acquired an arrangement of non-linear partial differential conditions and numerically fathomed them by a FV technique. Significant examinations on reproduction of natural convection sun based dehydrating of farming items and improvement have additionally been accounted [52-53]. Sanghi et al [7] have elaborated CFD model to simulate the corn dehydrating operation in a sun powered dryer. The execution of the dryer has been simulate at reasonable and over-cast climate conditions. Temperature, air speed profiles and humidity in the dryer have been showed. The outcome has been approved with numerical information. The simulated outcome over anticipate the factors, such as temperature and wetness.

Prakash and Kumar at el [54] have utilized ANSYS to anticipate the conduct of slightly modified greenhouse dryer under dynamic mode in no-load conditions. Climate condition have been taken has information, for example, Adaptive Neuro Fuzzy Inference framework (ANFIS). Warmth, surrounding temperature surrounding RH, solar radiation and period of experimentation. As yield greenhouse air temperature and relative moistness have been acquired. Outcomes from ANSYS model has been found to be in a decent understanding. Kumar and Tiwari [55] revealed thermal modelling of jiggery samples dehydrating in a natural convective greenhouse dryer. The miniature was customized in MATLAB, and anticipated data well. Khaldi et al [56] have empirical study of indirect sun based dryer incorporated with heat storage for dehydrating figs. Unsteady turbulent wind flow and thermal exchange utilizing two-dimensional model has been completed. The investigators have considered the impact of air channel estimate and thickness of backed bed on conduct of dryer. It is directed that the way of flow determined local parameters.

Janjai et al. [57] detailed an exploratory act of a Photovoltaic ventilated greenhouse dryer for dehydrating of chilies. Barnwal and Tiwari [58] revealed sun based dehydrating of grapes utilizing a hybrid PV-warm (PV/T) dryer and fabricated a multi linear equation to foresee wetness dissipation amid dehydrating. Kokate et al. [59] fabricated a sun powered dryer to help plastic industry as shown in Fig. 6. For preconditioning and lack of hydration of Nylon– 6 and Polypropylene (PP) about 3– 5% of all out power was required. For Polypropylene material the most extreme temperature rise up to 71°C henceforth with a similar dryer configuration preheating is possible. They formulated a numerical model of sun based dryer. Numerous parameters were in charge of predicting air temperature. Psychrometric curve was utilized to clarify it. The temperature of 55to 60°C was accomplished inside the dryer. The accompanying parameters to be specific sunlight based power and the outside wind speed of safeguard were in charge of inside air dehydrating temperature.

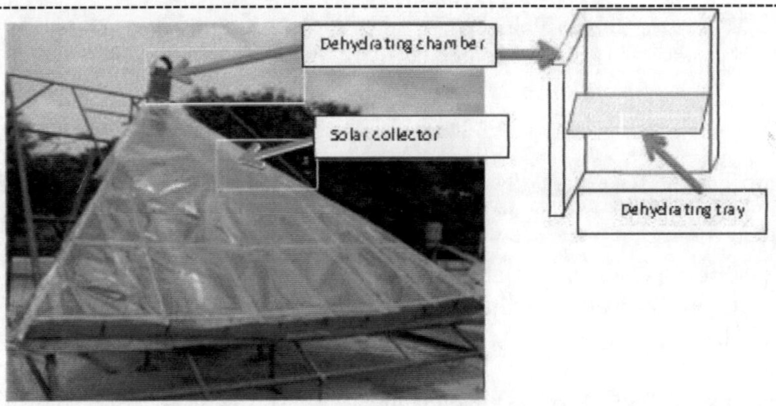

Figure 6. Pictorial view of solar dryer [59].

Dorouzi et al. [60] has fabricated liquid desiccant-helped sun based dryer for dehydrating of tomato samples. A photovoltaic warm sun powered collector has been utilized to supply the electrical power. Hue examination of dehydrated tomato samples was completed and it was incorporated that the expansion in the warmth of the dehydrating air expanded the gentility, and yellowness esteems, yet the expansion in the Relative humidity diminished the estimations of softness and redness. The temperature of 60°C & RH of 23 % were prescribed for tomato samples in the reported dryer.

3. Conclusion

In this article a survey on solar green house drying is presented and different approaches followed by the researchers are explained. Solar based Photo Voltaic fan usage as assistant connection to provide forced air circulation inside the dryer has been proposed to accomplish the higher dehydrating rates. For getting the better drying rate of dryer and investigation of dryer by using Computational Fluid Dynamics. The simulation output by CFD might be tried with actual drying tests and on the off chance that there is an understanding among predicted and test results, at that point the CFD code might be reapplied for other drying conditions to optimizing the drying performance of the unit. The CFD might be along these lines utilized as a drying optimization device.

Nomenclature

1.	FDM	Finite Difference Method
2.	TRNSYS	Simulation Program
3.	RMSD	Root-Mean-Square Deviation
4.	REA	Reacting Engineering Approach
5.	EF	Model Efficiency
6.	MAPE	Mean Absolute Percentage Error
7.	ANN	Artificial Neural Network
8.	SVM	Support Vector Machine
9.	ANFIS	Adaptive Neuro Fuzzy Inference framework
10.	LED	Light-Emitting Diode
11.	RH	Relative Humidity
12.	FVM	Finite Volume Method
13.	PV	Photo Voltaics
14.	TMA	Tridiagonal Matrix Algorithm

References

[1] D. K. Fidaros, C. A. Baxevanou, T. Bartzanas, and C. Kittas, "Numerical simulation of thermal behavior of a ventilated arc greenhouse during a solar day," *Renew. Energy*, vol. **35**, no. 7, pp. 1380–1386, 2010.

[2] M. Teitel, G. Ziskind, O. Liran, V. Dubovsky, and R. Letan, "Effect of wind direction on greenhouse ventilation rate, airflow patterns and temperature distributions," *Biosyst. Eng.*, vol. **101**, no. 3, pp. 351–369, 2008.

[3] C. Pontikakos, K. P. Ferentinos, T. A. Tsiligiridis, and A. B. Sideridis, "Natural ventilation efficiency in a twin-span greenhouse using 3D computational fluid dynamics," *Int. Conf. Inf. Syst. Sustain. Agric. Agroenvironment Food Technol.*, no. September, 2006.

[4] C. Baxevanou, D. Fidaros, T. Bartzanas, and C. Kittas, "Numerical simulation of solar radiation, air flow and temperature distribution in a naturally ventilated tunnel greenhouse," *Agric. Eng. Int. CIGR J.*, vol. **12**, no. 3–4, pp. 48–67, 2010.

[5] N. Malekjani and S. M. Jafari, "Simulation of food dehydrating processes by Computational Fluid Dynamics (CFD); recent advances and approaches," *Trends Food Sci. Technol.*, vol. **78**, pp. 206–223, 2018.

[6] N. Fernando, M. Narayana, and W. A. M. K. P. Wickramaarachchi, "The effects of air velocity, temperature and particle size on low-temperature bed dehydrating of wood chips," *Biomass Convers. Biorefinery*, vol. **8**, no. 1, pp. 211–223, 2018.

[7] A. Sanghi, R. P. K. Ambrose, and D. Maier, "CFD simulation of corn dehydrating in a natural

convection solar dryer," *Dry. Technol.*, vol. **36**, no. 7, pp. 859–870, 2018.

[8] H. T. Sabarez, "Airborne Ultrasound for Convective Dehydrating Intensification," *Innov. Food Process. Technol. Extr. Sep. Compon. Modif. Process Intensif.*, pp. 361–386, 2016.

[9] T. Boulard and S. Wang, "Radiactive and convective heterogeneity in a plastic tunnel: consequences on crop transpiration," *Plasticulture*, vol. **34**, pp. 22–35, 2002.

[10] S. A. Ould Khaoua, P. E. Bournet, C. Migeon, T. Boulard, and G. Chassériaux, "Analysis of Greenhouse Ventilation Efficiency based on Computational Fluid Dynamics," *Biosyst. Eng.*, vol. **95**, no. 1, pp. 83–98, 2006.

[11] R. Kumar, V. Gupta, and R. Varshney, "Numerical simulation of solar greenhouse dryer using computational fluid dynamics," *Int. J. Res. Sci. Innov.*, vol. **4**, no. June, pp. 111–115, 2017.

[12] Hao, Shen, and Zhang, "Three-dimensional steady simulation of microclimate patter inside single plastic greenhouse using computational fluid dynamics.," *Trans Chinese Soc Agric Mach.*, vol. **43**, pp. 222–228, 2014.

[13] D. W. Li YB, Zhou W, Li PF, Wang XC, "Temperature homogeneity control of greenhouse based on CFD simulation model.," *Trans Chinese Soc Agric Mach.*, vol. **43**, pp. 156–161, 2012.

[14] M. Manoj and A. Manivannan, "Simulation of Solar Dryer Utilizing Green House Effect for Cocoa Bean Dehydrating," *Int. J. Adv. Eng. Technol.*, vol. **IV**, no. II, pp. 24–27, 2013.

[15] S. Ben Mabrouk, B. Khiari, and M. Sassi, "Modelling of heat and mass transfer in a tunnel dryer," *Appl. Therm. Eng.*, vol. **26**, no. 17–18, pp. 2110–2118, 2006.

[16] M. Aghbashlo, J. Müller, H. Mobli, A. Madadlou, and S. Rafiee, "Modeling and Simulation of Deep-Bed Solar Greenhouse Dehydrating of Chamomile Flowers," *Dry. Technol.*, vol. **33**, no. 6, pp. 684–695, 2015.

[17] Chen XD and Chen NX, "Preliminary introduction to a unified approach to modeling dehydrating and equilibrium isotherms of moist porous solids," in *In: Proceedings of the Chemeca, Rotorua, New Zealand*, 1997, pp. 1–5.

[18] X. D. Chen and G. Z. Xie, "Fingerprints of the dehydrating behaviour of particulate or thin layer food materials established using a reaction engineering model," *Food Bioprod. Process. Trans. Inst. Chem. Eng. Part C*, vol. **75**, no. 4, pp. 213–222, 1997.

[19] A. Putranto, X. D. Chen, and P. A. Webley, "Infrared and convective dehydrating of thin layer of polyvinyl alcohol (PVA)/glycerol/water mixture-The reaction engineering approach (REA)," *Chem. Eng. Process. Process Intensif.*, vol. **49**, no. 4, pp. 348–357, 2010.

[20] A. Putranto, X. D. Chen, S. Devahastin, Z. Xiao, and P. A. Webley, "Application of the reaction engineering approach (REA) for modeling intermittent dehydrating under time-varying humidity and temperature," *Chem. Eng. Sci.*, vol. **66**, no. 10, pp. 2149–2156, 2011.

[21] B. Mohammadi, S. F. Ranjbar, and Y. Ajabshirchi, "Application of dynamic model to predict some inside environment variables in a semi-solar greenhouse," *Inf. Process. Agric.*, vol. **5**, no. 2, pp. 279–288, 2018.

[22] S. Misha, S. Mat, M. H. Ruslan, K. Sopian, and E. Salleh, "The Prediction of Dehydrating Uniformity in Tray Dryer System using CFD Simulation," *Int. J. Mach. Learn. Comput.*, vol. **3**, no. 5, pp. 419–423, 2014.

[23] Y. M. Yunus and H. H. Al-Kayiem, "Simulation of hybrid solar dryer," *IOP Conf. Ser. Earth Environ. Sci.*, vol. **16**, no. 1, 2013.

[24] M. Ranjbaran, B. Emadi, and D. Zare, "CFD Simulation of Deep-Bed Paddy Dehydrating Process and Performance," *Dry. Technol.*, vol. **32**, no. 8, pp. 919–934, 2014.

[25] V. M. Romero, E. Cerezo, M. I. Garcia, and M. H. Sanchez, "Simulation and validation of vanilla dehydrating process in an indirect solar dryer prototype using CFD Fluent program," *Energy Procedia*, vol. **57**, pp. 1651–1658, 2014.

IOP Conf. Series: Earth and Environmental Science **312** (2019) 012033 doi:10.1088/1755-1315/312/1/012033

[26] P. Somsila and U. Teeboonma, "Investigation of Temperature and Air Flow inside Para Rubber Greenhouse Solar Dryer Incline Roof Type by Using CFD Technique," *Adv. Mater. Res.*, vol. **931–932**, pp. 1238–1242, 2014.

[27] N. Tayyeb, J. Arezou, K. Mohammad Hossein, and A. Mortaza, "CFD simulation and optimization of factors affecting the performance of a fluidized bed dryer," *Iran. J. Chem. Chem. Eng.*, vol. **32**, no. 4, pp. 81–92, 2013.

[28] S. Misha, S. Mat, M. H. Ruslan, K. Sopian, and E. Salleh, "Comparison Between 2D and 3D Simulations of a Tray Dryer System Using CFD Software Faculty of Mechanical Engineering , Universiti Teknikal Malaysia Melaka ," vol. **29**, no. 10, pp. 1301–1309, 2014.

[29] M. Taki, S. Abdanan Mehdizadeh, A. Rohani, M. Rahnama, and M. Rahmati-Joneidabad, "Applied machine learning in greenhouse simulation; new application and analysis," *Inf. Process. Agric.*, vol. **5**, no. 2, pp. 253–268, 2018.

[30] G. Arunsandeep, A. Lingayat, V. P. Chandramohan, V. R. K. Raju, and K. S. Reddy, "A numerical model for dehydrating of spherical object in an indirect type solar dryer and estimating the dehydrating time at different moisture level and air temperature," *Int. J. Green Energy*, vol. **15**, no. 3, pp. 189–200, 2018.

[31] P. Singh Chauhan, A. Kumar, and P. Tekasakul, "Applications of software in solar dehydrating systems: A review," *Renew. Sustain. Energy Rev.*, vol. **51**, pp. 1326–1337, 2015.

[32] P. S. Chauhan, A. Kumar, and B. Gupta, "A review on thermal models for greenhouse dryers," *Renew. Sustain. Energy Rev.*, vol. **75**, no. February, pp. 548–558, 2017.

[33] O. Prakash, V. Laguri, A. Pandey, A. Kumar, and A. Kumar, "Review on various modelling techniques for the solar dryers," *Renew. Sustain. Energy Rev.*, vol. **62**, pp. 396–417, 2016.

[34] S. Sonthikun, P. Chairat, K. Fardsin, P. Kirirat, A. Kumar, and P. Tekasakul, "Computational fluid dynamic analysis of innovative design of solar-biomass hybrid dryer: An experimental validation," *Renew. Energy*, vol. **92**, pp. 185–191, 2016.

[35] L. K. Sahoo, M. K. Roul, and R. K. Swain, "CFD analysis of natural convection heat transfer augmentation from square conductive horizontal and inclined pin fin arrays," *Int. J. Ambient Energy*, vol. **39**, no. 8, pp. 840–851, 2018.

[36] O. Prakash and A. Kumar, "Application of artificial neural network for the prediction of jaggery mass during dehydrating inside the natural convection greenhouse dryer," *Int. J. Ambient Energy*, vol. **35**, no. 4, pp. 186–192, 2014.

[37] B. Nour-Eddine, Z. Belkacem, and K. Abdellah, "Experimental study and simulation of a solar dryer for spearmint leaves (Mentha spicata)," *Int. J. Ambient Energy*, vol. **36**, no. 2, pp. 50–61, 2015.

[38] D. Harjunowibowo, Y. Ding, S. Omer, and S. Riffat, "Recent active technologies of greenhouse systems – a comprehensive review," *Bulg. J. Agric. Sci.*, vol. **24**, no. 1, pp. 158–170, 2018.

[39] V. Gupta, A. Sharma, and K. S. Gupta, "NUMERICAL ANALYSIS OF DIRECT TYPE GREENHOUSE DRYER," in *Proceedings of the ASME 2017 Gas Turbine India Conference GTINDIA2017*, 2018, pp. 1–6.

[40] SermJanjai, "A greenhouse type solar dryer for small-scale dried food industries: Development and dissemination.," *Int. J. Energy Environ.*, vol. **3**, no. 3, pp. 383–398, 2012.

[41] A. R. H. Rigit and P. T. K. Low, "Heat and mass transfer in a solar dryer with biomass backup burner," *World Acad. Sci. Eng. Technol.*, vol. **38**, no. 2, pp. 105–108, 2010.

[42] P. S. Chauhan and A. Kumar, "Thermal modeling and dehydrating kinetics of gooseberry dehydrating inside north wall insulated greenhouse dryer," *Appl. Therm. Eng.*, vol. **130**, pp. 587–597, 2018.

[43] A. Mujumdar and W. Zhonghua, "Thermal dehydrating technologies - Cost-effective innovation aided by mathematical modeling approach," *Dry. Technol.*, vol. **26**, no. 1, pp. 146–154, 2008.

[44] A. Jyoti, D. P. Baredar, D. H. Kumar, and A. Kumar, "Design and Optimization of Solar Absorber Tube Using CFD Analysis," *Int. J. ONLINE Sci.*, vol. **4**, no. 3, pp. 1–6, 2018.

[45] A. Ficarella, A. Perago, G. Starace, and D. Laforgia, "Thermo-fluid-dynamic investigation of a dryer, using numerical and experimental approach," *J. Food Eng.*, vol. **59**, no. 4, pp. 413–420, 2003.

[46] S. Janjai, N. Srisittipokakun, and B. K. Bala, "Experimental and modelling performances of a roof-integrated solar dehydrating system for dehydrating herbs and spices," *Energy*, vol. **33**, no. 1, pp. 91–103, 2008.

[47] M. N. M. and E. A. R. A. M. S. Zuan, S. S. Hidayah S. Syahrullail, "Flow Analysis of Cooling Water in the Gas Turbine Fin Fan Cooler at Utilities Plant," *J. Adv. Res. Fluid Mech. Therm. Sci.*, vol. **14**, no. 1, pp. 8–23, 2015.

[48] J. Zhang, W. Johnson, and T. Plikas, "Application of computational fluid dynamics for solving ventilation problems in metallurgical industrial processes," *Int. J. Vent.*, vol. **16**, no. 3, pp. 200–212, 2017.

[49] K. Kant, A. Shukla, A. Sharma, A. Kumar, and A. Jain, "Thermal energy storage based solar dehydrating systems: A review," *Innov. Food Sci. Emerg. Technol.*, vol. **34**, pp. 86–99, 2016.

[50] M. Purusothaman and T. N. Valarmathi, "Computational fluid dynamics analysis of greenhouse solar dryer," *Int. J. Ambient Energy*, vol. **0**, no. 0, pp. 1–7, 2018.

[51] P. P. W. Ingle, A. A. Pawar, P. B. D. Deshmukh, and P. K. C. Bhosale, "CFD Analysis of Solar Flat Plate Collector," *Int. J. Emerg. Technol. Adv. Eng.*, vol. **3**, no. 4, pp. 337–342, 2013.

[52] F. K. Forson, M. A. A. Nazha, and H. Rajakaruna, "Modelling and experimental studies on a mixed-mode natural convection solar crop-dryer," *Sol. Energy*, vol. **81**, no. 3, pp. 346–357, 2007.

[53] B. K. Bala and J. L. Woods, "Simulation of the indirect natural convection solar dehydrating of rough rice," *Sol. Energy*, vol. **53**, no. 3, pp. 259–266, 1994.

[54] O. Prakash and A. Kumar, "ANFIS Prediction Model of a Modified Active Greenhouse Dryer in No-Load conditions in the Month of January," *Int. J. Adv. Comput. Res.*, vol. **3**, no. 1, pp. 220–223, 2013.

[55] A. Kumar and G. N. Tiwari, "Thermal modeling of a natural convection greenhouse dehydrating system for jaggery: An experimental validation," *Sol. Energy*, vol. **80**, no. 9, pp. 1135–1144, 2006.

[56] S. Khaldi, A. N. Korti, and S. Abboudi, "Applying CFD for Studying the Dynamic and Thermal Behavior of Solar Chimney Dehydrating System with Reversed Absorber," *Int. J. Food Eng.*, vol. **13**, no. 11, pp. 253–272, 2017.

[57] S. Janjai, V. Khamvongsa, and B. K. Bala, "Development, design, and performance of a PV-Ventilated greenhouse dryer," *Int. Energy J.*, vol. **8**, no. 4, pp. 249–258, 2007.

[58] P. Barnwal and G. N. Tiwari, "Grape dehydrating by using hybrid photovoltaic-thermal (PV/T) greenhouse dryer: An experimental study," *Sol. Energy*, vol. **82**, no. 12, pp. 1131–1144, 2008.

[59] D. H. Kokate *et al.*, "Energy conservation through solar energy assisted dryer for plastic processing industry," *Energy Procedia*, vol. **54**, no. 022, pp. 376–388, 2014.

[60] M. Dorouzi, H. Mortezapour, H. R. Akhavan, and A. G. Moghaddam, "Tomato slices dehydrating in a liquid desiccant-assisted solar dryer coupled with a photovoltaic-thermal regeneration system," *Sol. Energy*, vol. **162**, no. January, pp. 364–371, 2018.

AUTHOR INDEX

Abdullah, Syed Muhammad 213
Adharsh, K.S. Vishnu 55, 70
Anand, B 252
Anand, R. 128
Antonyraj, S 116
Arul, M 123
Babu, Safare Sudeep 213
Chandel, Sheshang S. 10
Chandran, G.R. Vijay 55, 70
Chary, S Shiva Kumar 206, 213
Chekuri, Bhargav Ganesh 206
Desai, Anuj Bimal 247
Dineshkumar, M 41
Elangovan, D 213
Elansezhian, R 92
Elavarasan, E 48, 173, 237, 275
Faizal, U. Mohamed 55
Harshavardhan, D Sri 163
JethMalani, C H Ram 237
Kalaisselvane, A 79
Kannan, K Bharathi 237
Kartheek, Gudapati 108
Karthikeyan, P 100
Karuppa, Raj R Thundil 213, 225, 265
Katkamwar, Vineet 225
Kumar, Alok 34
Kumar, Aman 1
Kumar, M Dinesh 123
kumar, S. Anish 55, 70
Kumar, Vandrasi Vinay 188
Kumaresan, G. 55, 70
Kumari, Sovana 237
Mathiarasu, A. 180
Mohamed, Raqeeb 1
Mondal, Swaraj 1
Murugavelh, K 257
Murugavelh, S 199, 252
Mutha, Mayuresh 225
Muthudineshkumar, R. 128
Nagarajan, R 199
Natarajan, Sendhil Kumar 1, 48, 100, 108, 152
Natarajan, Sendhil kumar 163
Natarajan, Sendhil Kumar 173, 247, 275
Patil, Aniket 265
Paulraj, Maheandera Prabu 10
Prasad, K Midhun 257
Prasanna, P. Lakshmi 21
Prasath, G Hari 163
Pugazhvadivu, M. 180
Purushothaman, Natarajan 92
Rajesh, Singuru 188
Ram, Jana Sai 188
Ramanathan, Anand 34, 41, 123, 135, 146
Rathna, R 1
Raviprakash, A V 92
Raviteja, B 163
Reddy, Dr. K Dharma 21
Reddy, Dr. K. Dharma 28
Reddy, Gummalla Venkata Sunil 108
Sahu, Santosh Kumar 10
Sahu, Susant Kumar 152
Samuel, G Giftson 116
Santhoshkumar, A. 135
Sathiya, Prakash P 79
Shrikar, Banagiri 41
Singh, K Arjun 48
Sivamurugan, P 225
Sravani, V. 28
Sudhagar, S. 70
Suraparaju, Subbarama Kousik 1, 48, 100, 108, 163
Takale, Mayuresh 265
Thangarasu, Vinoth 146
Thirunavukkarasu, Bageerathan 225, 265
Thundil, Raj R Karuppa 206
Vaskuri, P. Satya Subhasree 206
Verma, Ajay 10
Vijayakumar, P. 55, 70
Yashoda, Aavala 237